Current Topics in Microbiology and Immunology

258

Editors

R.W. Compans, Atlanta/Georgia
M. Cooper, Birmingham/Alabama · Y. Ito, Kyoto
H. Koprowski, Philadelphia/Pennsylvania · F. Melchers, Basel
M. Oldstone, La Jolla/California · S. Olsnes, Oslo
M. Potter, Bethesda/Maryland
P.K. Vogt, La Jolla/California · H. Wagner, Munich

Springer
Berlin
Heidelberg
New York
Barcelona
Hong Kong
London
Milan
Paris
Singapore
Tokyo

Epstein-Barr Virus and Human Cancer

Edited by K. Takada

With 38 Figures and 13 Tables

 Springer

Professor Dr. KENZO TAKADA
Hokkaido University
Institute for Genetic Medicine
Department of Tumor Virology
N15 W7, Kita-Ku
060-8638 Sapporo
Japan
e-mail: kentaka@med.hokudai.ac.jp.

Cover Illustration: BamHI fragment map for the EBV genome. The outer open boxes are viral genes expressed during latent infection in culture

ISSN 0070-217X
ISBN 3-540-41506-8 Springer-Verlag Berlin Heidelberg New York

This work is subject to copyright. All rights are reserved, whether the whole or part of the material is concerned, specifically the rights of translation, reprinting, reuse of illustrations, recitation, broadcasting, reproduction on microfilm or in any other way, and storage in data banks. Duplication of this publication or parts thereof is permitted only under the provisions of the German Copyright Law of September 9, 1965, in its current version, and permission for use must always be obtained from Springer-Verlag. Violations are liable for prosecution under the German Copyright Law.

Springer-Verlag Berlin Heidelberg New York
a member of BertelsmannSpringer Science + Business Media GmbH

http://www.springer.de

© Springer-Verlag Berlin Heidelberg 2001
Library of Congress Catalog Card Number 15-12910
Printed in Germany

The use of general descriptive names, registered names, trademarks, etc. in this publication does not imply, even in the absence of a specific statement, that such names are exempt from the relevant protective laws and regulations and therefore free for general use.

Product liability: The publishers cannot guarantee the accuracy of any information about dosage and application contained in this book. In every individual case the user must check such information by consulting other relevant literature.

Cover Design: *design & production GmbH*, Heidelberg
Typesetting: Scientific Publishing Services (P) Ltd, Madras
Printed on acid-free paper   SPIN: 10718320   27/3020/M 5 4 3 2 1 0

Preface

Epstein-Barr virus (EBV), a human herpesvirus, originally received much attention because of its associations with Burkitt's lymphoma and nasopharyngeal carcinoma. Subsequently it turned out that EBV is ubiquitous in the human population and most people carry the virus in memory B cells in a latent state. Now, many other malignancies such as T/NK cell lymphoma, AIDS-associated B-cell lymphoma, gastric carcinoma, and Hodgkin's disease have been causally linked to EBV.

The development of molecular biology techniques has allowed us to study the roles of individual EBV genes that act in the maintenance and disruption of EBV latency. The outbreak of AIDS revealed the oncogenic potential of EBV. AIDS-associated B-cell lymphoma is a proliferation of EBV-infected B cells in the absence of immune surveillance. This indicates that EBV-infected B cells have the ability to produce tumors if the host immune system does not work. The EBV-immortalized peripheral B cell is an in vitro model of AIDS lymphoma, and has been the focus of extensive studies. These studies revealed that latent membrane protein 1 (LMP1) is particularly important for the immortalization of B cells. On the other hand, As for the role of EBV in Burkitt's lymphoma and epithelioid malignancies including nasopharyngeal carcinoma and gastric carcinoma, we initially could not exclude the possibility that the virus was a passenger and had no role in their carcinogenesis. However, the recent establishment of in vitro models for Burkitt's lymphoma and epithelioid malignancies proved that EBV could contribute malignant conversion of these tumor cells, thus excluding a passenger scenario.

In this volume, outstanding researchers from the United States and Japan review recent progress in EBV research. I believe that this book will help readers to understand what has been done and what should be done in EBV research.

KENZO TAKADA

List of Contents

I Molecular Mechanisms of Maintenance and Disruption of Virus Latency

B. Sugden and E.R. Leight
EBV's Plasmid Replicon: An Enigma in *cis* and *trans* ... 3

K. Hirai[†] and M. Shirakata
Replication Licensing of the EBV *oriP* Minichromosome ... 13

S. Fujiwara
Epstein-Barr Virus Nuclear Protein 2-Induced Activation of the EBV-Replicative Cycle in Akata Cells: Analysis by Tetracycline-Regulated Expression ... 35

L.M. Hutt-Fletcher and C.M. Lake
Two Epstein-Barr Virus Glycoprotein Complexes ... 51

T. Tsurumi
EBV Replication Enzymes ... 65

II EBV-Associated Malignancies

M. Fukayama, J.-M. Chong, and H. Uozaki
Pathology and Molecular Pathology of Epstein-Barr Virus-Associated Gastric Carcinoma ... 91

K. Aozasa, H. Kanno, H. Miwa, and Y. Tomita
EBV and Malignant Lymphoma with Special Emphasis on Pyothorax-Associated Lymphoma ... 103

H. Katano, T. Sata, and S. Mori
AIDS Lymphoma: Its Virological Aspects ... 121

III Molecular Mechanisms of Oncogenesis

K. Takada
Role of Epstein-Barr Virus in Burkitt's Lymphoma ... 141

I.K. Ruf, P.W. Rhyne, H. Yang, C.M. Borza,
L.M. Hutt-Fletcher, J.L. Cleveland, and J.T. Sample
EBV Regulates c-MYC, Apoptosis,
and Tumorigenicity in Burkitt's Lymphoma 153

S. Imai, J. Nishikawa, M. Kuroda, and K. Takada
Epstein-Barr Virus Infection
of Human Epithelial Cells. 161

T. Sairenji, M. Tajima, M. Kanamori, N. Takasaka,
X. Gao, M. Murakami, K. Okinaga, Y. Satoh,
Y. Hoshikawa, H. Ito, Y. Miyazawa, and T. Kurata
Characterization of EBV-Infected Epithelial Cell Lines
from Gastric Cancer-Bearing Tissues 185

IV Animal Model and New Therapeutic Approach to Malignancy

F. Wang
A New Animal Model for Epstein-Barr
Virus Pathogenesis 201

C.M. Rooney, L.K. Aguilar, M.H. Huls,
M.K. Brenner, and H.E. Heslop
Adoptive Immunotherapy of EBV-Associated
Malignancies with EBV-Specific Cytotoxic
T-Cell Lines. 221

Subject Index. 231

List of Contributors

(Their addresses can be found at the beginning of their respective chapters.)

AGUILAR, L.K.   221

AOZASA, K.   103

BORZA, C.M.   153

BRENNER, M.K.   221

CHONG, J.-M.   91

CLEVELAND, J.L.   153

FUJIWARA, S.   35

FUKAYAMA, M.   91

GAO, X.   185

HESLOP, H.E.   221

HIRAI[†], K.   13

HOSHIKAWA, Y.   185

HULS, M.H.   221

HUTT-FLETCHER, L.M.   51, 153

IMAI, S.   161

ITO, H.   185

KANAMORI, M.   185

KANNO, H.   103

KATANO, H.   121

KURATA, T.   185

KURODA, M.   161

LAKE, C.M.   51

LEIGHT, E.R.   3

MIWA, H.   103

MIYAZAWA, Y.   185

MORI, S.   121

MURAKAMI, M.   185

NISHIKAWA, J.   161

OKINAGA, K.   185

RHYNE, P.W.   153

ROONEY, C.M.   221

RUF, I.K.   153

SAIRENJI, T.   185

SAMPLE, J.T.   153

SATA, T.   121

SATOH, Y.   185

SHIRAKATA, M.   13

SUGDEN, B.   3

TAJIMA, M.   185

TAKADA, K.   141, 161

TOMITA, Y.   103

TSURUMI, T.   65

TAKASAKA, N.   185

UOZAKI, H.   91

WANG, F.   201

YANG, H.   153

I
Molecular Mechanisms of Maintenance and Disruption of Virus Latency

EBV's Plasmid Replicon: An Enigma in *cis* and *trans*

B. SUGDEN and E.R. LEIGHT

References 9

Epstein-Barr virus (EBV) is a strikingly successful human parasite infecting more than 90% of humanity. It has adapted itself well to people; after primary infection, it usually remains latent for life and infrequently causes disease. The diseases it does cause, however, are often malignant. During latent infection, EBV maintains its duplex DNA extrachromosomally as a circular molecule in both nonproliferating and proliferating cells. The lytic phase of EBV's life-cycle can only be initiated in cells formerly supporting its latent phase. During the lytic phase, the plasmid replicon is the initial template for amplification of viral DNA. EBV's plasmid replicon is, therefore, an essential feature of both phases of EBV's life-cycle. The elements of this viral plasmid have been studied both to understand their roles in the life-cycle of this human pathogen and to gain any insights they might provide into the synthesis and segregation of other viral and cellular replicons. Much has been learned about this plasmid replicon, but much about it remains obscure.

Several defining features of the intact, viral, plasmid replicon have been established. The DNA of EBV was shown to be an extrachromosomal replicon in a Burkitt's lymphoma-derived cell line and in nasopharyngeal carcinoma biopsies by isolating large DNA from the cells and separating the DNAs as a function of their densities in CsCl gradients (LINDAHL et al. 1975; KASHKA-DIERICH et al. 1976) (Fig. 1a). EBV DNA contains 58% G+C while human DNA contains on the average 42% G+C. The intact viral DNA banded at equilibrium at a density appropriate for its being free of cellular sequences. The viral plasmid DNA was shown to be synthesized once per cell cycle by labeling cells with bromodeoxyuridine (BrdU) and finding that EBV first became heavy/light in its density at the same rate as did cellular DNA (ADAMS 1987). An origin of plasmid DNA synthesis within the B95-8 strain of EBV was determined via two-dimensional gel electrophoresis of replicative intermediates to map at or near its dyad symmetry (DS) element (GAHN and SCHILDKRAUT 1989), an element required for efficient replication of subgenomic, EBV-derived plasmids (YATES et al. 1984) (Fig. 1a). These

---

McArdle Laboratory for Cancer Research, University of Wisconsin, Madison, WI 53706, USA

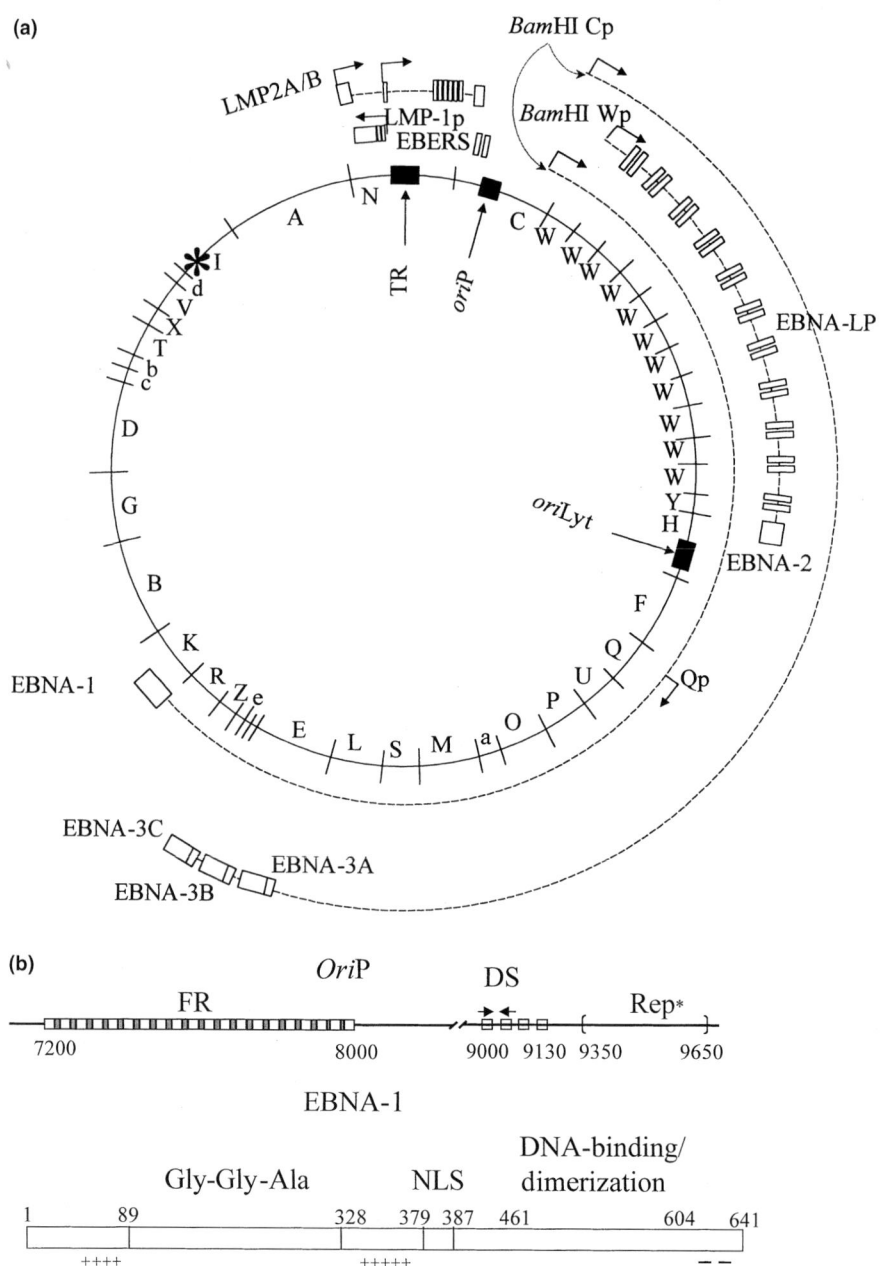

properties of the intact viral genome are defining characteristics of its plasmid replication. To understand them, mechanistically minimal elements of its plasmid replicon were identified.

Fig. 1a,b. Maps for the EBV genome and for *oriP* and EBNA-1 are shown. **a** The genome of the B95-8 strain of EBV is depicted with the fragments it yields upon digestion with *Bam*HI shown on the *inner circle* (BAER et al. 1984). (EBV's open reading frames are identified by their positions relative to the *Bam*HI-derived fragments.) TR represents the terminal repeats found on linear virion DNA, which are joined upon infection of cells. *oriP* and *oriLyt* are the identified origins of replication used during the plasmid or latent and the lytic phase of EBV's life-cycle. The *outer open boxes* connected by *dotted lines* denote the exons and introns, respectively, of the viral genes expressed during latency of cells in culture. Their RNAs are expressed from promoters denoted with *arrows*. These elements and genes are described in detail by ROWE (1999). The EBERs are nontranslated RNAs and Qp is a promoter used in other forms of latency. The star in the *Bam*HI I fragment identifies the site at which 13.6kbp of DNA absent in B95-8 are found in the Raji strain of EBV and which has been shown to contain an origin of DNA synthesis distinct from *oriP* (LITTLE and SCHILDKRAUT 1995). **b** Expanded representations of *oriP* and Rep* are shown *above* and the EBNA-1 protein *below*. The 20 related binding sites for EBNA-1 in FR (family of repeats) and the 4 in DS (dyad symmetry) are shown as *white boxes*. The 65bp dyad in DS is shown as *two inverted arrows*. Rep* maps approximately 300bp away from DS. The approximate map locations of *oriP* and Rep* are given as nucleotide positions from the B95-8 strain below the figure. The 641 amino acids of EBNA-1 are depicted as a *box* with the positively charged residues in the first linking region (residues 40–89) and in the second linking region (residues 328–379) noted along with nuclear localization signal (379–387). The last 40 residues of EBNA-1 are rich in aspartic and glutamic acid residues. The domain from 461 to 604 both dimerizes and binds DNA site-specifically

The required *cis*-acting elements of EBV's plasmid replicon were defined by first introducing into EBV-positive cells a library of DNAs consisting of overlapping fragments of EBV DNA cloned into a vector encoding resistance to G418. Those cells that maintained DNA were selected for their resistance to G418 and the input DNA was assessed for being integrated or being maintained extrachromosomally. These experiments defined the origin of plasmid replication of EBV, *oriP* (YATES et al. 1984). *OriP* consists of two elements including a family of repeated sequences (FR), and an element of DS (REISMAN et al. 1985) (Fig. 1b). The identification of *oriP* allowed the identification of those viral factors acting in *trans* to support the replication of *oriP*. Each member of the overlapping set of fragments of EBV DNA was individually integrated into an EBV-negative, thymidine kinase-negative human cell line. A vector consisting of *oriP* and the thymidine kinase gene of herpes simplex virus type I was introduced into the different cell clones, and cells capable of surviving in HAT medium were selected. Only the clones expressing one gene of EBV, the Epstein-Barr nuclear antigen (EBNA)-1, were found to support extrachromosomal replication of *oriP* (YATES et al. 1985) (Fig. 1b). A vector consisting of *oriP*, EBNA-1, and a selectable marker was found to replicate extrachromosomally in human, monkey, and dog but not rodent cells (LUPTON and LEVINE 1985; YATES et al. 1985), thus identifying EBNA-1 as the sole essential *trans*-acting viral gene of EBV's plasmid replicon. EBNA-1 was shown to bind each of the repeats within both the FR and DS elements that comprise *oriP* (RAWLINS et al. 1985), thereby also identifying EBNA-1 as a site-specific, DNA-binding protein.

Minimal *oriP* vectors behave similarly to intact EBV plasmids. Density labeling experiments akin to those showing that EBV DNA replicates once per cell cycle showed that *oriP* vectors also replicate in concert with cellular DNA (YATES and GUAN 1991). The origin of DNA synthesis in an *oriP*/EBNA-1 replicon was mapped at or close to DS as in the intact viral genome (GAHN and SCHILDKRAUT

1989). Thus, a foundation was laid to assess the contributions to EBV's plasmid replication made by *oriP* and EBNA-1 mechanistically. This assessment has revealed a series of unanticipated findings.

OriP was defined with a functional assay to screen all of the B95-8 strain of EBV DNA (YATES et al. 1984). However, two additional *cis*-acting elements within EBV have been found with different assays that map outside of *oriP* but have origin-like function. Two-dimensional gel analysis of replicative intermediates of EBV DNA from Raji cells has identified at least one locus in addition to *oriP* at which DNA synthesis initiates (LITTLE and SCHILDKRAUT 1995) (Fig. 1a). This additional origin would not have been detected in the experiments that identified *oriP* within the B95-8 strain of EBV DNA because the B95-8 strain lacks the DNA that contains the additional origin found in the Raji strain. Another origin-like element, Rep*, has been detected in a functional assay to identify DNA sequences that can substitute for DS within *oriP* to support extrachromosomal replication (KIRCHMAIER and SUGDEN 1998) (Fig. 1b). Rep* has approximately 50% of the activity of DS in supporting plasmid replication. It maps so closely to DS that any contribution from it to origin-like function would not have been detected in the original two-dimensional gel studies that defined DS as the site in *oriP* at which or near which DNA synthesis initiates. What is peculiar about the additional origin in the Raji viral genome and Rep* is that neither contains a DNA sequence bound directly by EBNA-1. The defining characteristic of DS, on the other hand, is two appropriately spaced binding sites for EBNA-1 (YATES et al. 2000). *OriP* is distinct, therefore, from other origin-like elements in EBV because its origin binds EBNA-1 and along with EBNA-1 *oriP* is sufficient to support extrachromosomal replication efficiently in human cells. The contributions of each of the origin-like elements to EBV's life-cycle have not been delineated; however, all available evidence indicates that at least the FR element and EBNA-1 can contribute to maintenance of the plasmid replicon.

EBNA-1 is unique among EBV-encoded proteins in that it alone has been found to be expressed in all EBV-infected, proliferating cells (ROWE et al. 1986). Its presence in all EBV-positive, proliferating cells is consistent with its providing one or more essential functions to these cells. It is likely that EBNA-1's contributions to the synthesis and/or maintenance of EBV's plasmid replicon are essential functions for the latent phase of EBV's life-cycle. These contributions, however, are uncertain mechanistically.

EBNA-1 is an origin-binding protein, as are multiple other DNA virus-encoded replication proteins. Polyoma-like viruses, papilloma viruses, and α-herpesviruses all encode origin-binding proteins with intrinsic DNA helicase activities (DEPAMPHILIS 1996). EBNA-1, however, lacks an intrinsic helicase activity (FRAPPIER and O'DONNELL 1991; MIDDLETON and SUGDEN 1992) and its contribution to DNA synthesis at *oriP* is obscure. EBNA-1 does have two activities shown genetically to correlate with replication of *oriP*. It binds DNA site-specifically (RAWLINS et al. 1985) and it loops or links DNAs to which it binds site-specifically (FRAPPIER and O'DONNELL 1991; Su et al. 1991; MIDDLETON and SUGDEN 1992). EBNA-1 on binding multiple sites on one DNA associates to "loop" out the DNAs between the binding sites; on binding sites on different DNAs EBNA-1 associates to "link" those DNAs together. Genetic analyses of EBNA-1 indicate that mutations within its DNA-binding and dimerization domain inhibit all of its functions (YATES and CAMIOLO 1988). In fact, derivatives of EBNA-1 that consist only of this domain are inhibitors of wild-type EBNA-1's functions (KIRCHMAIER and SUGDEN 1997). Studies of ten derivatives of EBNA-1, all of which bind DNA as does wild-type EBNA-1 but that vary in their abilities to link the bound DNAs, have shown that the ability of these derivatives to link DNAs correlates well with their support of *oriP*'s replication (MACKEY and SUGDEN 1999).

Although EBNA-1's DNA-looping/-linking is perplexing in its putative role in replication, it is an activity shared by other proteins, some of which may serve as models to understand EBNA-1. The E2 protein of papillomaviruses, for example, loops DNAs to which it binds (KNIGHT et al. 1991). It also, on binding to DNA, can associate with a DNA-bound Sp1 transcription factor to loop intervening DNA (LI et al. 1991). Both of these activities of E2 can be viewed as the result of E2's binding to one site on DNA, increasing the apparent affinity for a looping partner to bind to another site on that same DNA [such a mechanism has been documented for EBNA-1 (SU et al. 1991)]. The mechanism by which E2 loops DNAs to which it binds at multiple sites most likely reflects its ability to self-associate. E2's self-association has been revealed through determination of the structure of its amino-terminus in X-ray crystallographic studies (ANTSON et al. 2000). E2 binds DNA as a dimer, and three α-helices in each amino-terminal monomer can contact those α-helices in a monomer of another dimer to associate the dimers. It is not known if the α-helices of E2 that mediate its self-association are also involved in its binding Sp1. E2 can also associate with mitotic chromosomes through its aminoterminal one-half (SKIADOPOULOS and MCBRIDE 1998). Again, it is not known if the α-helices of E2 that mediate its self-association contribute to E2's association with mitotic chromosomes. However, E2's association with mitotic chromosomes occurs when E2 is bound to viral DNA and is likely to contribute to the association of papilloma viral DNAs with mitotic chromosomes (SKIADOPOULOS and MCBRIDE 1998). These latter observations on E2 might be explained by E2's binding papilloma DNA and linking it to E2 bound non-specifically to chromosomal DNA or linking it to other cellular proteins such as Sp1 which bind cellular DNA specifically.

EBNA-1 shares multiple features with the E2 protein of papillomaviruses. The structure of its dimerization and DNA-binding domain is quite similar to that of E2 (HEGDE et al. 1992; BOCHKAREV et al. 1996). It not only loops/links FR to DS within *oriP* as E2 loops sites it binds, but EBNA-1 also associates with mitotic chromosomes. Fusions of different domains of EBNA-1 to green fluorescent proteinGFP have shown that it is the linking regions of EBNA-1 that mediate the association of the fused GFP to mitotic chromosomes (MARECHAL et al. 1999). By analogy with E2 we hypothesize that EBNA-1 contributes to the distribution of *oriP* to daughter cells at mitosis by binding FR through its DNA-binding domain and associating the replicon with cellular proteins bound to chromatin through EBNA-1's linking regions. That both FR and EBNA-1 contribute to some step in maintenance is supported by findings that human DNA sequences can only support autonomous replication if FR is provided in *cis* and EBNA-1 is provided in *trans* (KRYSAN et al. 1989). The exact events during mitosis by which EBV's plasmid replicon is distributed to daughter cells are a fascinating mystery yet to be solved. How EBNA-1 might contribute to synthesis at the DS is not clear either. However, E2 helps to tether a viral helicase to the origin of viral DNA synthesis of bovine papilloma virus (SEDMAN and STENLUND 1995) and EBNA-1 may, by analogy, tether a cellular helicase such as the MCM complex to the DS (YOU et al. 1999).

One obvious route to resolve the enigmas clouding our understanding of EBV's plasmid replicon is to identify the cellular proteins that contribute to its synthesis and segregation and to define EBNA-1's possible role in recruiting those cellular proteins to FR and to DS. Multiple efforts to do so using yeast one- and two-hybrid assays have been reported but are most surprising for what they failed to find (FISCHER et al. 1997; KIM et al. 1997; WANG et al. 1997; AIYAR et al. 1998; SHIRE et al. 1999; ITO et al. 2000). No cellular proteins known to be part of a pre-replication complex such as the origin recognition complex (ORC) or minichromosome maintenance (MCM) proteins were found. No cellular proteins that are known candidates for mediating segregation of the plasmid replicon were found either. Yet several studies indicate that *oriP* plasmids are segregated faithfully in approximately 96% of mitoses (REISMAN et al. 1985; SUGDEN and WARREN 1989; KIRCHMAIER and SUGDEN 1995). These failures may be technical or may eventually be shown to reflect *oriP*'s and EBNA-1's employing unexpected cellular proteins to assemble a synthetic complex at DS and to mediate segregation of *oriP*.

One cellular protein found to bind EBNA-1, Rch1/importin-α (FISCHER et al. 1997; KIM et al. 1997; AIYAR et al. 1998) could clearly contribute to EBNA-1's functions by conducting it into the nucleus. EBNA-1 can efficiently shepherd DNAs it binds in the cytoplasm into the nucleus (LANGLE-ROUAULT et al. 1998) and this homing could be promoted by EBNA-1's binding Rch 1/importin-α. Two additional cellular proteins, P32/TAP/gClq-R (WANG et al. 1997; AIYAR et al. 1998) and EBP2 (SHIRE et al. 1999), have been identified in the yeast assays and proposed to contribute to EBNA-1's functions. Much less is known about these two proteins than about Rch 1/importin-α and only detailed genetic experiments will define possible roles for them in the replication of EBV's plasmid replicon.

One cellular protein essential for DNA synthesis but not found in the yeast assays has been shown to associate with EBNA-1 in in vitro assays. RPA, the single-stranded DNA-binding protein, binds to EBNA-1 as measured by surface plasmon resonance (ZHANG et al. 1998). RPA is a trimer and the 70-kDa subunit which binds single-stranded DNA is sufficient to bind EBNA-1 (ZHANG et al. 1998). The E2 protein of bovine papilloma virus also associates with RPA in vitro (LI and BOTCHAN 1993). If EBNA-1 binds RPA in vivo as it does in vitro, then EBNA-1 may recruit to DS not the pre-replication complex consisting in part of the ORC, CDC6, and MCM proteins, but the DNA synthetic complex of which RPA is an essential member. How DNA synthesis would be initiated at DS would remain a mystery.

Solving the mystery of how DS is identified as a site of initiation of DNA synthesis should provide a major mechanistic insight into EBV's plasmid replication. That EBV synthesizes its plasmid once per cell cycle indicates that it needs to be licensed to be synthesized (SHIRAKATA et al. 1999). In stark contrast, other DNA viruses that encode their own DNA helicases do not synthesize their genomes only once per S-phase and escape the cell's licensing control. Licensing of replication is poorly understood but involves a protein complex including MCM proteins (KUBOTA et al. 1997; PROKHOROVA and BLOW 2000). This understanding of the control of cellular DNA synthesis makes it likely that *oriP* should associate with licensing factors that would confer upon it "once per S-phase synthesis". *OriP* appears to be bound by EBNA-1 throughout the cell cycle (HSIEH et al. 1993), and it is therefore easiest to propose that EBNA-1 would bind some licensing factors, perhaps replacing the role of the ORC proteins and CDC6 to which the MCM proteins bind after mitosis. This proposition, however, lacks any direct evidence. The mystery remains.

Acknowledgements. This work was supported by grants from the National Institutes of Health: CA70723, CA22443, CA09135, and CA07175. B.S. is an American Cancer Society Research Professor.

References

Adams A (1987) Replication of latent Epstein-Barr virus genomes in Raji cells. J Virol 61:1743–1746

Aiyar A, Tyree C, Sugden B (1998) The plasmid replicon of EBV consists of multiple *cis*-acting elements that facilitate DNA synthesis by the cell and a viral maintenance element. EMBO J 17:6394–6403

Antson AA, Burns JE, Moroz OV, Scott DJ, Sanders CM, Bronstein IB, Dodson GG, Wilson KS, Maitland NJ (2000) Structure of the intact transactivation domain of the human papillomavirus E2 protein. Nature 403:805–809

Baer R, Bankier AT, Biggin MD, Deininger PL, Farrell PJ, Gibson TJ, Hatfull G, Hudson GS, Satchwell SC, Seguin C, Tuffnell PS, Barrell BG (1984) DNA sequence and expression of the B95-8 Epstein-Barr virus genome. Nature 310:207–211

Bochkarev A, Barwell JA, Pfuetzner RA, Bochkareva E, Frappier L, Edwards AM (1996) Crystal structure of the DNA-binding domain of the Epstein-Barr virus origin-binding protein, EBNA1, bound to DNA. Cell 84:791–800

DePamphilis M (ed) (1996) DNA Replication in Eukaryotic Cells. Cold Spring Harbor Laboratory Press

Fischer N, Kremmer E, Lautscham G, Mueller LN, Grasser FA (1997) Epstein-Barr virus nuclear antigen 1 forms a complex with the nuclear transporter karyopherin alpha2. J Biol Chem 272:3999–4005

Frappier L, O'Donnell M (1991) Epstein-Barr nuclear antigen 1 mediates a DNA loop within the latent replication origin of Epstein-Barr virus. Proc Natl Acad Sci USA 88:10875–10879

Gahn TA, Schildkraut CL (1989) The Epstein-Barr virus origin of plasmid replication, *oriP*, contains both the initiation and termination sites of DNA replication. Cell 58:527–535

Hegde RS, Grossman SR, Laimins LA, Sigler PB (1992) Crystal structure at 1.7A of the bovine papillomavirus-1 E2 DNA-binding domain bound to its DNA target. Nature 359:505–512

Hsieh DJ, Camiolo SM, Yates JL (1993) Constitutive binding of EBNA1 protein to the Epstein-Barr virus replication origin, *oriP*, with distortion of DNA structure during latent infection. EMBO J 12:4933–4944

Ito S, Ikeda M, Kato N, Matsumoto A, Ishikawa Y, Kumakubo S, Yanagi K (2000) Epstein-Barr virus nuclear antigen-1 binds to nuclear transporter karyopherin alpha1/NPI-1 in addition to karyopherin alpha2/Rch1. Virology 266:110–119

Kaschka-Dierich C, Adams A, Lindahl T, Bornkamm GW, Bjursell G, Klein G, Giovanella BC, Singh S (1976) Intracellular forms of Epstein-Barr virus DNA in human tumour cells in vivo. Nature 260:302–306

Kim AL, Maher M, Hayman JB, Ozer J, Zerby D, Yates JL, Lieberman PM (1997) An imperfect correlation between DNA replication activity of Epstein-Barr virus nuclear antigen 1 (EBNA1) and binding to the nuclear import receptor, Rch1/importin alpha. Virology 239:340–351

Kirchmaier AL, Sugden B (1995) Plasmid maintenance of derivatives of *oriP* of Epstein-Barr virus. J Virol 69:1280–1283

Kirchmaier AL, Sugden B (1997) Dominant-negative inhibitors of EBNA-1 of Epstein-Barr virus. J Virol 71:1766–1775

Kirchmaier AL, Sugden B (1998) Rep*: a viral element that can partially replace the origin of plasmid DNA synthesis of Epstein-Barr virus. J Virol 72:4657–4666

Knight JD, Li R, Botchan M (1991) The activation domain of the bovine papillomavirus E2 protein mediates association of DNA-bound dimers to form DNA loops. Proc Natl Acad Sci USA 88:3204–3208

Krysan PJ, Haase SB, Calos MP (1989) Isolation of human sequences that replicate autonomously in human cells. Mol Cell Biol 9:1026–1033

Kubota Y, Mimura S, Nishimoto S, Masuda T, Nojima H, Takisawa H (1997) Licensing of DNA replication by a multi-protein complex of MCM/P1 proteins in Xenopus eggs. EMBO J 16:3320–3331

Langle-Rouault F, Patzel V, Benavente A, Taillez M, Silvestre N, Bompard A, Sczakiel G, Jacobs E, Rittner K (1998) Up to 100-fold increase of apparent gene expression in the presence of Epstein-Barr virus *oriP* sequences and EBNA1: implications of the nuclear import of plasmids. J Virol 72:6181–6185

Li R, Botchan MR (1993) The acidic transcriptional activation domains of VP16 and p53 bind the cellular replication protein A and stimulate in vitro BPV-1 DNA replication. Cell 73:1207–1221

Li R, Knight JD, Jackson SP, Tjian R, Botchan MR (1991) Direct interaction between Sp1 and the BPV enhancer E2 protein mediates synergistic activation of transcription. Cell 65:493–505

Lindahl T, Adams A, Bjursell G, Bornkamm GW, Kaschka-Dierich C, Jehn U (1976) Covalently closed circular duplex DNA of Epstein-Barr virus in a human lymphoid cell line. J Mol Biol 102:511–530

Little RD, Schildkraut CL (1995) Initiation of latent DNA replication in the Epstein-Barr virus genome can occur at sites other than the genetically defined origin. Mol Cell Biol 15:2893–2903

Lupton S, Levine AJ (1985) Mapping genetic elements of Epstein-Barr virus that facilitate extrachromosomal persistence of Epstein-Barr virus-derived plasmids in human cells. Mol Cell Biol 5:2533–2542

Mackey D, Sugden B (1999) The linking regions of EBNA1 are essential for its support of replication and transcription. Mol Cell Biol 19:3349–3359

Marechal V, Dehee A, Chikhi BR, Piolot T, Coppey MM, Nicolas JC (1999) Mapping EBNA-1 domains involved in binding to metaphase chromosomes. J Virol 73:4385–4392

Middleton T, Sugden B (1992) EBNA1 can link the enhancer element to the initiator element of the Epstein-Barr virus plasmid origin of DNA replication. J Virol 66:489–495

Prokhorova TA, Blow JJ (2000) Sequential MCM/P1 subcomplex assembly is required to form a heterohexamer with replication licensing activity. J Biol Chem 275:2491–2498

Rawlins DR, Milman G, Hayward SD, Hayward GS (1985) Sequence-specific DNA binding of the Epstein-Barr virus nuclear antigen (EBNA-1) to clustered sites in the plasmid maintenance region. Cell 42:859–868

Reisman D, Yates J, Sugden B (1985) A putative origin of replication of plasmids derived from Epstein-Barr virus is composed of two *cis*-acting components. Mol Cell Biol 5:1822–1832

Rowe DT (1999) Epstein-Barr virus immortalization and latency. Frontiers in Bioscience 4:346–371

Rowe DT, Rowe M, Evan GI, Wallace LE, Farrell PJ, Rickinson AB (1986) Restricted expression of EBV latent genes and T-lymphocyte-detected membrane antigen in Burkitt's lymphoma cells. EMBO J 10:2599–2607

Sedman J, Stenlund A (1995) Co-operative interaction between the initiator E1 and the transcriptional activator E2 is required for replicator specific DNA replication of bovine papillomavirus in vivo and in vitro. EMBO J 14:6218–6228

Shirakata M, Imadome KI, Hirai K (1999) Requirement of replication licensing for the dyad symmetry element-dependent replication of the Epstein-Barr virus *oriP* minichromosome. Virology 263:42–54

Shire K, Ceccarelli DF, Avolio HT, Frappier L (1999) EBP2, a human protein that interacts with sequences of the Epstein-Barr virus nuclear antigen 1 important for plasmid maintenance. J Virol 73:2587–2595

Skiadopoulos MH, McBride AA (1998) Bovine papillomavirus type 1 genomes and the E2 transactivator protein are closely associated with mitotic chromatin. J Virol 72:2079–2088

Su W, Middleton T, Sugden B, Echols H (1991) DNA looping between the origin of replication of Epstein-Barr virus and its enhancer site: stabilization of an origin complex with Epstein-Barr nuclear antigen 1. Proc Natl Acad Sci USA 88:10870–10874

Sugden B, Warren N (1988) Plasmid origin of replication of Epstein-Barr virus, *oriP*, does not limit replication in cis. Mol Biol Med 5:85–94

Wang Y, Finan JE, Middeldorp JM, Hayward SD (1997) P32/TAP, a cellular protein that interacts with EBNA-1 of Epstein-Barr virus. Virology 236:18–29

Yates JL, Camiolo SM (1988) Dissection of DNA Replication and Enhancer Activation Functions of Epstein-Barr Virus Nuclear Antigen 1. Cancer Cells 6:197–205

Yates JL, Guan N (1991) Epstein-Barr virus-derived plasmids replicate only once per cell cycle and are not amplified after entry into cells. J Virol 65:483–488

Yates JL, Camiolo SM, Bashaw JM (2000) The minimal replicator of Epstein-Barr virus *oriP*. J Virol 74:4512–4522

Yates J, Warren N, Reisman D, Sugden B (1984) A *cis*-acting element from the Epstein-Barr viral genome that permits stable replication of recombinant plasmids in latently infected cells. Proc Natl Acad Sci USA 81:3806–3810

Yates JL, Warren N, Sugden B (1985) Stable replication of plasmids derived from Epstein-Barr virus in various mammalian cells. Nature 313:812–815

You Z, Komamura Y, Ishimi Y (1999) Biochemical analysis of the intrinsic MCM4-MCM6-MCM7 DNA helicase activity. Mol Cell Biol 19:8003–8015

Zhang D, Frappier L, Gibbs E, Hurwitz J, O'Donnell M (1998) Human RPA (hSSB) interacts with EBNA1, the latent origin binding protein of Epstein-Barr virus. Nucleic Acids Res 26:631–637

Replication Licensing of the EBV *oriP* Minichromosome

K. Hirai[†] and M. Shirakata

1	Introduction	13
2	Replication of Episomal EBV Genome from the Latent Viral Replication Origin *oriP*	14
2.1	Sequence Requirement in the DS Element for DNA Replication	15
2.2	Functions of the FR Element in DS-Dependent Replication	16
2.3	EBNA1 Domains Related to EBV Replication from *oriP*	16
3	Licensing Regulation of the DNA Replication from *oriP*	17
3.1	Requirement of Early G_1 Phase for DNA Replication from *oriP*	18
3.2	The DS Element Was Responsible for Cell Cycle-Regulated Replication	19
3.3	Structural Changes Occurred in the *oriP* Minichromosome Between G_2/M and Late G_1	19
3.4	Association of MCM2 with the *oriP* Minichromosome	20
3.5	Interaction of EBNA1 and MCM Proteins on the DS Element	22
3.6	A Possible Role of EBNA1 in the Initiation of DNA Replication	22
4	DS-Dependent Replication from *oriP* Is Suppressed in Several EBV-Infected Cell Lines	24
5	Stable Nuclear Retention and Partitioning of EBV Episomes During Cell Division	25
5.1	Requirement of the FR for Stable Retention of *oriP* Plasmid	26
5.2	EBNA1 Is Required for Nuclear Retention and Partitioning of *oriP* Plasmid	26
5.3	Analysis of Latent EBV State by Fluorescence In Situ Hybridization	27
6	Summary	28
References		29

1 Introduction

In human B lymphocytes latently infected in vitro and in vivo with Epstein-Barr virus (EBV), including Burkitt's lymphoma (BL) cells and various other EBV-related tumor cells, the viral DNA persists mostly in a circular episomal form with multiple copies and expresses a limited number of viral genes, while some of the latent EBV DNA may be integrated into the host chromosomal DNA (Kieff 1996; Hirai et al. 1998 for review). The original definition of the term "episome" coined by Jacob and Wollman (1961) is a genetic factor that can persist in both the integrated state and as an extrachromosomal replicon in host cells. However, the

Department of Tumor Virology, Division of Virology and Immunology, Medical Research Institute, Tokyo Medical and Dental University, Yushima 1-5-45, Bunkyo, Tokyo 113-8510, Japan

term episome has also been used to describe the circular, extrachromosomal plasmid state of the latent EBV genome in most recent papers related to the state of latent EBV. We also follow this definition.

EBV-infected peripheral blood cells are $CD19^+$, $CD23^-$, and $CD80^-$ resting B cells (MIYASHITA et al. 1995), and we harbor circular episomal EBV DNA molecules (BABCOCK et al. 1998). No viral replication is detectable in the peripheral blood (DECKER et al. 1996), indicating that latent EBV may persist in the resting B cell. In contrast to the latent state of EBV in vivo, it is still a matter of dispute where the EBV replicates in vivo. The tonsil was recently suggested to be one of sites of replication of EBV (BABCOCK et al. 1998; KOBAYASHI et al. 1998; IKEDA et al. 2000). EBV DNA replication may occur in the upper epithelial cell layers of the non-neoplastic tonsil (KOBAYASHI et al. 1998) and in the tonsillar lymphocytes of healthy, persistently infected individuals (BABCOCK et al. 1998; IKEDA et al. 2000).

The EBV genome replicates in two distinct modes during the viral life cycle (YATES 1996). In the lytic phase, the virus synthesizes the linear form of the double-stranded DNA genome by a rolling-circle type of replication and amplifies the genome for virus production. Replication factors that are required for this lytic replication are mostly encoded in the viral genome. The origins that initiate the lytic replication, *oriLyt*, are located in two separated regions of the viral genome (HAMMERSCHMIDT and SUGDEN 1988). Latent replication is initiated once in a single S phase so that the viral genomes are maintained stably in cells (HAMPER et al. 1974; ADAMS 1987; YATES and GUAN 1991). A 2.2-kb region in the *Bam*HI-C fragment of the EBV genome, *oriP*, was identified as the replication origin (YATES et al. 1984), and recently several replication initiation sites have also been identified in a region distant from *oriP* (LITTLE and SCHILDKRAUT 1995). In contrast to lytic replication, maintenance of the latent EBV genome requires only EBV nuclear antigen 1 (EBNA1) as a viral protein (LUPTON and LEVINE 1985; YATES et al. 1985), but the mechanism by which it initiates DNA replication is unknown. In this review, we summarize current knowledge regarding the latent replication of the EBV genome and describe some recent progress in our understanding of its cell-cycle-dependent regulation and maintenance of episomal EBV.

2 Replication of Episomal EBV Genome from the Latent Viral Replication Origin *oriP*

A study to search for autonomous replicating sequences (ARS) in the EBV genome identified two regions that were required for replication in mammalian cells (YATES et al. 1984). One region in the *Bam*HI-K fragment encodes the DNA-binding protein, EBNA1, and the other 2.2-kb region in the *Bam*HI-C fragment has several binding sites for EBNA1 and shows ARS activity in the presence of EBNA1 and was designated as *oriP*. The presence of ARS in the *oriP* region suggested that an actual replication origin is located in this region. EBNA1 is required for the ARS activity of *oriP* and it has two clusters of binding sites for EBNA1, the family of repeats (FR) and the dyad symmetry (DS) elements (LUPTON and LEVINE 1985; RAWLINS et al. 1985; REISMAN et al. 1985) (Fig. 1). The FR element contains twenty EBNA1-binding sites and the DS element consists of four EBNA1-binding sites. Mutagenesis studies of *oriP* have shown that the DS element is important for the ARS function of *oriP* (WYSOKENSKI and YATES 1989; HARRISON et al. 1994; NILLER et al. 1995). Further study of a transient replication assay using the HeLa cell stably expressing EBNA1 (HeLa/EB1) revealed that the DS element is actually the minimal sequence showing ARS activity that depends on EBNA1 (SHIRAKATA and HIRAI 1998). 2D-gel analysis of replicating plasmids determined the replication initiation sites of the small plasmid containing *oriP* (the *oriP* plasmid) and confirmed that bi-directional replication initiates at or very close to the DS element (GAHN and SCHILDKRAUT 1989). Thus, the DS element functions as a replication origin because it has ARS activity and the actual initiation site of DNA replication.

2.1 Sequence Requirement in the DS Element for DNA Replication

The DS element is 100bp in length and contains two repeats of a sequence arranged with dyad symmetry. Each repeat contains two EBNA1-binding sites (20bp) only 1bp apart: the DS element has four EBNA1-binding sites in total, from site 1 to site 4 (Fig. 1). EBNA1 forms a dimer and binds to a single binding site (BOCHKAREV et al. 1995, 1996). These EBNA1-binding sequences are essential for the ARS activity. The complete DS element containing four EBNA1 sites works alone as a functional ARS (SHIRAKATA and HIRAI 1998). The single repeat containing sites 1 and 2 (or sites 3 and 4) has only a background level of ARS activity that is 30-fold

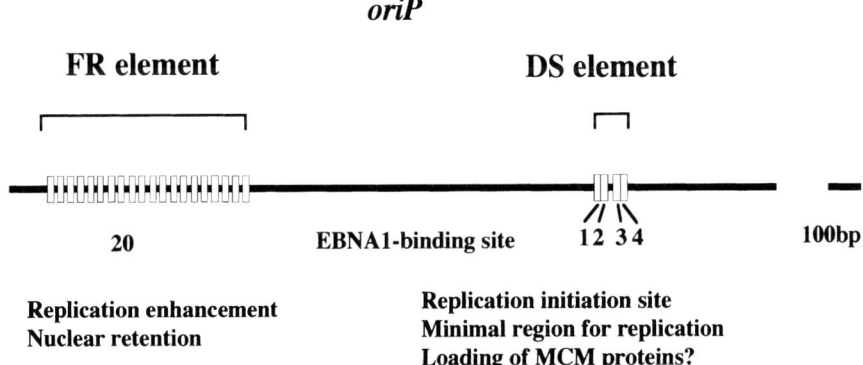

Fig. 1. Elements in the *oriP* region and their functions in the DS-dependent replication of the *oriP* plasmid. This figure summarizes the results reported in this and previous studies (GAHN and SCHILDKRAUT 1989; KRYSAN et al. 1989; WYSOKENSKI and YATES 1989; YATES and GUAN 1991; HARRISON et al. 1994; MIDDLETON and SUGDEN 1994; SHIRAKATA and HIRAI 1998; SHIRAKATA et al. 1999). An *open box* indicates an EBNA1 binding site recognized by an EBNA1 dimer

less than that of the DS element. However, in the context of the *oriP* sequence, the half of the element, a single repeat containing two EBNA1-binding sites, can also function as an ARS (HARRISON et al. 1994). Thus, the FR element in *oriP* may enhance the activity of the single repeat of the DS element and compensate for loss of the other half. Further mutagenesis of the DS element revealed that insertion of a 5-bp sequence between site 1 and site 2 or between site 3 and site 4 completely abolished the ARS activity (HARRISON et al. 1994). Insertion of a 5-bp sequence rotates site 2 (site 4) through about 180° along the DNA axis. EBNA1 can bind to all four sites in the mutated DS element but the relative positions of EBNA1 are completely different from those of the wild type. Thus, a particular spatial arrangement of EBNA1 on the DS element is important for the ARS activity. Crystal structure analysis of the complex consisting of EBNA1 DNA-binding domain and the target DNA suggested that the EBNA1 dimer bound to site 1 interacts with another bound to site 2 (BOCHKAREV et al. 1996). Formation of this tetramer complex consisting of two EBNA1 dimers on DNA may be important for the initiation of DNA replication from the DS element.

2.2 Functions of the FR Element in DS-Dependent Replication

The FR element also contains the EBNA1 binding sites, but this FR element has no more than 1% of the ARS activity of the DS element (SHIRAKATA and HIRAI 1998). As the spacing between the EBNA1-binding sites (bp) is longer than those in the DS element (1 bp), the FR element does not work as a replication origin but has distinct roles in maintenance of the EBV chromosome. Transient replication assay in HeLa/EB1 cells showed that ARS activity of the DS element is about 30% of that of *oriP* (SHIRAKATA and HIRAI 1998). This suggested that some sequence in *oriP* enhances replication of the plasmid. Deletion of the region between the DS and FR elements (900 bp) or the other region flanking to the DS element (300 bp) does not affect replication of the *oriP* plasmid so that the FR element enhances ARS activity of the DS element. The mechanism underlying this enhancement of replication is not known. A possible explanation is that the EBNA1 proteins bound to the FR element may stabilize the binding of EBNA1 on the DS element by EBNA1-EBNA1 interactions (FRAPPIER and O'DONNELL 1991b; SU et al. 1991; MACKEY et al. 1995). Besides this weak replication enhancement (threefold), the FR element has a major role in the nuclear retention of the EBV chromosome as described later in Sect. 5.1.

2.3 EBNA1 Domains Related to EBV Replication from *oriP*

As stated above, EBNA1 forms stable dimers to bind to *oriP* sequences. The dimerization domain of EBNA1 consisting of 641 amino acids was localized to amino acids (aa) 501–598 (AMBINDER et al. 1991; CHEN et al. 1993). Then, EBNA1 dimers bind to the DS and FR elements of *oriP* through aa 459–487. The dimerization and

DNA binding domains are essential for *oriP* replication (POLVINO-BODNAR et al. 1992). The DNA binding of EBNA1 dimers results in loop formation by multiple linking to the FR and DS elements (FRAPPIER and O'DONNELL 1991b; SU et al. 1991). The domains of EBNA1 that mediate DNA linking are located in at least three regions, aa 54–89, 331–361 and 372–391, by electrophoretic mobility shift assay (MACKEY et al. 1995). Glycerol gradient sedimentation assay also indicated that similar DNA linking regions were mapped to two regions, aa 40–100 and 327–377 (AVOLIO-HUNTER and FRAPPIER 1998). The DNA linking may facilitate initiation of *oriP* replication and contribute to nuclear retention and partitioning of the episomal EBV genome. The other characteristic structure of EBNA1 is the Gly-Ala repeat domain. The Gly-Ala repeats are encoded by the internal repeat 3 (IR3) of EBV DNA and affect the size of EBNA1 protein among EBV isolates (HENNESSY et al. 1983). The IR3 probe hybridizes to cell DNA sequences on all human chromosomes except the Y chromosome, suggesting that the viral DNA sequence may have been derived from cellular DNA (HELLER et al. 1982). Although the function of Gly-Ala repeats in *oriP* replication is not known, the repeats interfere with antigen processing and MHC class I-restricted presentation to escape from the host cytotoxic T-cell response (LEVITSKAYA et al. 1995).

3 Licensing Regulation of the DNA Replication from *oriP*

DNA replication of the latent EBV genome differs markedly from that of other DNA viruses such as polyomavirus, SV40, and papillomavirus in regulation during the cell cycle. The origin-binding factors of these viruses, the large T antigen of SV40 and the E1/E2 of papillomavirus, have DNA helicase activity and can initiate multiple rounds of DNA replication in a single S phase (STENLUND 1996; HASSELL and BRINTON 1996). This type of replication results in a large copy number of the genome in each cell. In contrast to these viruses, EBNA1 does not have DNA helicase activity (FRAPPIER and O'DONNELL 1991a; MIDDLETON and SUGDEN 1992) and the latent EBV replicates only once in a single S phase (HAMPER et al. 1974; ADAMS 1987). YATES and GUAN (1991) confirmed this by analyzing the replication of an *oriP*-containing plasmid in cultured cells. These results suggested that initiation of DNA replication from *oriP* is regulated to occur only once per S phase. Replication of the cellular genome is restricted to occur only once during a single S phase by the replication licensing mechanism (BLOW and LASKEY 1988). The same mechanism may restrict replication from *oriP*. Alternatively, *oriP* could be moderately regulated by a mechanism that is different from cellular replication licensing and specific for replication from *oriP*. It is possible, for example, that a protein kinase that is activated only at the G_1/S boundary, e.g., cdc7 kinase (KUMAGAI et al. 1999; ROBERTS et al. 1999), may activate EBNA1 transiently and initiate DNA replication from *oriP* once per S phase. As the regulatory mechanism of *oriP* should be part of the initiation process of DNA replication, it is important to determine whether *oriP* is regulated by the cellular regulatory mechanism or by an *oriP*-specific mechanism.

3.1 Requirement of Early G_1 Phase for DNA Replication from *oriP*

The replication licensing mechanism is conserved in yeast and higher eukaryotes. Replication origins are found in several autonomous replication sequences in yeast (BREWER and FANGMAN 1987) and the origin recognition complex (ORC1–6) binds to these origins constitutively during the cell cycle (BELL and STILLMAN 1992). In vertebrate cells, common sequences of replication origins have not yet been identified but the ORC-bound chromatin domains presumably function as replication origins. In early G_1 phase, replication licensing factors MCM2–7 were loaded onto chromatin (CHONG et al. 1995; KUBOTA et al. 1995; MADINE et al. 1995), and origins become competent at replication on formation of the pre-replicative complex, which consists at least of ORC1–6, MCM2–7, and Cdc6p (DIFFLEY et al. 1994; SANTOCANALE and DIFFLEY 1996; NEWLON 1997). These origins are not active in G_1 phase until Cdc45p is loaded to form the pre-initiation complex and the Cdc7p/Dbf4p kinase is activated at G_1/S boundary (BOUSSET and DIFFLEY 1998; DONALDSON et al. 1998; ZOU and STILLMAN 1998; KUMAGAI et al. 1999; ROBERTS et al. 1999; TAKEDA et al. 1999). Once origins fire in S phase, MCMs are released from chromatin and origins become incompetent again so that reinitiation of DNA replication is inhibited in the same S phase.

Loading of MCM proteins onto the cellular chromatin occurs in early G_1 phase, and this is essential for DNA replication in the next S phase. Thus, the cellular chromatin requires passage through early G_1 phase for the next round of DNA replication in S phase. If *oriP* is under control of the cellular licensing mechanism, *oriP* should have a similar requirement of early G_1 phase for DNA replication. To test this possibility, a series of transient replication experiments were performed using an *oriP*-containing plasmid (*oriP* plasmid) and HeLa/EB1 cells (SHIRAKATA et al. 1999). In the initial experiment, the *oriP* plasmid was transfected into HeLa/EB1 cells and then the cell-cycle progression was blocked at G_2/M phase with nocodazole immediately after transfection. With this nocodazole block protocol, G_1 and S cells proceeded through the initial S phase after transfection and stopped at G_2/M, but no cells were allowed to proceed through G_2/M and enter early G_1 phase after transfection. Therefore, if *oriP* required early G_1 phase for replication, the *oriP* plasmid transfected into G_1 and S cells would not replicate in the initial S phase, and if not, the plasmid could replicate in this initial S phase. The plasmids were effectively transfected into G_1 and S cells (41.0% of transfected cells were in G_0/G_1 and 34.5% were in S phase) by the calcium phosphate method, and most transfected cells (98%) were arrested at G_2/M phase within 24h by nocodazole. When replication was analyzed at 24h by the *Dpn*I/*Mbo*I method (DEPAMPHILIS 1995), replication of the *oriP* plasmid was not detected in the nocodazole-blocked cells, while replication of the plasmid was efficient in the unblocked cells. This result indicated that the *oriP* plasmid does not replicate in the initial S phase under G_2/M block. This was also confirmed by a bromodeoxyuridine (BRdU)-labeling experiment. Replication licensing does not regulate SV40 *ori* and the origin initiates multiple rounds of replication in a single S phase. SV40 *ori* does not require early G_1 phase for replication, and replicated in the initial S phase under G_2/M block in a similar nocodazole block experiment. When the cells were released from the G_2/M block, the *Dpn*I-resistant *oriP* plasmid (once-replicated) appeared by 24h after release. As the doubling time of these cells is about 24h, this result indicated that replication of the *oriP* plasmid occurred in the subsequent S phase after the cells proceeded though the initial G_2/M. In another experiment, the transfected cells were cultured under normal conditions for 24h and then the cell cycle was blocked at G_2/M. With this delayed G_2/M block protocol, most transfected cells underwent the initial G_2/M within 24h (one cell cycle) after transfection and then stopped at second G_2/M phase during the next 24h. If the *oriP* plasmid requires early G_1 phase for the next DNA replication, the plasmid should replicate in these cells only once. In fact, in this delayed G_2/M block experiment, the *Dpn*I-resistant *oriP* plasmid appeared within 24h after transfection and had increased in number at 48h, but no *Mbo*I-sensitive *oriP* plasmids (twice-replicated) were detected at 48h. Thus, like the cellular genome, the *oriP* plasmid requires the cell cycle window including early G_1 phase for the next round of replication in S phase.

3.2 The DS Element Was Responsible for Cell Cycle-Regulated Replication

The minimal region required for DNA replication is the DS element, but *oriP* could have additional regulatory elements that are important for cell cycle-regulated replication. To examine this possibility, the plasmid containing only the DS element, the DS plasmid, was analyzed by a similar transient replication assay under G_2/M block (SHIRAKATA et al. 1999). Like the *oriP* plasmid, the DS plasmid did not replicate in the initial S phase under G_2/M block. This result indicated that the DS region is responsible for the cell cycle-dependent regulation of replication. The replication-licensing event, if it is involved in the replication from *oriP*, should occur at the DS element.

3.3 Structural Changes Occurred in the *oriP* Minichromosome Between G_2/M and Late G_1

The requirement of early G_1 phase for replication from *oriP* suggested that some modifications or structural changes might occur in the circular minichromosome of the *oriP* plasmid (the *oriP* minichromosome) during this period of the cell cycle. An experiment was performed to detect any changes of the *oriP* minichromosome by sucrose density gradient centrifugation (SHIRAKATA et al. 1999). The *oriP* minichromosomes were prepared from cells transfected with the *oriP* plasmid. As an internal control, the SV40 *ori* plasmid was also transfected with the *oriP* plasmid and the SV40 minichromosomes were extracted from the same cells. The SV40 plasmid was chosen as the control because cellular replication licensing does not regulate the SV40 origin, as mentioned in Sect. 4.2. The mixture of the *oriP* minichromosome and the SV40 minichromosome was analyzed by sucrose density gradient centrifugation (15%–30%, 38,000rpm, 3h). Before extraction of minichromosomes, cells were arrested at G_2/M or late G_1 phase with nocodazole or mimosine so that the minichromosomes of G_2/M and those of late G_1 could be compared. When 11 fractions were collected in total, both the SV40 minichromosomes in G_2/M and those at late G_1 phases were collected mostly in fraction 6. Thus, there were no detectable differences in sedimentation velocity of the SV40 minichromosomes extracted from G_2/M and late G_1 cells. The *oriP* minichromosomes in G_2/M phase were collected in fraction 6, the same fraction in which the SV40 minichromosomes were detected. In contrast, the *oriP* minichromosomes in late G_1 migrated faster in a sucrose density gradient and were recovered mostly in fraction 4. The sedimentation velocity of the *oriP* minichromosome increased significantly between G_2/M and late G_1. Thus, this experiment revealed that the *oriP* minichromosome exists in two distinct states during the cell cycle, one formed in late G_1 and the other formed in G_2/M.

The result of sucrose density gradient centrifugation analysis suggested a large increase in molecular mass and/or structural change of the minichromosome. Nucleosome structure and supercoiling of the extracted minichromosomes of the *oriP* plasmid were analyzed, but there was no significant difference between G_2/M and late G_1. These results suggest that the conformation of the *oriP* minichromosome does not change, but molecular mass of the minichromosome increases during G_1 phase. Association of several replication factors, i.e., an MCM-hexamer complex (approx. 600kDa), may occur at early G_1 but it does not account for the difference in sedimentation velocity. Recent findings concerning eukaryotic DNA replication suggest that DNA replication may occur at particular locations in nuclei where the replication machinery forms a large factory, a replication factory (reviewed by COOK 1999). It is possible that in late G_1 phase several *oriP* minichromosomes may associate with a cellular replication factory in nuclei, and the extracted *oriP* minichromosome may contain a fragment of this large factory.

3.4 Association of MCM2 with the *oriP* Minichromosome

The requirement of early G_1 phase for the replication from *oriP* suggested that the MCM family of replication licensing factors may be loaded on the *oriP* minichromosome. Association of MCM2 and the *oriP* minichromosome was suggested by an experiment using an in situ cross-linking and chromatin immunoprecipitation method (SOLOMON et al. 1988, STRAHL-BOLSINGER et al. 1997) (Fig. 2). When the fixed chromatin of late G_1 cells was immunoprecipitated with anti-human MCM2 antibody, the *oriP* plasmid was precipitated. In contrast, the replication defective internal control, the *oriP*ΔDS plasmid, which was co-transfected with the *oriP* plasmid, was not precipitated by the antibody. This indicated

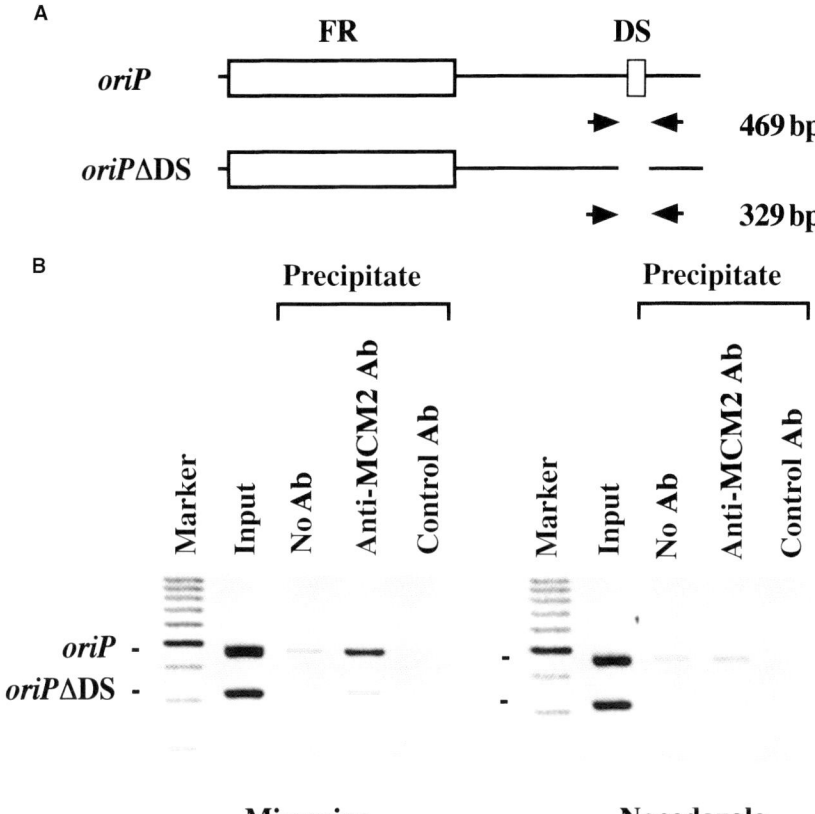

Fig. 2A,B. MCM2 associated with the *oriP* minichromosome in vivo. **A** The *oriP* plasmid and the *oriP*ΔDS plasmid. The same primer set used in the assay (*arrows*) produced the 469bp fragment from the *oriP* plasmid and the 329bp fragment from the *oriP*ΔDS plasmid. The *oriP*ΔDS plasmid did not replicate in cells because it lacked the DS region, the replication origin. **B** In situ cross-linking and chromatin immunoprecipitation analysis. The *oriP* plasmid and the *oriP*ΔDS plasmid were co-transfected into HeLa/EB1 cells and cultured for 2 days and then blocked with mimosine or nocodazole for 24h. The chromatin fragments (2μl) were prepared from the mimosine-treated (late G_1) and nocodazole-treated (G_2/M) cells and immunoprecipitated with anti-human MCM2 antibody or the control antibody (anti c-*fos*). DNA of the DS region in the precipitated chromatins was amplified by PCR using the primers described in **A** and electrophoresed in an agarose gel and stained with EtBr. DNA was also purified from the same amount of chromatin fragments without immunoprecipitation and amplified with PCR to show the amount of plasmid contained in the chromatin fragment sample (*input*). The gradation of gel images is inverted

that association of MCM2 with the minichromosome requires the DS element in the plasmid. Furthermore, the association of MCM2 with the *oriP* minichromosome is regulated by the cell cycle: it is detected at late G_1 but not at G_2/M phase. MCM proteins form a hexamer complex containing all of the MCM2–7 proteins and associate with the chromatin (THÖMMES et al. 1997; KUBOTA et al. 1995). However, associations of the other members of the MCM protein family have not yet been examined.

3.5 Interaction of EBNA1 and MCM Proteins on the DS Element

The requirement of the DS element for the loading of MCM2 on the *oriP* minichromosome suggests that the MCM proteins may associate with EBNA1. Yeast two-hybrid assay, however, did not reveal any interactions between EBNA1 and MCM proteins. As discussed in Sect. 2.2, a particular arrangement of EBNA1 proteins on the DS element is required for DNA replication (HARRISON et al. 1994). This suggests that MCM proteins may interact with EBNA1 only on the DS element. To analyze such interactions, a yeast one-hybrid system was used (Fig. 3). The reporter gene was the LacZ gene containing the DS element in the upstream regulatory sequence, pDSLacZi. When the GAL4AD-MCM5 hybrid protein and EBNA1 were co-expressed in yeast, the reporter gene was activated in a manner dependent on the DS element. Expression of either GAL4AD-MCM5 or EBNA1 alone did not activate the reporter gene. Co-expression of GAL4AD-MCM4 and EBNA1 also activates the reporter gene weakly. Although these interactions have to be confirmed by other methods, activation by MCM5 and MCM4 fusion proteins suggests that these MCM proteins may interact with EBNA1 on the DS element.

3.6 A Possible Role of EBNA1 in the Initiation of DNA Replication

The requirement of early G_1 phase for replication from *oriP*, association of MCM2 with the *oriP* minichromosome, and interaction of EBNA1 and MCM proteins on the DS element suggests that *oriP* is regulated by the cellular replication licensing mechanism. As loading of MCM2 on the *oriP* chromosome is dependent on the DS element and the MCM proteins associate with EBNA1 only on the DS element, EBNA1 may facilitate loading of MCMs onto the *oriP* minichromosome. For cellular chromatin, loading of MCM proteins occurs at the replication origins where the ORC proteins are bound. Like EBNA1 (HSIEH et al. 1993), the ORC proteins bind to a replication origin constitutively during the cell cycle and facilitate loading of the MCM proteins. EBNA1 may function as an ORC protein and form

Fig. 3A,B. EBNA1 associates with MCM5 and MCM4 on the DS element. Binding of EBNA1 to one of the human MCM proteins or human CDC6 protein was examined using the yeast one-hybrid assay. **A** Plasmids used for the one-hybrid assay. Plasmid pEBNA1d2 expressed the deletion mutant (aa 1–618) of EBNA1, and pGAD-MCM2–7 expressed the fusion protein of the GAL4 activator domain and one of the MCM proteins. Plasmid pGAD-CDC6 expressed the GAL4 fusion protein with human CDC6 protein. The pDSLacZi reporter plasmid contained the DS sequence upstream of LacZ gene. When the GAL4AD hybrid protein was bound to the EBNA1d2 protein in the DS region, it activated the LacZ gene. **B** Expression of β-galactosidase in the cells co-transfected with the EBNA1 expression plasmid (pEBNA1d2) and the MCM or CDC expression plasmid (pGAD-MCM, pGAD-CDC6). Plasmid pAS2–1 was the control vector without EBNAd2 coding sequence and expressed only the GAL4 DNA-binding domain. Plasmid pGAD10 was the control vector for pGAD plasmids. Levels of background expression of each reporter plasmids are indicated by *dotted lines*

Replication Licensing of the EBV *oriP* Minichromosome   23

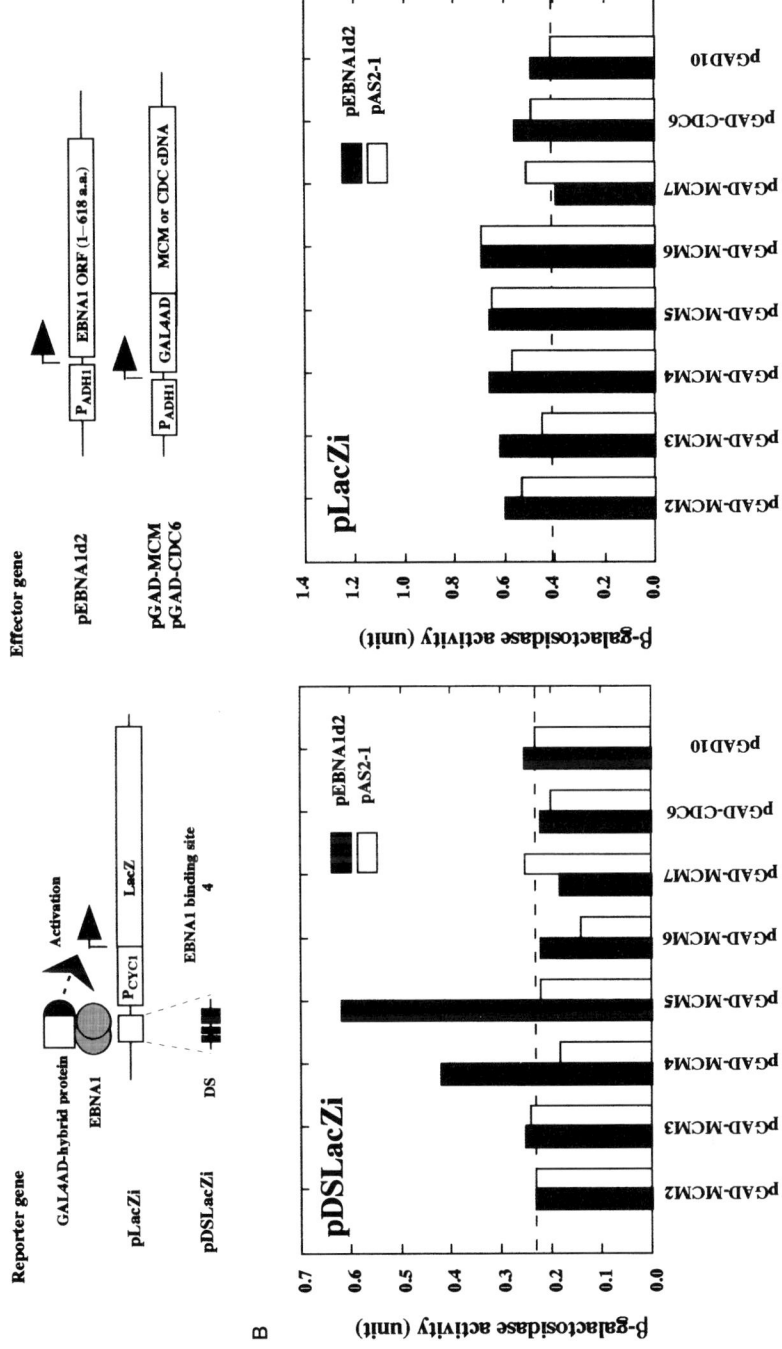

a pre-replicative complex at the DS element. It is also formally possible that the ORC binds at the DS element and EBNA1 facilitates its binding. However, this is unlikely to occur in cells, because genomic footprinting analysis did not reveal any strong protection near the DS element (NILLER et al. 1995), which is often observed at the ORC-binding site of the yeast cellular genome (BELL and STILLMAN 1992; DIFFLEY et al. 1994).

EBNA1 is essential for DNA replication from *oriP*, but it has none of the replication-related enzymatic activities, i.e. DNA helicase, polymerase, or ATPase (FRAPPIER and O'DONNELL 1991a; MIDDLETON and SUGDEN 1992). Besides its origin-binding activity, EBNA1 has been shown to induce distortion of the DNA structure at the binding site (FRAPPIER and O'DONNELL 1992; HSIEH et al. 1993) and form a DNA loop within *oriP* by the EBNA1-EBNA1 interaction (FRAPPIER and O'DONNELL 1991b; SU et al. 1991; MACKEY et al. 1995), and also bind to RNA (SNUDDEN et al. 1994) and EBP2, a cellular factor that may be important for chromosome segregation (SHIRE et al. 1999), but these functions are not associated directly with DNA replication. RP-A, a cellular single-stranded DNA binding protein, is a replication factor that associates with EBNA1 is (ZHANG et al. 1998). Like replication by large T antigen of SV40, this interaction may be involved in the process of DNA replication. However, loading of MCM proteins and formation of the pre-replicative complex could be a major function of EBNA1 in the initiation of DNA replication.

4 DS-Dependent Replication from *oriP* Is Suppressed in Several EBV-Infected Cell Lines

The *oriP* region was initially identified by screening for ARS activity (YATES et al. 1984), and the DS element was shown to function as a replication origin as discussed in Sect. 2. However, the DS element is not necessarily active as a replication origin in EBV-infected cells, and the DS-independent replication of the EBV genome was also reported. In the Burkitt's cell line Raji, DNA replication of the EBV genome initiates mainly at multiple sites in a region distant from *oriP*, meanwhile replication from the DS element is very rare (LITTLE and SCHILDKRAUT 1995). The occurrence of DS-independent replication of EBV DNA was also suggested by analysis of the EBV-transformed B-cell line X50-7, in which the EBV genome has a deletion of the DS element (YANDAVA and SPECK 1992). Furthermore, some plasmids containing *oriP* can replicate without the DS element in EBNA1-expressing 143B and 293 cells (KIRCHMAIER and SUGDEN 1998, AIYAR et al. 1998). Thus, in these cell lines, the DS element does not function as a replication origin and EBNA1 works only to prevent loss of the *oriP* plasmid (AIYAR et al. 1998). When the DS-independent replication occurs, some elements in the EBV genome and the plasmid work as a replication origin, and the origin function of the DS element is suppressed.

In contrast to these cell lines, DS-dependent replication did not occur above a detectable level in HeLa/EB1 cells when replication was analyzed with an *oriP* plasmid. Cell types in which the EBV genome or the *oriP* plasmid is present may determine the type of replication that can occur. Another possible explanation is that a specific size and structure of the plasmid may be required for a specific type of replication to occur. To examine these possibilities, long-term maintenance of an identical set of *oriP* plasmid and its deletion mutants was analyzed in HeLa and P3HR1 cells. About 20 copies of the *oriP* plasmid, p220.2 (provided by Dr. Sugden), were replicated and maintained in both HeLa and P3HR1. However, the mutated p220.2 plasmid containing only the FR element in the *oriP* region did not replicate in HeLa cells (< 1 copy per cell), but it replicated and was maintained efficiently in P3HR1 cells (170 copies per cell). The p220.2 plasmid was maintained by the DS-independent replication in P3HR1 but not in HeLa cells. These results suggested that DS-independent replication requires specific conditions in cells. This may involve expression of particular cellular factor(s) with or without mutations and the expression of viral factors. The involvement of EBNA1 in this DS-independent replication is not known, but the region that initiates this type of replication has no apparent EBNA-binding sites.

There are some possible explanations for the suppression of DS-dependent replication from *oriP* in several lymphoma cell lines. Transcription can interfere with DNA replication (HAASE et al. 1994; DESHPANDE and NEWLON 1996). Therefore, transcription occurring near the DS element may prevent DNA replication from the DS element. There are some TATA-box like sequences near the DS element that could act as transcriptional promoters. Formation of the transcriptional machinery near the DS element could interfere with formation of the replication machinery because of their very large size. Alternatively, EBNA1 may be modified, e.g. phosphorylated, in these cells and therefore may not initiate replication from the DS element. The DS-dependent replication is suppressed mostly in cultured cell lines derived from the EBV-positive Burkitt's lymphoma, in which the viral nuclear antigens (EBNA1–6) and the latent membrane proteins (LMP1, LMP2) are expressed in vitro. Therefore, it is likely that some of these virus-encoded proteins may be responsible for the suppression of DS-dependent replication. In contrast, the EBV-positive Burkitt's lymphoma and the circulating EBV-infected B cells express only EBNA1 and LMP2 in some cases (QU and ROWE 1992; TIERNEY et al. 1994; CHEN et al. 1995; THORLEY-LAWSON et al. 1996).

5 Stable Nuclear Retention and Partitioning of EBV Episomes During Cell Division

The copy number of latent EBV DNA may vary from one to several tens per cell in latently infected cell lines, whereas the number remains fairly constant in any specific cell line during multiple passage in culture. For example, the prototype BL cell line Raji harbors an average of 50 copies of latent EBV genome equivalents per cell, which are mostly circular plasmids. However, very little is known about the mechanism by which equal numbers of EBV episomes become segregated during mitosis and are deposited in each daughter cell. EBV minichromosomes associate with the cellular chromosomes during mitosis, possibly by EBNA1. Latent EBV DNA was shown by in situ hybridization to be randomly associated with metaphase chromosomes (HARRIS et al. 1985; SHIRAISHI et al. 1985; DELECLUSE et al. 1993). EBNA1 is also associated with metaphase chromosomes (REEDMAN and KLEIN 1973; GROGAN et al. 1983; PETTI et al. 1990).

5.1 Requirement of the FR for Stable Retention of *oriP* Plasmid

More than two EBNA1 binding sites on the FR are required for retention of *oriP*-containing plasmids (MIDDLETON and SUGDEN 1994). Plasmids carrying the FR but lacking the DS replicate very poorly but persist for a long time (more than 2 months) in cells expressing EBNA1 (REISMAN et al. 1985; KRYSAN et al. 1989). The plasmid containing only the DS element (the DS plasmid) replicates but is lost rapidly from HeLa/EB1 cells (28% loss per generation), meanwhile the *oriP* plasmid containing both DS and FR elements is retained in cells more efficiently (4% loss per generation) (SHIRAKATA et al. 1999). Rapid loss of plasmids was also observed in experiments using several cell lines (MIDDLETON and SUGDEN 1994). Thus, the FR element works to retain the EBV genome in nuclei, and EBNA1 is also required for this function. JANKELEVICH et al. (1992) demonstrated that the EBV chromosome is associated with nuclear matrix in the region of the *Bam*HI-C fragment that includes *oriP*. It is likely that the FR element is a matrix attachment region (MAR) and EBNA1 mediates binding between the FR element and the nuclear matrix structure.

5.2 EBNA1 Is Required for Nuclear Retention and Partitioning of *oriP* Plasmid

EBNA1 has been shown to play a role in efficient retention of the episomal EBV DNA as well as in its replication (RAWLINS et al. 1985; YATES et al. 1985) and EBNA1 binding to the FR element enables plasmids containing the element to segregate stably (REISMAN et al. 1985; KRYSAN et al. 1989). The requirement of EBNA1 for retention and segregation of latent EBV was also demonstrated using EBNA1-deficient mutants of EBV (LEE et al. 1999; SHIRE et al. 1999). Since EBNA1 is the sole viral gene product required for nuclear retention and segregation of the EBV plasmid, cellular factors that interact with EBNA1 may be required for plasmid maintenance. The human protein EBP2 that interacts with EBNA1 was isolated by a yeast two-hybrid system (SHIRE et al. 1999). Although the function of EBP2 is not known, it is a component of the nucleoli of proliferating human cells (CHATTERJEE et al. 1987). EBP2 was shown to interact with aa 330–386 of EBNA1.

This region includes the Gly-Ala repeats and DNA linking domains that mediate interactions between EBNA1 molecules bound to distant DNA binding sites for DNA looping (GOLDSMITH et al. 1993; LAINE et al. 1995; MACKEY et al. 1995). EBNA1 mutants that removed all of the Gly-Arg-rich region abrogated the plasmid maintenance function in human cells (WENDELBURG and VOS 1998; SHIRE et al. 1999).

Circular yeast artificial chromosomes containing *oriP* were also shown to bind to chromosomes in human cells expressing EBNA1 (SIMPSON et al. 1996). Therefore, EBNA1 may facilitate the interaction of episomal EBV on human cellular chromosomes. The domains that mediate binding to chromosomes have been identified in three independent regions of EBNA1 (Fig. 4, MARECHAL et al. 1999).

5.3 Analysis of Latent EBV State by Fluorescence In Situ Hybridization

Recently, the introduction of fluorescence in situ hybridization (FISH) for detection of latent EBV DNA has provided a unique tool for detection of the integrated and episomal forms. When the latent state of EBV DNA is examined by FISH, it is assumed that the integrated copies will be detected as the same symmetrical doublet signals at homologous sites of both chromatids. In contrast, the signals of the episomal copies are assumed to be distributed randomly over the host chromosomes. FISH demonstrated that several B-cell lines contained only integrated EBV DNA (HENDERSON et al. 1983; LAWRENCE et al. 1988; HURLEY et al. 1991; HIRAI et al. 1993, 1998), while coexistence of episomal and integrated EBV DNAs was observed in other B-cell lines (HURLEY et al. 1991; DELECLUSE et al. 1993). Integration of EBV DNA into the host chromosomal DNA may occur through the terminal repeat (TR) region (MATSUO et al. 1984; GULLEY et al. 1992; KLIPALANI-

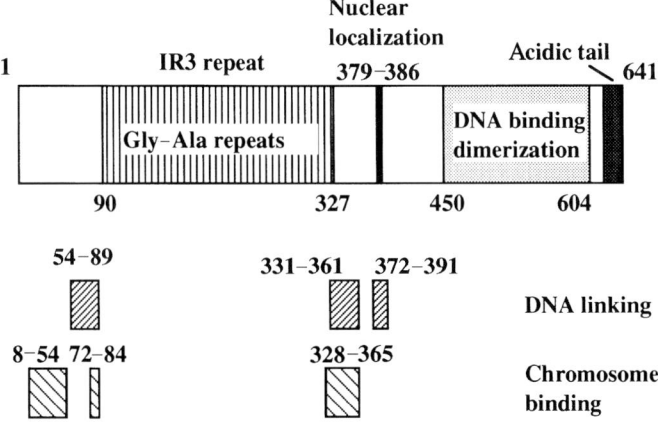

Fig. 4. Functional domains of EBNA1

JOSHI and LAW 1994) or the unique sequence region U1 adjacent to the left terminal end of the EBV genome (HURLEY et al. 1991; GUALANDI et al. 1992; YANDVA and SPECK 1992; HIRAI et al. 1993). The presence of full-sized integrated EBV DNA in the Namalwa cell line was demonstrated by molecular cloning of the cell-virus DNA junction fragments from restriction DNA fragments of cellular DNA (HENDERSON et al. 1983; MATSUO et al. 1984). Although the integration site does not appear to occur at a preferential site in human chromosomes, the integration may affect viral and cellular gene expression, possibly resulting in oncogenic transformation. However, the biological significance of the integrated form remains to be determined.

As stated above, the copy number of latent EBV DNA in the Burkitt's lymphoma cell line Raji remains constant during multiple passage in culture. However, the number by in situ hybridization was shown to be variable at the single-cell (DELECLUSE et al. 1993; HIRAI et al. 1998). Chromosomal denaturation required for in situ hybridization may affect the stability of episomal EBV DNA on chromosomes. Especially, the copy number at the single-cell level may be dependent on the denaturation temperature used for in situ hybridization, as suggested by HARRIS et al. (1985). Circular episomal DNA might be unstably associated with host chromosomes by the non-covalent linkage, and higher temperatures would be required for denaturation compared with the integrated DNA covalently linked to that of host cell. Thus, the increased temperature required for chromosomal denaturation resulted in release of the episomal molecules from host chromosomes, whereas the symmetrical doublet signals increased with decreases in temperature (HIRAI et al. 1998). Since the symmetrical doublet signals at the decreased temperature were weaker than the strong symmetrical doublet signals characteristic of the integrated form, the increased symmetrical doublet signals may represent the episomal form. This result indicated that the integrated form of latent EBV genome should not be determined only by FISH analysis. Molecular cloning of the cell–virus DNA junction fragments is also needed to define the integrated form. Thus, the episomal copy as well as the integrated EBV DNA would be present on both sister chromatids at the same location in every cell so that the replicated episomal copies as well as the integrated copies may be equally segregated to daughter cells. However, much more research is needed to further define the stability of the EBV episomal state on host chromosomes and to confirm the above hypothesis because interpretation of the FISH results is still very difficult.

6 Summary

The latent EBV genome may persist in the integrated form as well as the circular episomal form. However, most of the latent viral DNA molecules are known to exist in the circular episomal form, which binds to host chromosomes during mitosis. The DS element of *oriP* in the circular episomal DNA functions as a replication origin. As it replicates once in a single S phase, it is possible that *oriP* is regulated by the cellular replication licensing mechanism including the MCM family of replication licensing factors. Transient replication analysis using the *oriP* plasmid and HeLa/EB1 cells revealed that the DS element requires early G_1 phase for the next round of replication, the same cell-cycle window in which the replication licensing of cellular chromatin occurs. After this phase, the sedimentation velocity of the *oriP* minichromosome increases. MCM2 associates with the *oriP* minichromosome at late G_1 but not at G_2/M, and this association requires the DS element in the plasmid. The interaction of EBNA1 and the MCM proteins on the DS element was also suggested. These results suggested that the cellular licensing mechanism controls the replication from *oriP*. This also suggested a similarity in the replication machinery of the cellular chromatin and the latent EBV genome. In addition to DS-dependent replication, the EBV genome replicates in a manner independent of the DS element in several cultured cell lines. The DS-dependent replication is likely to be suppressed in these cell lines by the expression of other viral proteins. In contrast, EBV-positive Burkitt's lymphoma and circulating EBV-infected B cells express only EBNA1 or both EBNA1 and LMP2. DS-dependent replication may play a major role in these EBNA1-only cells, and the licensing regulation of *oriP* is important for maintenance of the EBV genome during this latent period of the viral life cycle. EBNA1 is required for efficient nuclear retention and partitioning of *oriP*-carrying plasmid by its binding to the FR element, thus providing stable persistence of the latent EBV genome during cell division. The copy number of latent EBV DNA molecules in B-cell lines remains fairly constant during multiple passage in culture. However, very little is known about the mechanism by which the viral DNA molecules are equally segregated into daughter cells. To understand the mechanisms responsible for stable nuclear retention and partitioning of the latent viral genome, it is essential to analyze the episomal and integrated viral DNAs at a single-cell level by FISH and other techniques.

Acknowledgements. This work was supported by a Grant-in-Aid for Scientific Research and a Grant-in-Aid for Scientific Research in Priority Areas from the Ministry of Education, Science, Sports, and Culture of Japan and by The Japan Health Science Foundation.

References

Adams A (1987) Replication of latent Epstein-Barr virus genomes in Raji cells. J Virol 61:1743–1746
Aiyar A, Tyree C, Sugden B (1998) The plasmid replicon of EBV consists of multiple cis-acting elements that facilitate DNA synthesis by the cell and a viral maintenance element. EMBO J 17:6394–6403
Ambinder RF, Mullen MA, Chang YN, Hayward GS, Hayward SD (1991) Functional domains of Epstein-Barr virus nuclear antigen EBNA-1. J Virol 65:1466–1478
Avolio-Hunter TM, Frappier L (1998) Mechanistic studies on the DNA linking activity of Epstein-Barr nuclear antigen 1. Nucleic Acids Res 26:4462–4470
Babcock GJ, Decker LL, Volk M, Thorley Lawson DA (1998) EBV persistence in memory B cell in vivo. Immunity 9:395–404

Bell SP, Stillman B (1992) Nucleotide dependent recognition of chromosomal origins of DNA replication by a multi-protein complex. Nature 357:128–134

Blow JJ, Laskey RA (1988) A role for the nuclear envelope in controlling DNA replication within the cell cycle. Nature 332:546–548

Bochkarev A, Barwell JA, Pfuetzner RA, Furey WJr, Edwards AM, Frappier L (1995) Crystal structure of the DNA-binding domain of the Epstein-Barr virus origin-binding protein EBNA1. Cell 83:39–46

Bochkarev A, Barwell JA, Pfuetzner RA, Bochkareva E, Frappier L, Edwards AM (1996) Crystal structure of the DNA-binding domain of the Epstein-Barr virus origin-binding protein, EBNA1, bound to DNA. Cell 84:791–800

Bousset K, Diffley JF (1998) The Cdc7 protein kinase is required for origin firing during S phase. Genes Dev 12:480–490

Brewer BJ, Fangman WL (1987) The localization of replication origins on ARS plasmids in S. cerevisiae. Cell 51:463–471

Chatterjee A, Freeman JW, Busch H (1987) Identification and partial characterization of a Mr 40,000 nucleolar antigen associated with cell proliferation. Cancer Res 47:1123–1129

Chen MR, Middeldorp JM, Hayward SD (1993) Separation of the complex DNA binding domain of EBNA-1 into DNA recognition and dimerization subdomains of novel structure. J Virol 67:4875–4885

Chen F, Zou JZ, di Renzo L, Winberg G, Hu LF, Klein E, Klein G, Ernberg I (1995) A subpopulation of normal B cells latently infected with Epstein-Barr virus resembles Burkitt lymphoma cells in expressing EBNA-1 but not EBNA-2 or LMP1. J Virol 69:3752–3758

Chong JPJ, Mahbubani HM, Khoo CY, Blow JJ (1995) Purification of an MCM-containing complex as a component of the DNA replication licensing system. Nature 375:418–421

Cook PR (1999) The organization of replication and transcription. Science 284:1790–1795

Decker LL, Klaman LD, Thorley-Lawson DA (1996) Detection of the latent form of Epstein-Barr virus DNA in the peripheral blood of healthy individulas. J Virol 70:3286–3289

Delecluse EH, Bartnike S, Hammerschmidt W, Bullerdiek J, Bornkamm GW (1993) Episomal and integrated copies of Epstein-Barr virus coexist in Burkitt lymphoma cell lines. J Virol 67:1292–1299

DePamphilis ML (1995) Specific labeling of newly replicated DNA. Methods in Enzymology 262:628–669

Deshpande AM, Newlon CS (1996) DNA replication fork pause sites dependent on transcription. Science 272:1030–1033

Diffley JF, Cocker JH, Dowell SJ, Rowley A (1994) Two steps in the assembly of complexes at yeast replication origins in vivo. Cell 78:303–316

Donaldson AD, FangmanWL, Brewer BJ (1998) Cdc7 is required throughout the yeast S phase to activate replication origins. Genes Dev 12:491–501

Frappier L, O'Donnell M (1991a) Overproduction, purification, and characterization of EBNA-1, the origin binding protein of Epstein-Barr virus. J Biol Chem 266:7819–7826

Frappier L, O'Donnell M (1991b) Epstein-Barr nuclear antigen-1 mediates a DNA loop within the latent replication origin of Epstein-Barr virus. Proc Natl Acad Sci USA 88:10875–10879

Frappier L, O'Donnell M (1992) EBNA1 distorts oriP, the Epstein-Barr virus latent replication origin. J Virol 66:1786–1790

Gahn TA, Schildkraut CL (1989) The Epstein-Barr virus origin of plasmid replication, oriP, contains both the initiation and termination sites of DNA replication. Cell 58:527–535

Goldsmith K, Bendell L, Frappier L (1993) Identification of EBNA1 amino acid sequences required for the interaction of the functional elements of the Epstein-Barr virus latent origin of DNA replication. J Virol 67:3418–3426

Grogan EA, Summers WP, Dowling S, Shedd D, Gradoville L, Miller G (1983) Two Epstein-Barr viral nuclear neoantigens distinguished by gene transfer, serology, and chromosome binding. Proc Natl Acad Sci USA 80:7650–7653

Gualandi G, Santolini E, Calef E (1992) Epstein-Barr virus DNA recombines via latent origin of replication with the human genome in the lymphoblastoid cell line RGN1. J Virol 66:5677–5681

Gulley ML, Raphael M, Lutz CT, Ross DW, Raab-Traub N (1992) Epstein-Barr virus integration in human lymphomas and lymphoid cell lines. Cancer 70:185–191

Haase SB, Heinzel SS, Calos MP (1994) Transcription inhibits the replication of autonomously replicating plasmids in human cells. Mol Cell Biol 14:2516–2524

Hammerschmidt W, Sugden B (1988) Identification and characterization of orilyt, a lytic origin of DNA replication of Epstein-Barr virus. Cell 55:427–433

Hamper B, Tanaka A, Nonoyama M, Derge JG (1974) Replication of the resident repressed Epstein-Barr virus genome during the early S-phase (S-1 period) of nonproducer Raji cells. Proc Natl Acad Sci USA 71:631–635

Harris A, Young BD, Griffin BE (1985) Random association of Epstein-Barr virus genomes with host cell metapahase chromosomes in Burkitt's lymphoma-derived cell lines. J Virol 56:328–332

Harrison S, Fisenne K, Hearing J (1994) Sequence requirements of the Epstein-Barr virus latent origin of DNA replication. J Virol 68:1913–1925

Hassell JA, Briton BT (1996) SV40 and polyomavirus DNA replication. In: DePamphilis ML (ed) DNA replication in eukaryotic cells. Cold Spring Harbor Laboratory Press, Cold Spring Harbor

Heller M, Henderson A, Kieff E (1982) Repeat array in Epstein-Barr virus DNA is related to cell DNA sequences interspersed on human chromosomes. Proc Natl Acad Sci USA 79:5916–5920

Henderson A, Ripley S, Heller M, Kieff E (1983) Chromosome site for Epstein-Barr virus in a Burkitt tumor cell line and in lymphocytes growth-transformed in vitro. Proc Natl Acad Sci USA 80:1987–1991

Hennessy K, Heller M, van Santan V, Kieff E (1983) simple repeat array in Epstein-Barr virus DNA encodes part of the Epstein-Barr nuclear antigen. Science 220:1396–1398

Hirai K, Hironaka T, Yamamoto T (1993) Structure and expression of standard Epstein-Barr virus DNA in a P3HR-1 subclone containing a single viral genome equivalent per cell. In: Tursz T, Pagano JS, Albashi DV, de The G, Lenoir G, Pearson GR (eds) The Epstein-Barr virus and associated diseases. Colloque INSERM/John Libbey Eurotext Ltd, Vol 225, pp 289–294

Hirai K, Yamamoto T, Hironaka T, Arai T, Takeuchi H, Kobayashi R, Hojo I, Oishi T (1998) Epstein-Barr virus genome in infected cells and cancer. In: Osato T, Takada K, Tokunaga M (eds) Epstein-Barr virus and human cancer. Monograph on cancer research No.45, Japan Scientific Societies Press Karger, pp 29–39

Hsieh DJ, Camiolo SM, Yates J (1993) Constitutive binding of EBNA1 protein to the Epstein-Barr virus replication origin, $oriP$, with distortion of DNA structure during latent infection. EMBO J 12:4933–4944

Hurley E, Klaman LD, Agger S, Lawrence JB, Thorley-Lawson DA (1991) The prototypical Epstein-Barr virus-transformed lymphoblastoid cell line IB4 is an unusual variant containing integrated but no episomal viral DNA. J Virol 65:3958–5963

Ikeda T, Kobayashi R, Horiuchi M, Nagata Y, Hasegawa M, Mizuno F, Hirai K (2000) Detection of lymphocytes productively infected with Epstein-Barr Virus in non-neoplastic tonsils. J Gen Virol in press

Jacob F, Wollman E (1961) Sexuality and the genetics of bacteria. Academic Press

Jankelevich S, Kolman JL, Bodnar JW, Miller G (1992) A nuclear matrix attachment region organizes the Epstein-Barr viral plasmid in Raji cells into a single DNA domain. EMBO J 11:1165–1176

Kirchmaier AL, Sugden B (1998) Rep*: a viral element that can partially replace the origin of plasmid DNA synthesis of Epstein-Barr virus. J Virol 72:4657–4666

Klipalani-Joshi S, Law HY (1994) Identification of integrated Epstein-Barr virus in nasopharyngeal carcinoma using pulse field gel electrophoresis. Int J Cancer 56:187–192

Kobayashi R, Takeuchi H, Sasaki M, Hasegawa M, Hirai K (1998) Detection of Epstein-Barr Virus infection in the epithelial cells and lymphocytes of non-neoplastic tonsils by in situ hybridization and in situ PCR. Arch Virol 143:803–813

Krysan PJ, Haase SB, Calos MP (1989) Isolation of human sequences that replicate autonomously in human cells. Mol Cell Biol 9:1026–1033

Kubota Y, Mimura S, Nishimoto SI, Takisawa H, Nojima H (1995) Identification of the yeast MCM3-related protein as a component of Xenopus DNA replication licensing factor. Cell 81:601–609

Kumagai H, Sato N, Yamada M, Mahony D, Seghezzi W, Lees E, Arai K, Masai H (1999) A novel growth- and cell cycle-regulated protein, ASK, activates human Cdc7-related kinase and is essential for G1/S transition in mammalian cells. Mol Cell Biol 19:5083–5095

Laine A, Frappier L (1995) Identification of Epstein-Barr nuclear antigen 1 protein domains that direct interactions at a distance between DNA-bound proteins J Biol Chem 270:30914–30918

Lawrence JB, Villnave CA, Singer RH (1988) Sensitive, high-resolution chromatin and chromosome mapping in situ: presence and orientation of two closely integrated copies of EBV in a lymphoma line. Cell 52:51–61

Lee MA, Diamond ME, Yates JL (1999) Genetic evidence that EBNA-1 is needed for efficient, stable latent infection by Epstein-Barr virus. J Virol 73:2974–2982

Levitskaya J, Coram M, Levitsky V, Imreh S, Steigerwald-Mullen PM, Klein G, Kurilla MG, Masucci MG (1995) Inhibition of antigen processing by the repeat region of the Epstein-Barr nuclear antigen 1. Nature 375:685–688

Little R, Schildkraut C (1995) Initiation of latent DNA replication in the Epstein-Barr virus genome can occur at sites other than the genetically defined origin. Mol Biol Cell 5:2893–2903

Lupton S, Levine AJ (1985) Mapping genetic elements of Epstein-Barr virus that facilitate extrachromosomal persistence of Epstein-Barr virus derived plasmids in human cells. Mol Cel Biol 5: 2533–2542

Mackey D, Middleton T, Sugden B (1995) Multiple regions within EBNA1 can link DNAs. J Virol 69:6199–6208

Madine MA, Khoo CY, Milles AD, Laskey RA (1995) MCM3 complex required for cell cycle regulation of DNA replication in vertebrate cells. Nature 375:421–424

Marechal V, Dehee A, Chikhi-Brachet R, Piolot T, Coppey-Moisan M, Nicolas JC (1999) Mapping EBNA-1 domains involved in binding to metaphase chromosomes. J Virol 73:4385–4392

Matsuo T, Heller M, Petti L, Oshiro E, Kieff E (1984) Persistence of the entire EBV genome integrated into human lymphocyte DNA. Science 226:1322–1325

Middleton T, Sugden B (1992) EBNA1 can link the enhancer element to the initiator element of the Epstein-Barr virus plasmid origin of DNA replication. J Virol 66:489–495

Middleton T, Sugden B (1994) Retention of plasmid DNA in mammalian cells is enhanced by binding of the Epstein-Barr virus replication protein. J Virol 68:4067–4071

Miyashita EM, Yang B, Lam KMC, Crawford DC, Thorley-Lawson DA (1995) A novel form of Epstein-Barr virus latency in normal B cells in vivo. Cell 80:593–601

Newlon CS (1997) Putting it all together: building a rereplicative complex. Cell 91:717–720

Niller HH, Glaser G, Knuchel R, Wolf H (1995) Nucleoprotein complexes and DNA 5′-ends at *oriP* of Epstein-Barr virus. J Biol Chem 270:12864–12868

Petti L, Sample C, Kieff E (1990) Subnucler localization and phosphorylation or Epstein-Barr virus latent infection nuclear proteins. Virology 176:563–574

Polvino-Bodnar M, Schaffer PA (1992) Mutational analysis of Epstein-Barr virus nuclear antigen (EBNA-1). Nucleic Acids Res 16:3415–3435

Qu L, Rowe DT (1992) Epstein-Barr virus latent gene expression in cultured peripheral blood lymphocytes. J Virol 66:3715–3724

Rawlins DR, Milman G, Hayward SD, Hayward GS (1985) Sequence-specific DNA binding of the Epstein-Barr virus nuclear antigen (EBNA-1) to clustered sites in the plasmid maintenance region Cell 42:859–868

Reedman BM, Klein G (1973) Cellular localization of an Epstein-Barr virus (EBV)-associated complement-fixing antigen in producer and non-producer lymphoblastoid cell lines. Int J Cancer 11:499–520

Reisman D, Yates J, Sugden B (1985) A putative origin of replication of plasmids derived from Epstein-Barr virus is composed of two cis-acting components. Mol Biol Cell 5:1822–1832

Roberts BT, Ying CY, Gautier J, Maller JL (1999) DNA replication in vertebrates requires a homolog of the Cdc7 protein kinase. Proc Natl Acad Sci USA 96:2800–2804

Santocanale C, Diffley JF (1996) ORC- and Cdc6-dependent complexes at active and inactive chromosomal replication origins in Saccharomyces cerevisiae. EMBO J 15:6671–6679

Shiraishi Y, Taguchi T, Ohta Y, Hirai K (1985) Chromosomal localization of the Epstein-Barr virus (EBV) genome in Bloom's syndrome B-lymphoblastoid cell lines transformed with EBV. Chromosoma 93:157–164

Shirakata M, Hirai K (1998) Identification of minimal *oriP* of Epstein-Barr virus required for DNA replication. J Biochem 123:175–181

Shirakata M, Imadome K, Hirai K (1999) Requirement of replication licensing for the dyad symmetry element-dependent replication of the EBV *oriP* minichromosome. Virology 263:42–54

Shire K, Ceccarelli DFJ, Avolio-Hunter TM, Frappier L (1999) EBP2, a human protein that interacts with sequences of the Epstein-Barr virus nuclear antigen 1 important for plasmid maintenance. J Virol 73:2587–2595

Simpson K, McGuigan A, Huxley C (1996) Stable episome maintenance of yeast chromosomes in human cells. Mol Cell Biol 16:5117–5126

Snudden DE, Hearing J, Smith PR, Grasser FA, Griffin BE (1994) EBNA1, the major nuclear antigen of Epstein-Barr virus, resembles 'RGG' RNA binding proteins. EMBO J 13:4840–4847

Solomon MJ, Larsen PL, Varshavsky A (1988) Mapping protein-DNA interactions in vivo with formaldehyde: evidence that histone H4 is retained on a highly transcribed gene. Cell 53:937–947

Stenlund A (1996) Papillomavirus DNA replication. In: DePamphilis ML (ed) DNA replication in eukaryotic cells. Cold Spring Harbor Laboratory Press, Cold Spring Harbor

Strahl-Bolsinger S, Hecht A, Luo K, Grunstein M (1997) SIR2 and SIR4 interactions differ in core and extended telomeric heterochromatin in yeast. Gene Dev 11:83–93

Su W, Middleton T, Sugden B, Echols H (1991) DNA looping between the origin of replication of Epstein-Barr virus and its enhancer site: Stabilization of an origin complex with Epstein-Barr nuclear antigen 1. Proc Natl Acad Sci USA 88:10870–10874

Takeda T, Ogino K, Matsui E, Cho MK, Kumagai H, Miyake T, Arai K, Masai H (1999) A fission yeast gene, him1(+)/dfp1(+), encoding a regulatory subunit for Hsk1 kinase, plays essential roles in S-phase initiation as well as in S-phase checkpoint control and recovery from DNA damage. Mol Cell Biol 19:5535–5547

Thömmes P, Kubota Y, Takisawa H, Blow JJ (1997) The RLF-M component of the replication licensing system forms complexes containing all six MCM/P1 polypeptides. EMBO J 16:3312–3319

Thorley-Lawson DA, Miyashita EM, Kahn G (1996) Epstein-Barr virus and the B cell: that all it takes. Trends Microbiol 4:204–208

Tierney RJ, Steven N, Young LS, Rickinson AB (1994) Epstein-Barr virus latency in blood mononuclear cells: Analysis of viral gene transcription during primary infection and in the carrier state. J Virol 68:7374–7385

Wendelburg BJ, Vos JM (1998) An enhanced EBNA1 variant with reduced IR3 domain for long-term episomal maintenance and transgene expression of oriP-based plasmids in human cells. Gene Ther 5:1389–1399

Wysokenski DA, Yates J (1989) Multiple EBNA-1-binding sites are required to form an EBNA1-dependent enhancer and to activate a minimal replicative origin within *oriP* of Epstein-Barr virus. J Virol 63:2657–2666

Yandava CN, Speck SH (1992) Characterization of the deletion and rearrangement in the BamHI C region of the X50-7 Epstein-Barr virus genome, a mutant viral strain which exhibits constitutive BamHI W promoter activity. J Virol 66:5646–5650

Yates JL, Warren N, Reisman D, Sugden B (1984) A cis-acting element from the Epstein-Barr viral genome that permits stable replication of recombinant plasmids in latently infected cells. Proc Natl Acad Sci USA 81:3806–3810

Yates JL, Wallen N, Sugden B (1985) Stable replication of plasmids derived from Epstein-Barr virus in various mammalian cells. Nature 313:812–815

Yates JL, Guan N (1991) Epstein-Barr virus-derived plasmids replicate only once per cell cycle and are not amplified after entry into cells. J Virol 65:483–488

Yates JL (1996) Epstein-Barr virus DNA replication. In: DePamphilis ML (ed) DNA replication in eukaryotic cells. Cold Spring Harbor Laboratory Press, Cold Spring Harbor

Zhang D, Frappier L, Gibbs E, Hurwitz J, O'Donnell M (1998) Human RPA (hSSB) interacts with EBNA1, the latent origin binding protein of Epstein-Barr virus. Nuc Acids Res 26:631–637

Zou L, Stillman B (1998) Formation of a preinitiation complex by S-phase cyclin-dependent loading of Cdc45p onto chromatin. Science 280:593–596

Epstein-Barr Virus Nuclear Protein 2-Induced Activation of the EBV-Replicative Cycle in Akata Cells: Analysis by Tetracycline-Regulated Expression

S. FUJIWARA

1	Introduction.	35
2	EBNA2 Is a Transcription Factor Essential for EBV-Induced Immortalization of B Lymphocytes	36
3	EBNA2 Functions as a Transcription Factor Through Interaction with Cellular Proteins	37
4	The Effects of EBNA2 on Cellular Phenotype.	39
5	Expression of EBNA2 in Akata Cells by Gene Transfer	40
5.1	Burkitt's Lymphoma Line Akata	40
5.2	Expression of EBNA2 by a Constitutive Expression Vector	41
5.3	Expression of EBNA2 by an Inducible Expression Vector	42
5.4	Discussion.	45
6	Concluding Remarks.	47
	References	47

1 Introduction

Epstein-Barr virus has a unique ability to transform human B cells into activated lymphoblasts and establish lymphoblastoid cell lines (LCLs) with infinite proliferative potential in vitro (for review, see KIEFF 1996). This process, termed growth transformation or immortalization, is supposed to be a prerequisite for the virus' ability to establish life-long persistent infection in humans. Since similar signal pathways are activated and similar genes are induced in both EBV-mediated immortalization and antigen-induced activation, it is generally considered that EBV utilizes the normal mechanism of antigen-induced activation to immortalize B lymphocytes. In EBV-immortalized lymphoblastoid cells, viral DNA persists as circular episomes, and 11 viral genes are expressed. This program of EBV gene expression is called latency III (for review, see RICKINSON and KIEFF 1996). Six of the eleven genes code for EBV nuclear proteins (EBNAs 1, 2, 3A, 3B, 3C, and LP) and three for latent membrane proteins (LMPs 1, 2A, and 2B). The remaining two

---

Department of Immunology and Microbiology, Nihon University School of Medicine, Oyaguchikami-machi, Itabashi-ku, Tokyo 173-8610, Japan

genes code for the EBV-encoded small RNAs (EBERs 1 and 2), that are not translated into proteins. In addition, a family of extensively spliced messenger (m)RNAs containing the BARF0 open reading frame are transcribed from a region including the *Bam*HI A fragment, but their protein products have not been definitely identified. Genetic analyses using recombinant EBVs so far showed that EBNA2, EBNA3A, EBNA3C, and LMP1 are essential for lymphocyte immortalization (KIEFF 1996). EBNA1 is required for intracellular persistence of EBV DNA in a plasmid form.

EBV nuclear protein 2 (EBNA2) is one of the six nuclear proteins expressed in LCLs and has a pivotal role in lymphocyte immortalization (HAMMERSCHMIDT and SUGDEN 1989; COHEN et al. 1989). EBNA2 transactivates certain viral and cellular genes that, in concert, cause the uncontrolled proliferation and the characteristic activated phenotype of LCLs (AMAN et al. 1990; CORDIER et al. 1990; PENG and LUNDGREN 1992; KEMPKES et al. 1995). In the following discussion, recent advances in the analysis of EBNA2 function are reviewed first, and then the author's own results of EBNA2 gene transfer into Burkitt's lymphoma (BL) Akata cells, revealing new aspects of EBNA2 function, are described.

EBV isolates can be divided into two types, 1 and 2, by sequence divergence in several genes including EBNA2. Only 53% identity in the predicted amino acid sequence is found between type-1 and type-2 EBNA2s. Type-2 EBV has less efficiency of lymphocyte immortalization and this is ascribed to the functional differences in EBNA2 (COHEN et al. 1989; WANG et al. 1990a). The properties of EBNA2 reviewed below will apply to that encoded by the type-1 EBV.

2 EBNA2 Is a Transcription Factor Essential for EBV-Induced Immortalization of B Lymphocytes

EBNA2 encoded by a prototype EBV strain, B95-8, is composed of 487 amino acid residues and has an apparent molecular weight of 85kDa in sodium dodecyl sulfate (SDS) -polyacrylamide gel electrophoresis. EBNA2 is encoded by the open reading frame BYRF1 residing in the *Bam*HI YH region of the EBV genomic DNA. The importance of this region of EBV DNA in lymphocyte immortalization was initially suggested by properties of a laboratory EBV mutant, P3HR-1. The P3HR-1 virus lacked the ability to immortalize lymphocytes and had a deletion in this particular region of the genome. Subsequent analyses identified two open reading frames, coding for EBNA-LP and EBNA2 (HENNESSY and KIEFF 1985; MUELLER-LANTZSCH et al. 1985; ROWE et al. 1985; RYMO et al. 1985), in this region and showed further that the deficiency in the EBNA2 coding capacity was responsible for the loss of immortalizing ability (COHEN et al. 1989; HAMMERSCHMIDT and SUGDEN 1989). Although EBNA-LP was found dispensable for immortalization, the efficiency of immortalization was severely impaired in EBNA-LP-deficient recombinant viruses, indicating that this protein also plays an important role in immortalization (MANNICK et al. 1991). EBNA2, together with EBNA-LP, is one of the first EBV proteins to be synthesized after EBV infection of B lymphocytes (ALLDAY et al. 1989; ALFIERI et al. 1991).

EBNA2 acts as a key regulator of latent EBV gene expression in LCLs through transactivation of the Cp EBNA promoter (SUNG et al. 1991; WALLS and PERRICAUDET 1991; JIN et al. 1992) and the LMP promoters (ABBOT et al. 1990; FAHRAEUS et al. 1990; WANG et al. 1990b; ZIMBER-STROBL et al. 1991) (Fig. 1). Cp is responsible for transcription of a family of extensively spliced mRNAs coding for all six EBNAs, whereas the LMP promoters direct transcription of mRNAs coding for all three LMPs. EBNA2 also transactivates cellular genes, including CD21 (WANG et al. 1990a), CD23 (WANG et al. 1987), c-*fgr* (KNUTSON 1990), and c-*myc* (KAISER et al. 1999). Because CD21, CD23, and c-*myc* are supposed to have important roles in antigen-induced B-cell activation, they may have critical functions also in EBV-induced immortalization. However, the exact roles of EBNA2 in immortalization have not been fully elucidated, and it is very likely that EBNA2 transactivates more cellular genes than is known at present.

3 EBNA2 Functions as a Transcription Factor Through Interaction with Cellular Proteins

EBNA2 does not bind DNA directly. In several promoters including Cp, the LMP promoters, the CD23 promoter, and probably also in the CD21 promoter, its

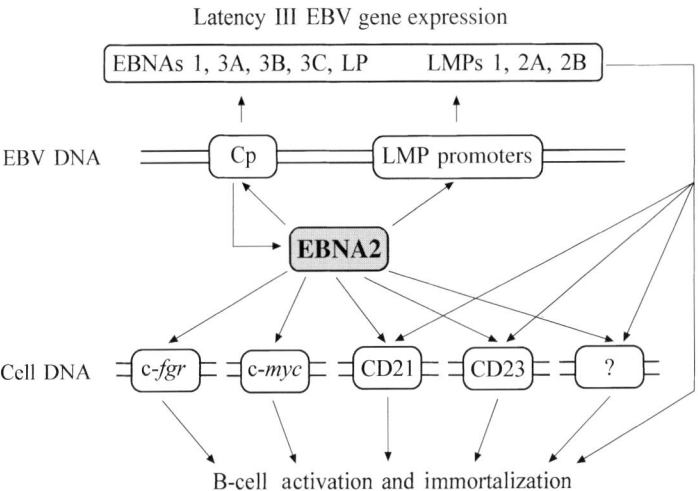

Fig. 1. Schematic representation of the roles of EBNA2 in the gene regulation in EBV-immortalized lymphoblastoid cells

transcriptional activation function is dependent on association with a DNA-binding protein, recombination signal-binding protein (RBP)-Jκ (GROSSMAN et al. 1994; HENKEL et al. 1994; ZIMBER-STROBL et al. 1994; HSIEH et al. 1995). RBP-Jκ is a cellular protein initially identified as a protein binding to the recombination signal Jκ (MATSUNAMI et al. 1989). Later studies, however, indicated that it does not bind to the recombination signal itself, but recognizes the related sequence GTGGGAA. RBP-Jκ is normally a transcriptional repressor protein, but when associated with EBNA2, its transcriptional repressor domain is masked by EBNA2 (HSIEH et al. 1995) and the protein complex acts as a transcriptional activator by virtue of the acidic transcriptional activation domain of EBNA2. RBP-Jκ is a component of the Notch signal pathway that has critical roles in vertebrate embryogenesis (for review, see ARTAVANIS-TSAKONAS et al. 1995). Notch is a transmembrane receptor protein and activation by its ligand Delta or Serrate leads to its proteolytic cleavage. The intracellular portion of the protein (Notch-IC) is then translocated to the nucleus and transactivates genes previously repressed by RBP-Jκ in the same way as EBNA2 exerts its function (SCHROETER et al. 1998; STRUHL and ADACHI 1998). By binding to the common target RBP-Jκ, EBNA2 and Notch transactivate similar genes (HOFELMAYR et al. 1999). Furthermore, EBNA2 has the ability to functionally replace the intracellular region of the mouse Notch1 protein and suppresses differentiation of myeloblast progenitor cells (SAKAI et al. 1998). Do EBNA2 and Notch have more functional similarity than just transcriptional activation assisted by RBP-Jκ? This may not be very likely since these two proteins are not structurally related outside the RBP-Jκ-binding motif.

In contrast to other EBNA2-regulated genes, c-*myc* does not contain the RBP-Jκ-binding motif and the RBP-Jκ-binding domain of EBNA2 is dispensable for transactivation of this gene. An interferon-stimulated response element termed the plasmacytoma repressor factor (PRF) element was found to be involved in the c-*myc* transactivation by EBNA2 (JAYACHANDRA et al. 1999).

Besides RBP-Jκ, several cellular proteins are involved in the EBNA2-mediated transactivation of the LMP-1 promoter. The *ets* family protein PU.1 interacts with EBNA2 and takes part in the EBNA2-mediated transactivation (JOHANNSEN et al. 1995). Because PU.1 has important roles in gene regulation in B cells and myeloid cells, it may also be involved in the EBNA2-induced transactivation of cellular genes important for immortalization. The EBNA2 domain required for association with PU.1 overlaps with that for binding to RBP-Jκ (JOHANNSEN et al. 1995). The LMP1 promoter contains an ATF/CRE [activating transcription factor/cyclic adenosine monophosphate (cAMP) response element] having a role in EBNA2-induced transactivation, and direct interaction of EBNA2 with an ATF-2/c-Jun heterodimer was demonstrated (SJOBLOM et al. 1998). In addition, a POU domain protein distinct from Oct-1 and Oct-2 may mediate EBNA2-induced transactivation of the latent membrane protein (LMP)1 promoter (SJOBLOM et al. 1995).

EBNA2 also interacts with hSNF/Ini1, the human homolog of the yeast transcription factor SNF5 (WU et al. 1996). SNF5 is a component of the SNF-SWI complex that is supposed to mediate chromatin structure alteration associated with transcriptional regulation. The association of EBNA2 with hSNF/Ini1 therefore suggests that it has a role in alteration of nucleosome structure and thereby promotes transcription from certain genes. Only the phosphorylated form of EBNA2 is associated with hSNF/Ini1 in cells. A recent report described the association of EBNA2 with a novel DEAD box protein, DP103 (GRUNDHOFF et al. 1999). The DEAD box family proteins are thought to be ATP-dependent RNA helicases, but the functional significance of this association is at present not known. EBNA2 has an acidic transactivation domain near its C terminus and general transcription factors such as TFIIB, TFIIH, and TAF40 are recruited to the RNA polymerase II transcription complex by interaction with this domain (TONG et al. 1995).

4 The Effects of EBNA2 on Cellular Phenotype

Gene transfer into EBV-negative cell lines as well as into primary-culture B cells provided evidence that EBNA2 can produce certain phenotypes characteristic to both EBV-immortalized lymphoblastoid cells and antigen-activated B cells. As described above, EBNA2 induces expression of CD21, CD23, c-*myc*, and c-*fgr*. CD21 and CD23 are both B-cell activation surface markers and supposed to have important roles in the process of antigen-induced activation. The soluble form of CD23 may act as a growth factor. c-*Myc* has an essential role in the regulation of cell proliferation in general and therefore should be important in proliferation of activated B cells and EBV-immortalized LCLs. Although little is known about the behavior of c-*fgr* in B-cell activation, there is evidence that murine c-*fgr* expression is induced by B-cell-activating stimuli and that the c-Fgr kinase activity is activated following cross-linking of the B-cell antigen receptor (BCR) (WECHSLER and MONROE 1995; HATAKEYAMA et al. 1996). EBNA2 and EBNA-LP cooperatively induce expression of the G1 cyclin, cyclin D2 (SINCLAIR et al. 1994). EBNA2 also changes cellular phenotype and growth characteristics of non-lymphoid cells. EBNA2 reduces the serum requirement for the growth of the Rat-1 cells, although it did not confer malignant characteristics such as transplantability to nude mice and anchorage independence (DAMBAUGH et al. 1986). Transgenic mice possessing EBNA2 gene under control of the SV40 early promoter fused to the endogenous Wp EBNA promoter developed renal adenocarcinoma (TORNELL et al. 1996).

Kempkes and others engineered EBNA2-estrogen receptor chimera proteins and established an experimental system in which EBNA2 functions could be controlled by estrogen (KEMPKES et al. 1995). Lymphoblastoid cell lines were established by co-infection of P3HR-1 EBV and mini EBV (HAMMERSCHMIDT and SUGDEN 1989) encoding this chimeric protein. When estrogen was removed from culture and EBNA2 function was blocked, cells went into either growth arrest or apoptosis. Growth arrest occurred at G1 and G2 stages of the cell cycle, suggesting that EBNA2 regulates the cell cycle at multiple stages. Growth arrest at the G1 stage was reversible and could be released by re-addition of estrogen, but that in G2 was irreversible. These results clearly demonstrated that EBNA2 is required not only for initiation of lymphocyte immortalization but also for its maintenance. Interestingly, the role of EBNA2 in continued proliferation of these cells could be substituted by constitutive expression of c-*myc* by gene transfer, suggesting that c-*myc* transactivation is a critical step in the growth promotion by EBNA2 (POLACK et al. 1996). Another interesting finding in this work was that EBNA2 can transcriptionally downregulate the μ immunoglobulin heavy chain gene (JOCHNER et al. 1996). This downregulation is quick and evident within 30min of activation of EBNA2. It is at present not known whether RBP-Jκ is involved in this function of EBNA2. The downregulation of the Ig heavy chain gene by EBNA2 is consistent with the growth suppression of BL cell lines by this protein (KEMPKES et al. 1996). In these cell lines, c-*myc* on chromosome 8 is juxtaposed to the Ig heavy chain locus on chromosome 14 by translocation, and c-*myc* expression is constitutively activated. This c-*myc* deregulation, resulting in its overexpression, is thought to be critical in the uncontrolled proliferation of BL cells. When EBNA2 suppresses transcription of Ig heavy chain gene, c-*myc* expression under control of the heavy chain locus, will be also downregulated, and this should lead to growth suppression. This explanation, however, does not apply to the cell line BJAB that also went into growth arrest and cell death after induction of EBNA2 (JOCHNER et al. 1996; KEMPKES et al. 1996). BJAB cells do not have chromosomal translocation involving c-*myc* and the Ig heavy chain gene. Thus, EBNA2 may have intrinsic anti-proliferative and cytocidal effect when expressed in established cell lines. These effects may have been overlooked in previous gene transfer experiments using constitutive expression vector, because cells expressing EBNA2 should not survive the selection process after transfection.

5 Expression of EBNA2 in Akata Cells by Gene Transfer

This section deals with the author's own results of EBNA2 gene transfer into Burkitt's lymphoma Akata cells. Here experiments with a constitutive expression vector gave a suggestion of deleterious effect of EBNA2 on cells, and subsequent analysis with an inducible vector revealed the unexpected EBNA2-induced activation of EBV-replicative cycle in Akata cells.

5.1 Burkitt's Lymphoma Line Akata

EBV gene expression in Burkitt's lymphoma cells is more restricted than LCLs, and only EBNA1, BARF0, and occasionally LMP2A are expressed as proteins (for review, see RICKINSON and KIEFF 1996). In this form of EBV latency, termed latency I, Cp is not active and instead Qp is used for transcription of the EBNA1 mRNAs. When BL cells are transferred in vitro and established as a cell line, this restricted pattern of EBV gene expression is not usually maintained and eventually replaced by latency III.

The Burkitt's lymphoma line Akata, derived from a Japanese case of BL, is unique in that it maintains the latency I phenotype after long-term culture (TAKADA et al. 1991). Another unique property of Akata cells is that they have a tendency to lose EBV genomes spontaneously and to give rise to virus-negative sublines (SHIMIZU et al. 1994). Akata cells express surface IgG molecules and their cross-linking by antibodies results in activation of EBV replication, through signal transduction pathways involving Ca^{2+} mobilization and activation of protein kinase C (TAKADA and ONO 1989; DAIBATA et al. 1990). In contrast, EBV genomes in LCLs with the latency III phenotype are not significantly activated by ligation of surface immunoglobulin molecules.

To obtain a possible explanation for these unique properties of Akata cells, EBNA2 gene transfer experiments were carried out to answer two primary questions: (1) Does EBNA2 induce the latency III phenotype in Akata cells? (2) regardless of whether latency III is induced, does EBNA2 have any influence on the anti-IgG-induced EBV activation in Akata cells?

5.2 Expression of EBNA2 by a Constitutive Expression Vector

For efficient stable expression of EBNA2 by gene transfer, an EBNA2 expression vector, pOH-SGE2, was constructed. This vector contains the following elements in the pBluescript SK(–) backbone; SV–40 early promoter, β-globin intron, EBNA2 open reading frame, poly-A signal, EBV Ori-P, and a hygromycin resistance gene. The detail of the structure of this plasmid was given elsewhere (FUJIWARA et al. 1999). When EBNA1 is provided in *trans*, Ori-P is the only *cis* element required for episomal persistence of a plasmid. Since Akata cells produce EBNA1, expressed from their endogenous EBV genomes, it was expected that pOH-SGE2 would be maintained in Akata cells as multiple copies of episomes, thereby facilitating efficient expression of EBNA2.

Akata cells were transfected with pOH-SGE2 and hygromycin-resistant clones were examined for expression of EBNA2. In most of these clones, however, either no EBNA2-positive cells, or only a few, were found. After screening 220 hygromycin-resistant clones, only two clones were found, in which a majority of the cells were EBNA2-positive. This inefficient selection of EBNA2-positive cells suggested that expression of this protein was disadvantageous for the cells. Concurrently, two hygromycin-resistant Akata clones were isolated after transfection with pOH-SG2E, containing the EBNA2 gene in the reverse orientation with respect to the SV40 promoter, and used as negative controls.

To examine whether these EBNA2-positive Akata cells have any traits of latency III, activities of latent EBV promoters as well as transcription of some latent EBV genes characteristic to latency III were examined by the reverse transcription-polymerase chain reaction (RT-PCR) method. The results indicated that Cp was not detectably activated by expression of EBNA2, and consistent to this, no detectable level of EBNA3A and EBNA3B mRNAs was found. The EBNA2 mRNAs transcribed from the endogenous Akata EBV genome was not detected either. Both EBNA2-positive and control Akata cells used Qp for expression of EBNA1. LMP1 was detected by immunoblotting in one of the two EBNA2-positive Akata clones, but its level was much lower than in B95-8 and LCLs. Thus these EBNA2-positive Akata clones did not have characteristics of the latency III in spite of expression of EBNA2. There were, however, two significant changes induced by EBNA2; RT-PCR showed that expression of LMP2A and transcription from the Fp promoter were induced by EBNA2. Induction of LMP2A was consistent to previous results (ZIMBER-STROBL et al. 1991). Activation of Fp, on the other hand, was unexpected and suggested that EBNA2 has a potential to activate EBV replication, since this promoter is activated in lytic EBV infection (LEAR et al. 1991).

To test whether expression of EBNA2 has any influence on spontaneous and anti-IgG-induced activation of the EBV cycle, the two Akata transfectant clones expressing EBNA2 and the two control clones were cultured with or without goat affinity-purified antibodies to human IgG and activation of the viral cycle was assessed by indirect immunofluorescence. Without anti-IgG, less than 0.1% of the control transfectant cells expressed the early antigens (EA). Upon stimulation with anti-IgG, EA was induced in 15–29% of these control cells. In contrast, an unexpectedly large fraction (1.1–4.8%) of the EBNA2-expressing cells were already positive with EA even without treatment with anti-IgG. After addition of anti-IgG, however, increase in the number of EA-positive cells was significantly smaller than control cells (1.4–8.1%).

These results strongly suggested that EBNA2 has a potential to induce the EBV cycle. Since EBV replication is generally presumed to have deleterious effects on cells, it was suspected that significant number of EBNA2-positive cells were lost after transfection with pOH-SGE2. The unexpectedly low rate (2/220) of EBNA2-expressing clones among hygromycin-resistant clones (see above), despite the presence of both the EBNA2 gene and the hygromycin resistance gene on the same plasmid, supported this notion.

5.3 Expression of EBNA2 by an Inducible Expression Vector

More direct evidence of EBNA2-induced activation of EBV replication in Akata cells was obtained by experiments with a tetracycline-regulated expression vector. Two plasmids, pTet-SGE2 and pTAk-Hyg, derived from a modified tetracycline-regulated expression system described by Schokett and others were constructed (GOSSEN and BUJARD 1992; SCHOKETT et al. 1995). pTet-SGE2 contains the EBNA2 gene under the control of the tetracycline-responsive promoter, whereas PTAk-Hyg contains tetracycline-responsive transactivator and a hygromycin-resistance gene. Akata cells were co-transfected with these two plasmids and a number of Akata clones expressing EBNA2 in a tetracycline-regulated manner were isolated (Fig. 2) (FUJIWARA et al. 1999). In presence of 0.5µg/ml tetracycline, EBNA2 was not detected either by immunoblot analysis or immunoenzymatic staining. Upon removal of tetracycline, EBNA2 expression was efficiently induced

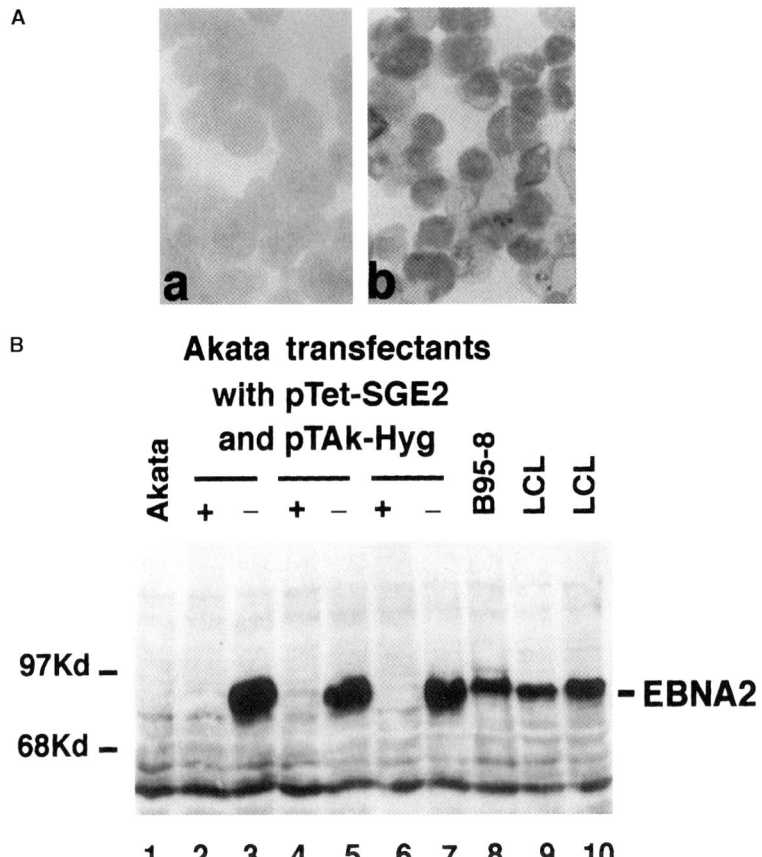

Fig. 2A,B. Tetracycline-regulated expression of EBNA2 in Akata cells. **A** Immunoenzymatic staining. Cell smears of an Akata transfectant clone harboring pTet-SGE2 and pTAk-Hyg were cultured with (*a*) or without (*b*) tetracycline and examined with the anti-EBNA2 monoclonal antibody PE2. **B** Immunoblot analysis. Cells of three Akata transfectant clones (lanes 2–7) were maintained with (lanes 2, 4, and 6) or without (lanes 3, 5, and 7) tetracycline and cell lysates were analyzed by immunoblotting with the PE2 antibody. As a reference, untransfected Akata cells (lane 1), B95–8 cells (lane 8) and two LCLs (lanes 9 and 10) were also analyzed

and the level of its expression was higher than the average among EBV-immortalized lymphoblastoid cell lines. When the cells were examined at various times after removal of tetracycline, EBNA2 was first detected at 12h and reached the plateau by 48h.

To examine whether EBNA2 expression in these cells induces the EBV-replicative cycle, smears of an Akata transfectant clone were prepared at 72h after removing tetracycline and examined for expression of EA and viral capsid antigens (VCA) by indirect immunofluorescence. The results indicate that EA was induced in more than 50% of the cells and VCA around 25% (Fig. 3A). Similar levels of EA

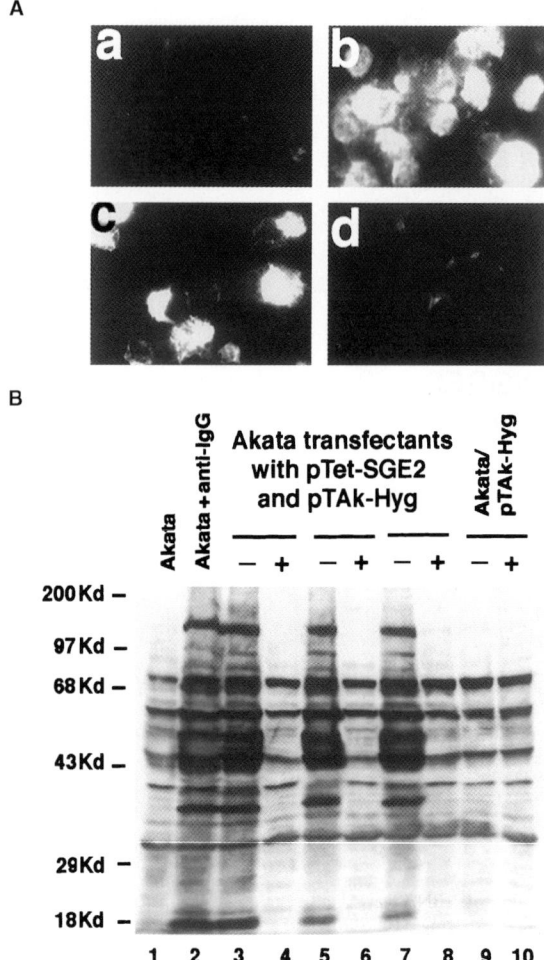

Fig. 3A,B. Expression of EBV-replicative cycle proteins following induced expression of EBNA2. **A** Immunofluorescence. An Akata transfectant clone harboring both pTet-SGE2 and pTAk-Hyg were maintained in culture medium with (*a*) or without (*b* and *c*) tetracycline and examined by indirect immunofluorescence. Serum from an NPC patient was used to detect EA (*a* and *b*) and that from a healthy EBV carrier was used to detect VCA (*c*). As a reference, a control Akata transfectant clone harboring pTAk-Hyg alone was maintained in tetracycline-free medium and examined with serum from a patient with NPC (*d*). **B** Immunoblot analysis. Three Akata transfectant clones harboring pTet-SGE2 and PTAk-Hyg (lanes 3–8) and a control Akata clone harboring pTAk-Hyg alone (lanes 9 and 10) were cultured with (lanes 4, 6, 8, and 10) or without (lanes 3, 5, 7, and 9) tetracycline and examined by immunoblot analysis with pooled sera from patients with NPC. As a reference, Akata cells before (lane 1) and after (lane 2) treatment with anti-IgG antibodies were also examined

and VCA were detected in three other Akata transfectant clones after removing tetracycline. Expression of EBV lytic-cycle proteins similar to those induced by anti-IgG antibodies were also detected by immunoblot analysis of three Akata transfectant clones (Fig. 3B). As expected, no such replication-cycle proteins were induced in control Akata transfectants harboring pTAk-Hyg alone. These lytic-cycle proteins were first detected at 18h after removing tetracycline in a low level and increased until 36h and remained at plateau until 96h. The BZLF1 protein, an immediate early protein critical to activation of EBV replication from latent state, was also shown induced. Immunoblot analysis using a monoclonal antibody to the BFRF3 protein, an immunodominant component of the EBV capsid, indicated that the 18-Kd band seen in Fig. 3B corresponds to this protein.

Simultaneous to EBV activation, proliferation of Akata cells were suppressed and cell viability declined after several days of induction of EBNA2 (S. Fujiwara et al., submitted for publication). Because the EBV-replicative cycle is harmful to cells, it was first suspected that these effects of EBNA2 were consequences of EBV activation. To test this hypothesis, the same tetracycline-regulated EBNA2 expression vector was introduced into cells of an EBV-negative Akata subline. Interestingly, growth suppression and cell death were also seen in these EBV-negative Akata cells that were induced to express EBNA2. Thus, growth retardation and cell death are not dependent on EBV-replicative cycle and could be ascribed to intrinsic functions of EBNA2. Because Akata cells contain the t(8;14) translocation and c-*myc* expression is suppressed upon induction of EBNA2, these EBNA2 effects may be mediated by c-*myc* downregulation.

To test whether RBP-Jκ is involved in the EBNA2-induced EBV-replicative cycle, a deletion mutant lacking the RBP-Jκ -binding domain was investigated (S. Fujiwara et al., submitted for publication). By deleting a *Sph*I fragment within BYRF1, an EBNA2 mutant lacking the amino acid residues 248–382 was prepared. This mutant barely, if ever, induced the EBV cycle, suggesting that RBP-Jκ has an important role in the EBNA2-induced EBV cycle, although critical involvement of other proteins such as PU.1, also binding to this portion of EBNA2, cannot be excluded.

5.4 Discussion

Gene transfer experiments with pOH-SGE2 indicated that in Akata cells constitutively expressing EBNA2 (1) the efficiency of the anti-IgG-induced viral cycle was decreased, (2) the rate of spontaneous activation of EBV-replicative cycle was increased, and (3) the Cp promoter was not activated and therefore the latency III program was not induced. The decreased rated of anti-IgG-induced viral cycle may be explained by the EBNA2-induced upregulation of LMP2A. LMP2A blocks Ca^{2+} mobilization and thereby impedes activation of the EBV cycle triggered by cross-linking of surface immunoglobulins (MILLER et al. 1993). The spontaneous EBV replicative cycle in Akata transfectants, despite expression of LMP2A, therefore, suggests that EBNA2 promotes a certain step downstream of Ca^{2+} mobilization, in the signal pathway of EBV activation. The Cp promoter, a hallmark of latency III, was not induced by EBNA2 and this is consistent with the finding that DNA methylation around Cp is a decisive factor in the maintenance and possibly establishment of latency I (SCHAEFER et al. 1997).

Gene transfer with a tetracycline-regulated EBNA2 expression vector gave decisive evidence for the EBNA2-induced EBV activation. EBV-replicative cycle was more efficiently induced in this experiment ($>50\%$) than with pOH-SGE2 (1.1–4.8%). Considering deleterious effects of EBV-replicative cycle on cells, constitutive expression vectors like pOH-SGE2 may not be suitable for the analysis of stable effects of EBNA2 on Akata cells, since most cells expressing the protein are likely to be lost soon after transfection. Because the level of EBNA2 expression in individual cells was much higher in transient transfection of pOH-SGE2 than in stable transfection with the same plasmid (S. Fujiwara, unpublished observation), it is suspected that only cells with low EBNA2 expression survived. This implies that high-level EBNA2 expression is required to activate the EBV cycle.

Akata is a rare exception among BL cell lines in that its latency I phenotype was not replaced by latency III after long-term culture. The EBNA2-induced disruption of the EBV cycle may provide an explanation for this unique property of the cell line. Similar to other BL-derived cell lines, occasional Akata cells may express EBNA2 spontaneously, yet instead of inducing the latency III phenotype, EBNA2 expression will result in activation of the EBV replicative cycle and cell death. This scenario may be also relevant to the mechanism of spontaneous loss of EBV genomes from Akata cells. If the rate of spontaneous EBNA2 expression and consequent viral replication is beyond a certain level, cells that have lost EBV genomes should have advantage for survival. It will be interesting to test whether EBNA2 activates viral cycle in other few BL cell lines that retain the latency I phenotype.

The EBV-activating, growth-suppressing, and cell-death-inducing potentials of EBNA2 are apparently suppressed in latency III, possibly by the products of other latent EBV genes, such as EBNA3C and LMP1. It is also conceivable that high-level expression is required for EBNA2 to exert these effects, because gene transfer experiments, especially that by tetracycline-regulated system, tend to cause overexpression. The molecular mechanism involved in EBNA2-induced the EBV cycle remains as an open question. Although the RBP-Jκ -binding motif (GTGGGAA) is not found close to immediate-early EBV genes such as BZLF1 and BRLF1, there still remains possibility that other cellular factors may mediate the action of EBNA2. A more likely mechanism may be that EBNA2 primarily transactivates a certain cellular gene(s) and its product(s) induces an immediate-early EBV gene.

As reviewed in the earlier part of this chapter, EBNA2 can induce cellular genes responsible for the uncontrolled proliferation and activated-B-cell phenotype of LCLs. The author's work described here, together with that presented by Kempkes and others, not only confirmed these previous findings but also showed that EBNA2 can suppress lymphoid cell proliferation, induce cell death, and activate EBV-replicative cycle. When B lymphocytes are stimulated through their antigen receptor (BCR), either by antigen or antibody, the cellular response varies depending on a number of factors, including the strength of stimulation, the stage of B-cell differentiation, and accessory signals from other cells and cytokines. A variety of responses ranging from B-cell activation and proliferation to apoptosis can be elicited by stimulation of BCR. In Burkitt's lymphoma cell lines, anti-immunoglobulin antibodies can either suppress proliferation, induce apoptosis, or activate EBV replication (reviewed in THOMPSON 1998). Thus, it should be noted that a striking similarity exists between the cellular responses to stimulation of BCR and those to expression of EBNA2. A possible explanation for this interesting parallelism is that EBNA2 may transactivate a certain cellular gene(s) that is involved in gene regulation coupled with the signal pathway from BCR.

6 Concluding Remarks

Previously, effects of EBNA2 on EBV and cellular gene regulation, as well as on cellular phenotype, has been mainly analyzed by gene transfer with constitutive expression vectors. These experiments provided evidence that EBNA2 plays a central role in EBV gene regulation in virus-immortalized LCLs, and also that, in cooperation with other latent EBV proteins, EBNA2 is responsible for initiation and maintenance of cell cycle and activated B-cell phenotype in these cells. By designing experimental systems in which either expression or function of EBNA2 can be readily controlled, the author's work described here and that by Kempkes and others revealed unexpected effects of EBNA2, such as growth suppression, cell death, and activation of the EBV cycle. These novel effects of EBNA2 may provide valuable information in elucidating the exact roles of EBNA2 in lymphocyte immortalization.

Acknowledgements. The experiments of EBNA2 gene transfer into Akata cells described here were carried out in collaboration with Yoshikazu Nitadori, Hiroyuki Nakamura, Takashi Nagaishi, and Yasushi Ono. This work was financially supported in part by Grant-in Aid for the High-Tech Research Center from the Japanese Ministry of Education, Science, Sports, and Culture to Nihon University.

References

Abbot SD, Rowe M, Cadwallader K, Rickinson A, Gordon J, Wang F, Rymo L, Rickinson AB (1990) Epstein-Barr virus nuclear antigen 2 induces expression of the virus-encoded latent membrane protein. J Virol 64:2126–2134
Alfieri C, Birkenbach M, Kieff E (1991) Early events in Epstein-Barr virus infection of human B lymphocytes. Virology 181:595–608
Allday MJ, Crawford DH, Griffin BE (1989) Epstein–Barr virus latent gene expression during the initiation of B cell immortalization. J Gen Virol 70:1755–1764
Aman P, Rowe M, Kai C, Finke J, Rymo L, Klein E, Klein G (1990) Effect of the EBNA2 gene on the surface antigen phenotype of transfected EBV-negative B-lymphoma lines. Int J Cancer 45:77–82
Artvanis-Tsakonas S, Matsuno K, Fortini ME (1995) Notch signaling. Science 268:225–232
Cohen J, Wang F, Mannick J, Kieff E (1989) Epstein-Barr virus nuclear protein 2 is a key determinant of lymphocyte transformation. Proc Natl Acad Sci USA 86:9558–9562

Cordier M, Calender A, Billaud M, Zimber U, Rousselet G, Pavlish O, Banchereau J, Tursz T, Bornkamm G, Lenoir GM (1990) Stable transfection of Epstein-Barr virus (EBV) nuclear antigen 2 in lymphoma cells containing the EBV P3HR-1 genome induces expression of B-cell activation molecules CD21 and CD23. J Virol 64:1002–1013

Daibata M, Humphreys RE, Takada K, Sairenji T (1990) Activation of latent Epstein-Barr virus via anti-IgG-triggered, second messenger pathways in the Burkitt's lymphoma cell line Akata. J Immunol 144:4788–4793

Dambaugh T, Wang F, Hennessy K, Woodland E, Rickinson A, Kieff E (1986) Expression of the Epstein-Barr virus nuclear protein 2 in rodent cells. J Virol 59:453–462

Fåhraeus R, Jansson A, Ricksten A, Sjöblom A, Rymo L (1990) Epstein-Barr virus-encoded nuclear antigen 2 activates the viral latent membrane protein promoter by modulating the activity of a negative regulatory element. Proc Natl Acad Sci USA 87:7390–7394

Fujiwara S, Nitadori Y, Nakamura H, Nagaishi T, Ono Y (1999) Epstein-Barr virus (EBV) nuclear protein 2-induced disruption of EBV latency in the Burkitt's lymphoma cell line Akata: analysis by tetracycline-regulated expression. J Virol 73:5214–5219

Gossen M, Bujard H (1992) Tight control of gene expression in mammalian cells by tetracycline-responsive promoters. Proc Natl Acad Sci USA 89:5547–5551

Grossman SR, Johannsen E, Tong X, Yalamanchili R, Kieff E (1994). The Epstein-Barr virus nuclear antigen 2 transactivator is directed to response elements by the Jκ recombination signal binding protein. Proc Natl Acad Sci USA 91:7568–7572

Grundhoff AT, Kremmer E, Türeci Ö, Glieden A, Gindorf C, Atz J, Mueller-Lantzsch N, Schubach WH, Grässer FA (1999) Characterization of DP103, a novel DEAD box protein that binds to the Epstein-Barr virus nuclear proteins EBNA2 and EBNA3C. J Biol Chem 274:19136–19144

Hammerschmidt W, Sugden B (1989) Genetic analysis of immortalizing functions of Epstein-Barr virus in human B lymphocytes. Nature 340:393–397

Hatakeyama S, Iwabuchi K, Ato M, Iwabuchi C, Kajino K, Takami K, Katoh M, Ogasawara K, Good RA, Onoe K (1996) Fgr expression restricted to subpopulation of monocyte/macrophage lineage in resting conditions is induced in various hematopoietic cells after activation or transformation. Microbiol Immunol 40:223–231

Henkel T, Ling PD, Hayward SD, Petersen MG (1994). Mediation of Epstein-Barr virus EBNA2 transactivation by recombination signal-binding protein Jκ. Science 265:92–95

Hennessy K, Kieff E (1985) A second nuclear protein is encoded by Epstein-Barr virus in latent infection. Science 227:1238–1240

Höfelmayr H, Strobl LJ, Stein C, Laux G, Marschall G, Bornkamm GW, Zimber-Strobl U (1999) Activated mouse Notch1 transactivates Epstein-Barr virus nuclear antigen 2-regulated viral promoters. J Virol 73:2770–2780

Hsieh JJ-D, Hayward SD (1995) Masking of the CBF1/RBPJκ transcriptional repression domain by Epstein-Barr virus EBNA2. Science 268:560–563

Jayachandra S, Low KG, Thlick A-E, Yu J, Ling PD, Chang Y, Moore PS (1999) Three unrelated viral transforming proteins (vIRF, EBNA2, and E1A) induce the *MYC* oncogene through the interferon-responsive PRF element by using different transcription coadaptors. Proc Natl Acad Sci USA 96:11566–11571

Jin XW, Speck SH (1992) Identification of critical cis elements involved in mediating Epstein-Barr virus nuclear antigen 2-dependent activity of an enhancer located upstream of the viral *Bam* HI C promoter. J Virol 66:2846–2852

Jochner N, Eick D, Zimber-Strobl U, Pawlita M, Bornkamm GW, Kempkes B (1996) Epstein-Barr virus nuclear antigen 2 is a transcriptional suppressor of the immunoglobulin μ gene: implications for the expression of the translocated c-*myc* gene in Burkitt's lymphoma cells. EMBO J 15:375–382

Johannsen E, Koh E, Mosialos G, Tong X, Kieff E, Grossman SR (1995) Epstein-Barr virus nuclear protein 2 transactivation of the latent membrane protein 1 promoter is mediated by J kappa and PU.1. J Virol 69:253–262

Kaiser C, Laux G, Eick D, Jochner N, Bornkamm GW, Kempkes B (1999) The proto-oncogene c-*myc* is a direct target gene of Epstein-Barr virus nuclear antigen 2. J Virol 73:4481–4484

Kempkes B, Spitkovsky D, Jansen-Dürr P, Ellwart JW, Kremmer E, Delecluse H-J, Rottenberger C, Bornkamm GW, Hammerschmidt W (1995) B-cell proliferation and induction of early G_1-regulating proteins by Epstein-Barr virus mutants conditional for EBNA2. EMBO J 14:88–96

Kempkes B, Zimber-Strobl U, Eissner G, Pawlita M, Falk M, Hammerschmidt W, Bornkamm GW (1996) Epstein-Barr virus nuclear antigen 2 (EBNA2)-oestrogen receptor fusion proteins complement the EBNA2-deficient Epstein-Barr virus strain P3HR1 in transformation of primary B cells but suppress growth of human B cell lymphoma lines. J Gen Virol 77:227–237

Kieff E (1996) Epstein–Barr virus and its replication. In: Fields BN, Knipe DM, Howley PM, Chanock RM, Melnick JL, Monath TP, Roizman B, Straus SE (eds) Virology, 2nd edn. Lipppincott-Raven, Philadelphia, pp 2343–2396

Knutson JC (1990) The level of c-*fgr* RNA is increased by EBNA2, an Epstein-Barr virus gene required for B-cell immortalization. J Virol 64:2530–2536

Lear AL, Rowe M, Kurilla MG, Lee S, Henderson S, Kieff E, Rickinson AB (1992) The Epstein-Barr virus (EBV) nuclear antigen 1 *Bam*HI F promoter is activated on entry of EBV-transformed B cells into the lytic cycle. J Virol 66:7461–7468

Mannick JB, Cohen JI, Birkenbach M, Marchini A, Kieff E (1991) The Epstein-Barr virus nuclear protein encoded by the leader of the EBNA RNAs is important in B-lymphocyte transformation. J Virol 65:6826–6837

Matsunami N, Hamaguchi Y, Yamamoto Y, Kuze K, Kangawa K, Matsuo H, Kawaichi M, Honjo T (1989) A protein binding to the Jκ recombination sequence of immunoglobulin genes contains a sequence related to the integrase motif. Nature 342:934–937

Miller CL, Longnecker R, Kieff E (1993) Epstein-Barr virus latent membrane protein 2A blocks calcium mobilization in B lymphocytes. J Virol 67:3087–3094

Mueller-Lantzsch N, Lenoir GM, Sauter M, Takaki K, Bechet J-M, Kuklik-Roos C, Wunderlich D, Bornkamm GW (1985) Identification of the coding region for a second Epstein-Barr virus nuclear antigen (EBNA2) by transfection of cloned DNA fragments. EMBO J 4:1805–1811

Peng M, Lundgren E (1992) Transient expression of the Epstein-Barr virus *LMP1* gene in human primary B cells induces cellular activation and DNA synthesis. Oncogene 7:1775–1782

Polack A, Hortnagel K, Pajic A, Christoph B, Baier B, Falk M, Mautner J, Geltinger C, Bornkamm GW, Kempkes B (1996) c-*myc* activation renders proliferation of Epstein-Barr virus (EBV)-transformed cells independent of EBV nuclear antigen 2 and latent membrane protein 1. Proc Natl Acad Sci USA 93:10411–10416

Rickinson AB, Kieff E (1996) Epstein–Barr virus. In: Fields BN, Knipe DM, Howley PM, Chanock RM, Melnick JL, Monath TP, Roizman B, Straus SE (eds) Virology, 2nd edn. Lipppincott-Raven, Philadelphia, pp 2397–2446

Rowe D, Heston L, Metlay J, Miller G (1985) Identification and expression of a nuclear antigen from the genomic region of the Jijoe strain of Epstein-Barr virus that is missing in its nonimmortalizing deletion mutant, P3HR-1. Proc Natl Acad Sci USA 82:7429–7433

Rymo L, Klein G, Ricksten A (1985) Expression of a second Epstein-Barr virus determined nuclear antigen in mouse cells after gene transfer with a cloned fragment of the viral genome. Proc Natl Acad Sci USA 82:3435–3439

Sakai T, Taniguchi Y, Tamura K, Minoguchi S, Fukuhara T, Strobl LJ, Zimber-Strobl U, Bornkamm GW, Honjo T (1998) Functional replacement of the intracellular region of the Notch1 receptor by Epstein-Barr virus nuclear antigen 2. J Virol 72:6034–6039

Schaefer BC, Strominger JL, Speck SH (1997) Host-cell-determined methylation of specific Epstein-Barr virus promoters regulates the choice between distinct viral latency programs. Mol Cell Biol 17: 364–377

Schroeter EH, Kisslinger JA, Kopan R (1998) Notch-1 signalling requires ligand-induced proteolytic release of intracellular domain. Nature 393:382–386

Shimizu N, Tanabe-Tochikura A, Kuroiwa Y, Takada K (1994) Isolation of Epstein-Barr virus-negative cell clones from the EBV-positive Burkitt's lymphoma line Akata: malignant phenotypes of BL cells are dependent on EBV. J Virol 68:6069–6073

Shokett P, Difilippantonio M, Hellman N, Schatz DG (1995) A modified tetracycline-regulated system provides autoregulatory, inducible expression in cultured cells and transgenic mice. Proc Natl Acad Sci USA 92:6522–6526

Sinclair AJ, Palmero I, Peters G, Farrell PJ (1994) EBNA2 and EBNA-LP cooperate to cause G_0 to G_1 transition during immortalization of resting human B lymphocytes by Epstein-Barr virus. EMBO J 13:3321–3328

Sjöblom A, Jansson A, Yang W, Laín S, Nilsson T, Rymo L (1985) PU box-binding transcription factors and a POU domain protein cooperate in the Epstein-Barr virus (EBV) nuclear antigen 2-induced transactivation of the EBV latent membrane protein 1 promoter. J Gen Virol 76:2679–2692

Sjöblom A, Yang W, Palmqvist L, Jannson A, Rymo L (1998) An ATF/CRE element mediates both EBNA2-dependent and EBNA2-independent activation of the Epstein-Barr virus LMP1 gene promoter. J Virol 72:1365–1376

Struhl G, Adachi A (1998) Nuclear access and action of notch in vivo. Cell 93:649–660
Sung NS, Kenney S, Gutsch D, Pagano YS (1991) EBNA2 transactivates a lymphoid-specific enhancer in the *Bam*HI C promoter of Epstein-Barr virus. J Virol 65:2164–2169
Takada K, Horinouchi K, Ono Y, Aya T, Osato T, Takahashi M, Hayasaka S (1991) An Epstein-Barr virus producer line Akata: establishment of the cell line and analysis of viral DNA. Virus Genes 5:147–156
Takada K, Ono Y (1989) Synchronous and sequential activation of the latently infected Epstein-Barr virus genomes. J Virol 63:445–449
Thompson EB (1998) The many roles of c-Myc in apoptosis. Annu Rev Physiol 60:575–500
Tong X, Wang F, Thut CJ, Kieff E (1995) The Epstein-Barr virus nuclear protein 2 acidic domain can interact with TFIIB, TAF40, and RPA70, but not with TBP. J Virol 69:585–588
Törnell J, Farzad S, Espander-Jansson A, Matejka G, Isaksson O, Rymo L (1996) Expression of Epstein-Barr nuclear antigen 2 in kidney tubule cells induces tumors in transgenic mice. Oncogene 12:1521–1528
Walls D, Perricaudet M (1991) Novel downstream elements upregulate transcription initiated from an Epstein-Barr virus latent promoter. EMBO J 10:143–151
Wang F, Gregory CD, Rowe M, Rickinson AB, Wang D, Birkenbach M, Kikutani H, Kishimoto T, Kieff E (1987) Epstein-Barr virus nuclear antigen 2 specifically induces expression of the B-cell activation antigen CD23. Proc Natl Acad Sci USA 84:3452–3456
Wang F, Gregory C, Sample C, Row M, Liebowitz D, Murray R, Rickinson A, Kieff E (1990a) Epstein-Barr virus latent membrane protein (LMP1) and Nuclear proteins 2 and 3 C are effectors of phenotypic changes in B lymphocytes: EBNA2 and LMP1 cooperatively induce CD23. J Virol 64: 2309–2318
Wang F, Tsang SF, Kurilla MG, Cohen JI, Kieff E (1990b) Epstein-Barr virus nuclear antigen 2 transactivates latent membrane protein LMP1. J Virol 64:3407–3416
Wechsler RJ, Monroe JG (1995) *src*-family tyrosine kinase p55^{fgr} is expressed in murine splenic B cells and is activated in response to antigen receptor cross-linking. J Immunol 154:3234–3244
Wu DY, Kalpana GV, Goff SP, Schubach WH (1996) Epstein-Barr virus nuclear protein 2 (EBNA2) binds to a component of the human SNF-SWI complex, hSNF5/Ini1. J Virol 70:6020–6028
Zimber-Strobl U, Suentzenich KO, Laux G, Eick D, Cordier M, Calender A, Billaud M, Lenoir GM, Bornkamm GW (1991) Epstein-Barr virus nuclear antigen 2 activates transcription of the terminal protein gene. J Virol 65:415–423
Zimber-Strobl U, Strobl LJ, Meitinger C, Hinrichs R, Sakai T, Furukawa T, Honjo T, Bornkamm GW (1994) Epstein-Barr virus nuclear antigen 2 exerts its transactivating function through interaction with recombination signal binding protein RBP-Jκ, the homologue of Drosophila suppressor of Hairless. EMBO J 13:4973–4982

Two Epstein-Barr Virus Glycoprotein Complexes

L.M. HUTT-FLETCHER and C.M. LAKE

1	Introduction	51
2	The gH/gL/gp42 Complex	53
2.1	Identifying the Components	53
2.2	The Role of the gH/gL/gp42 Complex in Entry into B Lymphocytes	54
2.3	The Role of the gH/gL/gp42 Complex in Entry into Epithelial Cells	56
2.4	Implications for Pathogenesis	57
3	The gN/gM Complex	58
3.1	Identifying the Components	58
3.2	The Role of the gN/gM Complex in Virus Exit and Entry	59
4	Unanswered Questions	61
References		62

1 Introduction

The envelopes of all herpesviruses contain multiple glycoprotein species, each one of which is potentially important to virus entry, to virus egress and to trafficking of virus-producing cells throughout the body. In addition, membrane-associated proteins or glycoproteins that are found in the infected cell, but not in the virion, may influence virus assembly and yield. Thus, as a class these molecules have a major impact on virus tropism and virus load and contribute significantly to the outcome of infection.

The total complement of membrane proteins encoded by Epstein-Barr virus (EBV) is not yet certain. Currently, eleven unique species are known to be expressed (Table 1), although the information available about several is still quite meager. Least is know about four gene products that have no known homologs in alpha and beta herpesviruses. These are the BDLF3, BILF2, BILF1, and BMRF2 gene products (BAER et al. 1984). The BDLF3 open reading frame (ORF) encodes gp150 (KURILLA et al. 1995; NOLAN and MORGAN 1995), a mucin-like molecule to which N- and O-linked sugars contribute more than 50% of the mass. Recombinant

School of Biological Sciences, University of Missouri-Kansas City, 5007 Rockhill Road, Kansas City, MO 64110, USA

Table 1. Open reading frames currently known to express membrane proteins

Open reading frame	Number of amino acids in primary translation product	Protein	Homolog conserved in all known herpesviruses
BDLF3	234	gp150	None
BILF2	248	gp78	None
BILF1	312	gp60	None
BMRF2	356	?p55	None
BLLF1a/b	907/710	gp350/220	None
BALF4	857	gp110/125	gB
BXLF2	706	gp85	gH
BKRF2	137	gp25	gL
BZLF2	223	gp42	None
BLRF1	102	gp15	gN
BBRF3	405	gp48/84/113	gM

viruses lacking gp150 are not impaired for growth in tissue culture (BORZA and HUTT-FLETCHER 1998). The BILF2 ORF encodes gp78 (MACKETT et al. 1990), another highly glycosylated protein that carries primarily N-linked sugars. Both gp150 and gp78, which are predicted to be type 1 membrane proteins with relatively short cytoplasmic tails, are known to be present in the virion. However, the distribution of the poorly studied products of the BILF1 and BMRF2 ORFs remains uncertain. Preliminary data suggest that BILF1 encodes a glycosylated protein of approximately 60kDa, which aggregates upon boiling (KENYON and HUTT-FLETCHER 1999). This is consistent with its predicted structure as protein that spans the membrane multiple times. The BMRF2 gene product has a similar predicted structure, but although RNA transcripts mapping to this ORF have been shown to be present in abundance in the virus producing cells of oral hairy leukoplakia (PENARANDA et al. 1997), the only information available concerning its protein product comes from a study with an anti-peptide antibody that places its mass at approximately 55kDa (MODROW et al. 1992).

Considerably more information is available about the remaining gene products and there is increasing understanding of the roles they play in infection. The products of the BALF4 and BLLF1 open reading frames are perhaps the most abundant gene products although they have complementary distributions. BLLF1 encodes the virion envelope proteins gp350 and gp220, which are often referred to as one, gp350/220, since the smaller form is derived by in a frame splice which results in the loss of residues 500–757 of the 907-amino-acid gp350 protein (BEISEL et al. 1985). Both molecules carry a large amount of N- and O-linked sugar and they are responsible for attachment of virus to the B-cell receptor for EBV, the complement receptor type 2, CR2, or CD21 (FINGEROTH et al. 1984; NEMEROW et al. 1985). The binding site in gp350/220 includes a short sequence, 21 amino acids from the N-terminus of the protein, which is homologous to the binding site in the natural ligand, the C3dg fragment of complement (TANNER et al. 1988; NEMEROW et al. 1989). CR2 is part of a signal transduction complex and the interaction of gp350/220 with CR2 interaction may do more than simply tether virus to the cell surface. It apparently triggers endocytosis of the virion (TANNER et al.

1987) and several observations have suggested that it may alter the phenotype of the B cell and facilitate expression of virus genes (SINCLAIR and FARRELL 1995; SUGANO et al. 1997).

In contrast, the BALF4 gene product, gp110, is primarily a cell-associated glycoprotein that carries only endoglycosidase-sensitive high mannose sugars (GONG and KIEFF 1990; PAPWORTH et al. 1997; LEE 1999). This is consistent with its retention in the endoplasmic reticulum and nuclear membrane (GONG et al. 1987; GONG and KIEFF 1990) via a string of four arginine residues in its cytoplasmic tail (LEE 1999). The distribution is somewhat unexpected given that gp110 is the homolog of gB, a highly conserved glycoprotein which in most herpesviruses is abundant in the virion envelope where it plays a major role in virus-cell fusion (SPEAR 1993). The only BALF4 gene product that may be found in the virion in very small amounts (L.M. Hutt-Fletcher, unpublished data) is a differentially glycosylated form, gp125 (EMINI et al. 1987; GONG and KIEFF 1990; LEE 1999), which carries complex sugars. Studies of the relative distributions of gp350/220 and gp110, the former found in the plasma membrane and Golgi and the latter in the nuclear membrane, have suggested a model for assembly and egress of EBV that involves two sites of envelopment. The model, which is compatible with those proposed for other herpesviruses, suggests that virus first acquires an envelope as it buds through the nuclear membrane. At some point this enveloped form fuses back out into the cytoplasm. Unenveloped virions then acquire a second envelope rich in gp350/220 either by budding back into the Golgi and leaving the cell by exocytosis, or by budding through the plasma membrane (GONG and KIEFF 1990). The model proposes that gp110 is important to the exit of virus from the nucleus, and is consistent with the observation that recombinant virus lacking gp110 is not released from the cell (HERROLD et al. 1995; LEE and LONGNECKER 1997).

In addition to those molecules that appear to exist either as single proteins or as homopolymers, EBV encodes at least two heteromeric, multi-protein complexes, the members of which have inter-dependent functions. One, the gH/gL/gp42 complex, plays a major role in virus entry and tropism, the other, the gN/gM complex, appears to be important to assembly and disassembly of the enveloped particle. The remainder of this chapter focuses on these two important protein complexes.

2 The gH/gL/gp42 Complex

2.1 Identifying the Components

The largest of the three components of the gH/gL/gp42 complex is the 85-kDa gH molecule itself, described in the early literature as gp85, and for many years its association with the other two components went undetected. gp85 was identified as the EBV homolog of the conserved herpesvirus gH protein when it was mapped as the product of the BXLF2 ORF (HEINEMAN et al. 1988; OBA and HUTT-FLETCHER 1988). It is a 706-amino-acid protein which carries N-linked sugar and is predicted to be a type I membrane protein with a sequence that is overall quite hydrophobic. Like its counterparts in other herpesviruses, the EBV gH is improperly folded and retained in the endoplasmic reticulum if it is expressed alone as a recombinant protein (YASWEN et al. 1993). Coexpression with a viral chaperone encoded by the BKRF2 ORF is essential for its authentic processing (LI et al. 1995). The BKRF2 ORF is the EBV homolog of the conserved glycoprotein gL and encodes a 137-amino-acid protein which, after addition of N-linked sugars, is expressed as a molecule of 25kDa sometimes referred to as gp25 (YASWEN et al. 1993). It is predicted to have only one hydrophobic domain, close to the amino terminus, which is long enough to span a membrane, and since all attempts to demonstrate that the protein is secreted have failed (Q.X. Li and L.M. Hutt-Fletcher, unpublished data), it is assumed to be a type II membrane protein.

Formation of a complex between gH and its chaperone gL is a common theme in herpesviruses. However, unlike the prototype gH/gL complex of the alpha herpesviruses, there is a third component in the EBV gH/gL complex, gp42. This 223-amino-acid glycoprotein has been mapped to the BZLF2 ORF (LI et al. 1995). Like gL, it is a type II membrane protein anchored by an uncleaved signal sequence and carries N-linked sugars. Neither gp42 nor gL are dependent on gH for processing if they are expressed as recombinant proteins.

2.2 The Role of the gH/gL/gp42 Complex in Entry into B Lymphocytes

One of the original monoclonal antibodies (Mabs) made to study EBV glycoproteins was one called F-2-1 which immunoprecipitated gH and neutralized infection of B cells. The antibody had no effect on virus binding, but inhibited virus/cell fusion as judged by a fluorescence dequenching assay (MILLER and HUTT-FLETCHER 1988). In addition, the same assay was used to show that virosomes made from EBV proteins could bind specifically to CR2 positive cells, but could not fuse if the gH/gL/gp42 complex was first removed by affinity chromatography (HADDAD and HUTT-FLETCHER 1989). Both observations are compatible with a role for gH in virus cell fusion, which is proposed for its homologs in other herpesviruses. The identification and expression of gL and gp42, however, revealed that Mab F-2-1 did not recognize an epitope on gH as had originally been proposed (STRNAD et al. 1982), but instead reacted with gp42 (LI et al. 1995).

Computer-assisted analysis performed by investigators at Immunex indicated that gp42 has some features characteristic of the C-type lectin gene family and raised the possibility that there might be a cellular partner with which it could interact. The predicted extracellular domain of the protein (amino acids 34–223) was expressed as a soluble chimeric molecule, in which the putative signal peptide was replaced by the Fc portion of human immunoglobulin G_1. Probing of a complementary (c)DNA expression library made from activated T cells and analysis of monkey kidney cells transfected with wild-type or mutant HLA-DR chains demonstrated that the chimeric protein bound to the β_1 domain of the HLA class II protein, HLA-DR (SPRIGGS et al. 1996).

Put in the context of our own observations that gp42 carries an epitope critical to penetration of B cells, this suggested that an interaction between gp42 and HLA class II might have important implications for B-cell infection. In collaboration with Immunex, four additional observations were then made. First it was shown that the soluble form of gp42, gp42.Fc, could inhibit B-cell infection, second, that the interaction between gp42.Fc and HLA class II could be inhibited by the neutralizing Mab F-2-1, third, that a Mab called Alva 42 that reacts with HLA-DR (GAYLE et al. 1994) could inhibit both gp42.Fc binding to class II and virus infection, and, finally, that B cells that lacked HLA class II could only be superinfected with EBV if class II expression were restored (LI et al. 1997). To confirm the importance of gp42 to B-cell infection, a virus was made in which the BZLF2 ORF was interrupted with a neomycin resistance cassette. This virus, which lacked gp42, could bind to B cells but could not infect them unless cells and bound virus were treated with the exogenous fusogen polyethylene glycol (WANG and HUTT-FLETCHER 1998).

The interpretation of these data was that penetration of B lymphocytes by EBV involves use of HLA class II as a virus coreceptor. The model we propose for B-cell infection, based on results to this point, is that EBV first uses gp350/220 to bind to CR2. An essential interaction between gp42 and HLA class II then occurs, perhaps bringing virus closer to the cell surface. Finally, based on analogy with other herpesvirus, the gH/gL complex mediates virus/cell fusion, either alone or in cooperation with additional virus proteins.

One unresolved question was whether or not a conformational change has to occur in the gH-gL-gp42 complex before an interaction with class II can take place. Such a conformational change occurs when the HIV attachment protein gp120 binds to CD4 and creates a new recognition site for one of the chemokine receptors used to facilitate penetration (TRKOLA et al. 1996; WU et al. 1996). Immunoprecipitation of HLA class II from Akata cells in which lytic replication of virus has been induced has never revealed any association between gp42 or the gH/gL/gp42 complex and HLA class II within the cell. However, an interaction between gp42 and HLA class II in the endoplasmic reticulum might lead to targeting of gp42 to endosomal compartments where it could be the substrate for serine proteases. The HLA class II complex in the endoplasmic reticulum consists of three dimers of alpha and beta chains and three invariant chains. The invariant chain occupies the peptide binding groove, and targets HLA class II to endosomal compartments. Here it is partially degraded and displaced so that exogenously derived peptides can be loaded for antigen presentation (MELLMAN et al. 1995). Were gp42 to replace one or more of the invariant chains in the nine-component class II complex, it too might be targeted to an endosome and degraded.

To test this possibility, an association between HLA class II and the gH/gL/gp42 complex was sought in cells that had been treated with the serine protease inhibitor leupeptin. Drug treatment revealed such an association and suggested that no conformational changes would be required for the gH/gL/gp42 complex to interact with HLA class II at the cell surface. The observations also have some implications for the relative behavior of virus released from B cells and epithelial cells as discussed below.

2.3 The Role of the gH/gL/gp42 Complex in Entry into Epithelial Cells

Although the model described above could account for experimental observations made with B lymphocytes, direct application of the same model of infection to epithelial cells was complicated by three observations made with an epithelial cell line called SVKCR2, and subsequently confirmed with the gastric carcinoma cell line AGS. The SVKCR2 cell line is derived from skin epithelium that has been transformed with simian virus 40 and stably transfected with the B-cell receptor CR2 (LI et al. 1992), whereas the AGS line can be infected in a CR2-independent manner (YOSHIYAMA et al. 1997). The observations that were made with these lines were (1) that they do not constitutively express HLA class II, (2) that infection cannot be neutralized by Mab F-2-1 which reacts with gp42, and (3) that another Mab called E1D1 that reacts with gH, neutralizes infection although it has no effect on infection of B lymphocytes (LI et al. 1995). This suggested that HLA class II is not required for epithelial cell infection, that gp42 may not be required either, and that usage of gH is different.

To explore these issues further (WANG et al. 1998) we first confirmed the lack of interaction with HLA class II by showing that the Mab Alva 42 to HLA-DR that neutralized B-cell infection had no effect on infection of SVKCR2 cells. To confirm that gp42 was not involved in the process a comparison was made of the infectability of wild-type and gp42 minus virus. If the amounts of wild-type and recombinant virus were normalized in terms of virion DNA, gp42 minus virus could infect epithelial cells equally well as wild-type.

However, the soluble form of gp42, gp42.Fc, was if anything, better able to inhibit infection of SVKCR2 cells that infection of B cells. Previous workers have demonstrated the presence of HLA class II in the EBV virion (KNOX and YOUNG 1995) so it was possible that gp42.Fc inhibited SVKCR2 cells infection by binding to class II in the virus and interfering, perhaps sterically, with the function of EBV glycoproteins. To eliminate this possibility, infections were repeated with the P3HR1 strain of virus. P3HR1 virus is derived from cells that express only an allele of HLA-DQ to which gp42.Fc cannot bind (Li et al.) and although infection of SVKCR2 cells with this strain is not very efficient, it was nevertheless inhibited by gp42.Fc.

If inhibition of epithelial cell infection could not be attributed to gp42.Fc binding to class II in the virion, the next most likely possibility seemed to be that it might result from binding to gH or gL. The fact that such an interaction could occur was demonstrated by showing that soluble gp42.Fc in conjunction with protein A-agarose beads, which bind to the Fc domain of the construct, could precipitate not only HLA class II, but also gH and gL from lysates of virus-producing Akata cells. A series of amino terminal deletion mutants of gp42.Fc made at Immunex could still interact with class II, but not with gH and gL. This allowed the gH/gL binding domain to be mapped to amino acids 34–58; the class II binding domain was already known to be in the carboxyterminal 110 amino acids of the 223-amino-acid protein (SPRIGGS et al. 1996)

Since gp42.Fc retained both the class II binding domain and the gH-gL interactive domain it was also possible that it could substitute for the native protein in *trans* and rescue the ability of the gp42 minus virus to transform normal B cells. B-cell transformation by recombinant virus was, in fact, rescued by gp42.Fc at concentrations below those needed to inhibit infection of B cells, but only as long as the full-length construct was used. None of the constructs that lacked the gH/gL binding domain could rescue recombinant virus. Furthermore, only the construct that retained the gH/gL binding domain, and not those that had lost it, could inhibit epithelial cell infection, although, to varying extents depending on the extent of the deletion, they could still inhibit B-cell infection. It was already clear that a virus that contained only a bipartite complex of gH and gL could not infect B cells. It now appeared that a virus that contained only three part complexes, or at least one in which the stoichiometry of the complex was altered to contain much larger amounts of gp42, was unable to infect epithelial cells. This was consistent with the observation that more than four times as much gp42.Fc was needed to inhibit gp42 minus virus than wild-type virus. Stoichiometric analysis of proteins in wild-type virus then indicated that both three-part gH/gL/gp42 and two-part gH-gL complexes coexist (WANG et al. 1998).

The interpretation of these data (WANG et al. 1998) was that in contrast to infection of B cells, where an interaction between gp42 and HLA class II is required, infection of epithelial cells requires direct interaction of gH with a yet unidentified novel coreceptor which serves an analogous function to HLA class II. Conversion of all gH/gL complexes in virus to the three-part form gH/gL/gp42 by addition of soluble gp42.Fc, or addition of Mab EID1 blocks the interaction of gH and this novel coreceptor. The possibility also exists that in the absence of CR2, for example on AGS cells, the coreceptor functions as a primary receptor as well.

2.4 Implications for Pathogenesis

The addition of gp42 to the EBV gH/gL complex is an event that from an evolutionary standpoint might be thought of as occurring as divergence of EBV as a gamma herpesvirus took place allowing expansion of its host range from epithelial cells to lymphocytes. The retention of two forms of the complex in virus, however, has interesting potential implications for both cell types.

In the case of the B cell, the use of a highly polymorphic molecule as a coreceptor for entry raises the question of whether or not there might be differences in the efficiency with which alleles expressed in different individuals might function. Expression of alleles with higher binding affinities for gp42 might influence the ease with which lymphocytes can be infected and thus potentially affect the size of the virus load that is established in an individual during primary infection. In a similar way, mutations in gp42 that increase or decrease its affinity for HLA class II could affect the fitness and spread of a given virus isolate.

There is some controversy in the field as to whether or not infection of epithelial cells normally occurs during infection with the virus. Indeed it has been cogently argued that there is no need to evoke a role for any cell other than the B lymphocyte in the persistence of EBV in vivo (THORLEY-LAWSON et al. 1996). However, although infection of epithelial cells is not very efficient, a model that uniquely involves EBV with B cells provides little insight into the development of nasopharyngeal carcinoma, its presence in carcinomas of the gastrointestinal tract (GULLEY et al. 1996; OSATO and IMAI 1996), and its replication in the lesions of oral hairy leukoplakia (GREENSPAN et al. 1985). Certainly access to B cells at mucosal surfaces during primary infection would be easier if at least a small amount of replication were possible in the epithelium, even if it were not a persistent phenomenon. A transient, but nevertheless important interaction between EBV and epithelial cells might explain the conflicting results from several searches for replicating virus in this cell type (LEMON et al. 1977; SIXBEY et al. 1984; NIEDOBITEK et al. 1992; ANAGNASTOPOULOS et al. 1995). The finding that a proportion of the gp42 that is made in B cells is degraded as a result of association with HLA class II implies that virus replicated in an HLA class II-negative epithelial cell might contain a higher proportion of gH complexes in the three part gH/gL/gp42 form. This should increase its ability to infect B cells and decrease its tropism for epithelial cells, driving it directionally through one round of lytic replication in an epithelial cell into latency in a B cell.

3 The gN/gM Complex

3.1 Identifying the Components

The second heteromeric complex of proteins detected in EBV is the gN/gM complex, so named because it also represents two molecules that are conserved throughout the herpesvirus family. The EBV gN homolog is encoded by the BLRF1 ORF (BARNETT et al. 1992; JÖNS et al. 1996). The sequence predicts a type 1 membrane protein of 102 amino acids with a 9-amino-acid cytoplasmic tail and a cleavable signal peptide of 32 amino acids. An anti-peptide antibody made to a sequence in the predicted extracellular domain immunoprecipitated an 8-kDa doublet from cells expressing the ORF as a recombinant protein (LAKE et al. 1998). However, the same antibody immunoprecipitated the 8-kDa doublet, and four additional glycosylated proteins of approximately 113, 84, 48, and 15kDa from virus-producing Akata cells. The 15-kDa species proved to be the mature form of gN. After cleavage of the signal peptide, O-linked sugar is added to an approximately 8-kDa molecule to produce the upper component of the 8-kDa doublet. This sugar is further processed by addition of 2,6-linked sialic acid, rendering the mature protein readily visible with radiolabeled sugar.

The remaining three higher-molecular-weight glycoprotein species proved to be products of the BBRF3 ORF, which encodes the EBV gM homolog. The sequence predicts a 405-amino-acid protein with multiple membrane spanning domains and a 78-amino-acid carboxyterminus, which is unusually rich in proline residues and is predicted to be on the cytoplasmic face of the membrane. An anti-peptide antibody made to a sequence in this predicted cytoplasmic tail immunoprecipitated the same complex of 113-, 84-, and 48-kDa glycoproteins as the antibody to gN. Long exposures of the autoradiograms provided evidence of mature gN as well. Expression of gM as a recombinant molecule recapitulated the expression of the three largest glycoproteins and coexpression of gN and gM as recombinant molecules restored the authentic processing of gN (LAKE et al. 1998). Thus expression of mature gN was dependent on coexpression with gM whereas gM expressed alone, at least as a recombinant protein expressed at high levels under control of a heterologous T7 promoter, was processed in the same way as gM made in the context of all other virus proteins. The complex thus appeared to behave in a manner very similar to that of gH and gL, where gH expression is dependent on gL, but gL is processed in the same way in the presence or absence of gH.

3.2 The Role of the gN/gM Complex in Virus Exit and Entry

The role of the gN/gM complex in the virus life cycle was initially explored by making recombinant viruses in which the BLRF1 ORF encoding gN was interrupted by a neomycin resistance cassette. Somewhat surprisingly in view of the findings described above with respect to the expression of the complex from plasmids, these viruses apparently lacked both gN and its partner gM. This might reflect a very rapid turnover of the smaller amounts of virally produced gM in the absence of its partner or a failure of the protein to be glycosylated, perhaps as a result of retention by another, unknown, virus protein. It does, however, argue that a virus that lacks gN is deficient in both components of the complex.

Further analysis of the phenotype of the virus lacking gN and gM revealed several defects. First, a comparison of the amount of the total amount of virion DNA associated with induced cells and the DNAse-resistant, encapsidated DNA that was released from cells indicated that consistently less virion DNA was released in the absence of the complex. This stimulated an electron microscopic examination of virus-producing cells. Cells making wild-type virus contained the expected condensed and marginated chromatin and capsids in various stages of maturation in the nucleus. These included capsids surrounded by what appeared to be rings of precursors. Virus particles could also be seen in vesicles in the cytoplasm and enveloped virus was visible outside cell. In contrast in cells making virus that lacks gN, there was a marked accumulation of capsids within the condensed chromatin itself. Although clearly not all particles were within chromatin, a majority were. Subjectively few viruses were seen in vesicles in the cytoplasm and the number of cells with extracellular virus was at too low a frequency to detect.

Next it was found that although the specificity of binding of recombinant virus to B cells was unchanged, more than tenfold less of the encapsidated virus DNA harvested from cells making recombinant virus bound to receptor positive, EBV negative cells than did similarly isolated wild-type virus DNA. Since the specificity of virus binding was unaltered this suggested either that preparations of recombinant virus contained a large number of empty virions with empty capsids that competed for binding or that the virus was damaged in some way. To test the first possibility, stocks of recombinant and wild-type virus that contained the same amount of virion DNA were added singly or as mixtures to receptor-positive cells and the amount of DNA that bound was measured. The amount of virion DNA bound in the mixtures was the same as would have been expected from simple addition of each virus. Thus, although less recombinant virus was bound to cells, there was no evidence that this was because the recombinant stocks contained empty virions that could compete for binding. To test the second possibility, namely, that virus was damaged in some way, equal amounts of recombinant and wild-type virus, as judged by virion DNA content, were sedimented through a 20–40% Nycodenz gradient. Fractions were collected and analyzed by slot blot and scanned to measure virus DNA content. Wild-type virus sedimented as a large peak consistent with enveloped virus and a second smaller peak that sedimented with a higher density. In contrast, little of the recombinant virus sedimented with the first lower density peak and more was found distributed throughout the higher density region of the gradient. This was consistent with the idea that more of the recombinant virus was either lacking an envelope or was incompletely enveloped. The viability of cells producing recombinant virus was typically about 50% less than those making wild-type virus at 72h after induction, even though similar numbers of cells were induced. This suggested that improperly enveloped particles might be being released from the recombinant cells as a result of cell death and lysis.

Finally, a defect was also found in those recombinant viruses that were able to bind to receptor positive cells. If the amounts of wild-type and recombinant virus added to EBV-negative Akata cells were adjusted so that equal amounts were bound, infectivity of the bound recombinant was significantly lower. To determine whether this represented a block in virus–cell fusion, wild-type virus, a recombinant virus that lacks gp42 and is known to be deficient in fusion, and the gN minus virus were bound to normal B cells. Cells and bound virus were then treated with the exogenous fusogen polyethylene glycol. Infectivity of virus lacking gp42 was restored at a low level and polyethylene glycol treatment even increased infectivity of wild-type virus. In repeated attempts, however, it was never possible to increase the infectivity of virus lacking gN, suggesting that the defect was in a step that occurred after the fusion of virus and cell. The combination of the binding defect and the post-fusion defect resulted in a virus that, when adjusted for DNA content, was greater than 3 logs lower in infectivity than wild-type virus.

Thus, in summary, viruses that lack both gN and at least a mature form of its partner gM are deficient at several points in the replication cycle. As described above, wild-type virus is thought to bud through the inner nuclear membrane and either follow the default exocytic pathway to the cell surface, or perhaps, more likely, undergo a second step of de-envelopment and re-envelopment. Enveloped particles bind to a new cell, fuse with the cell membrane, and the capsid moves away from the membrane to the nucleus. In contrast, it appears that in the absence of the gN/gM complex, capsids associate with condensed chromatin. Fewer of them appear as enveloped particles in vesicles and a significant amount of the virus that is released is damaged, perhaps lacking an envelope completely. Even those viruses that do bind to new cells are impaired in infectivity, likely in an event that occurs after fusion, perhaps involving movement of capsids away from the cell membrane to the nucleus. Thus, both association of capsids with envelopes in preparation for exit from the cells and dissociation of capsids from an envelope that has fused with a cell membrane to allow penetration into a cell appear to be impaired.

The hypothesis that is currently being explored to explain these data is first that loss of gN affects function of gM, and second that the long, potentially cytoplasmic tail of gM plays a role in association and dissociation of capsids with membranes. Although the phenotypes of other viruses lacking gN or gM have not been dramatic in vitro, Mettenleiter and colleagues have recently suggested that in pseudorabies virus gE, gI, and gM share overlapping functions that include maturation of enveloped virions (BRACK et al. 1999). In EBV, which lacks a gE or gI homolog, somewhat similar functions may have been concentrated in the gN/gM complex itself.

4 Unanswered Questions

The two heteromeric complexes of EBV that are described here provide several different examples of the ways in which membrane proteins impact virus replication, virus tropism, and thus ultimately virus pathogenesis. The gH/gL/gp42 complex clearly plays an important part in virus penetration and an unexpectedly important role in virus tropism. Neither has yet been fully explored, either for the B cell, where at least one of its cellular partners has been identified, or the epithelial cell, on which the coreceptor remains unidentified. Basic questions remain about whether the components of this complex are the only virus proteins required to mediate fusion of the virus envelope with the cell membrane and about how fusion is triggered; more subtle issues concerning potential polymorphism of either virus or cell proteins involved in the process also need to be addressed. There is even greater uncertainty about how proteins such as gN and gM influence the critical steps of virus assembly and disassembly and how they may complement the activities of gp110, the EBV gB homolog. Although these proteins do not pertain directly to the unique transforming properties of EBV, their influences on growth and spread of the virus have a far-reaching impact on the diseases that the virus causes.

References

Anagnastopoulos IMH, Kreschel C, Stein H (1995) Morphology, immunophenotype, and distribution of latently and/or productively Epstein-Barr virus-infected cells in acute infectious mononucleosis: implications for the interindividual infection route of Epstein-Barr virus. Blood 85:744–750

Baer R, Bankier AT, Biggin MD, Deininger PL, Farrell PJ, Gibson TJ, Hatfull G, Hudson GS, Satchwell SC, Seguin C, Tuffnell PS, Barrell BG (1984) DNA sequence and expression of the B95-8 Epstein-Barr virus genome. Nature 310:207–211

Barnett BC, Dolan A, Telford EAR, Davison AJ, McGeoch DJ (1992) A novel herpes simplex virus gene (UL49A) encodes a putative membrane protein with counterparts in other herpesviruses. J Gen Virol 73:2167–2171

Beisel C, Tanner J, Matsuo T, Thorley-Lawson D, Kezdy F, Kieff E (1985) Two major outer envelope glycoproteins of Epstein-Barr virus are encoded by the same gene. J Virol 54:665–674

Borza C, Hutt-Fletcher LM (1998) Epstein-Barr virus recombinant lacking expression of glycoprotein gp150 infects B cells normally but is enhanced for infection of the epithelial line SVKCR2. J Virol 72:7577–7582

Brack AR, Dijkstra JM, Granzow H, Klupp BG, Mettenleiter TC (1999) Inhibition of virion maturation by simultaneous deletion of glycoproteins E, I, and M of pseudorabies virus. J Virol 73: 5364–5372

Emini EA, Luka J, Armstrong ME, Keller PM, Ellis RW, Pearson GR (1987) Identification of an Epstein-Barr virus glycoprotein which is antigenically homologous to the varicella-zoster glycoprotein II and the herpes simplex virus glycoprotein B. Virology 157:552–555

Fingeroth JD, Weis JJ, Tedder TF, Strominger JL, Biro PA, Fearon DT (1984) Epstein-Barr virus receptor of human B lymphocytes is the C3d complement CR2. Proc Natl Acad Sci USA 81:4510–4516

Gayle MA, Sims JE, Dower SK, Slack JL (1994) Monoclonal antibody 1994–2001 (also known as ALVA 42) reported to recognize type II IL-1 receptor is specific for HLA-DR alpha and beta chains. Cytokine 6:83–86

Gong M, Kieff E (1990) Intracellular trafficking of two major Epstein-Barr virus glycoproteins, gp350/220 and gp110. J Virol 64:1507–1516

Gong M, Ooka T, Matsuo T, Kieff E (1987) Epstein-Barr virus glycoprotein homologous to herpes simplex virus gB. J Virol 61:499–508

Greenspan JS, Greenspan D, Lennette ET, Abrams DI, Conant MA, Petersen V, Freese UK (1985) Replication of Epstein-Barr virus within the epithelial cells of oral "hairy" leukoplakia, an AIDS-associated lesion. New Engl J Med 313:1564–1571

Gulley ML, Pulitzer DR, Eagan PA, Schneider BG (1996) Epstein-Barr virus infection is an early event in gastric carcinogenesis and is independent of bcl-2 expression and p53 accumulation. Hum Pathol 27:20–27

Haddad RS, Hutt-Fletcher LM (1989) Depletion of glycoprotein gp85 from virosomes made with Epstein-Barr virus proteins abolishes their ability to fuse with virus receptor-bearing cells. J Virol 63:4998–5005

Heineman T, Gong M, Sample J, Kieff E (1988) Identification of the Epstein-Barr virus gp85 gene. J Virol 62:1101–1107

Herrold RE, Marchini A, Frueling S, Longnecker R (1995) Glycoprotein 110, the Epstein-Barr virus homolog of herpes simplex virus glycoprotein B, is essential for Epstein-Barr virus replication in vivo. J Virol 70:2049–2054

Jöns A, Granzlow H, Kuchling R, Mettenleiter TC (1996) The UL49.5 gene of pseudorabies virus codes for an O-glycosylated structural protein of the virus envelope. J Virol 70:1237–1241

Knox PG, Young LS (1995) Epstein-Barr virus infection of CR2-transfected epithelial cells reveals the presence of MHC class II on the virion. Virology 213:147–157

Kurilla MG, Heineman T, Davenport LC, Kieff E, Hutt-Fletcher LM (1995) A novel Epstein-Barr virus glycoprotein gp150 expressed from the BDLF3 open reading frame. Virology 209:108–121

Lake CM, Molesworth SJ, Hutt-Fletcher LM (1998) The Epstein-Barr virus (EBV) gN homolog BLRF1 encodes a 15kDa glycoprotein that cannot be authentically processed unless it is co-expressed with the EBV gM homolog BBRF3. J Virol 72:5559–5564

Lee SK (1999) Four consecutive arginine residues at positions 836–839 of EBV gp110 determine intracellular localization of gp110. Virology 264:350–358

Lee SK, Longnecker R (1997) The Epstein-Barr virus glycoprotein 110 carboxy-terminal tail domain is essential for lytic virus replication. J Virol 71:4092–4097

Lemon SM, Hutt LM, Shaw JE, Li J-LH, Pagano JS (1977) Replication of EBV in epithelial cells during infectious mononucleosis. Nature 268:268–270

Li QX, Spriggs MK, Kovats S, Turk SM, Comeau MR, Nepom B, Hutt-Fletcher LM (1997) Epstein-Barr virus uses HLA class II as a cofactor for infection of B lymphocytes. J Virol 71:4657–4662

Li QX, Turk SM, Hutt-Fletcher LM (1995) The Epstein-Barr virus (EBV) BZLF2 gene product associates with the gH and gL homologs of EBV and carries an epitope critical to infection of B cells but not of epithelial cells. J Virol 69:3987–3994

Li QX, Young LS, Niedobitek G, Dawson CW, Birkenbach M, Wang F, Rickinson AB (1992) Epstein-Barr virus infection and replication in a human epithelial system. Nature 356:347–350

Mackett M, Conway MJ, Arrand JR, Haddad RS, Hutt-Fletcher LM (1990) Characterization and expression of a glycoprotein encoded by the Epstein-Barr virus BamHI 1 fragment. J Virol 64:2545–2552

Mellman I, Pierre P, Amigorena S (1995) Lonely MHC molecules seeking immunogenic peptides for meaningful relationships. Current Opinion in Cell Biology 7:564–572

Miller N, Hutt-Fletcher LM (1988) A monoclonal antibody to glycoprotein gp85 inhibits fusion but not attachment of Epstein-Barr virus. Journal of Virology 62:2366–2372

Modrow S, Hoflacker B, Wolf H (1992) Identification of a protein encoded in the EB-viral open reading frame BMRF2. Arch Virol 127:379–386

Nemerow GR, Houghton RA, Moore MD, Cooper NR (1989) Identification of the epitope in the major envelope proteins of Epstein-Barr virus that mediates viral binding to the B lymphocyte EBV receptor (CR2). Cell 56:369–377

Nemerow GR, Wolfert R, McNaughton M, Cooper NR (1985) Identification and characterization of the Epstein-Barr virus receptor on human B lymphocytes and its relationship to the C3d complement receptor (CR2). J Virol 55:347–351

Niedobitek G, Herbst H, Young LS, Brooks L, Masucci MG, Crooker J, Rickinson A, Stein H (1992) Patterns of Epstein-Barr virus infection in non-neoplastic lymphoid tissue. Blood 79: 2520–2526

Nolan LA, Morgan AJ (1995) The Epstein-Barr virus open reading frame BDLF3 codes for a 100–150kDa glycoprotein. J Gen Virol 76:1381–1392

Oba DE, Hutt-Fletcher LM (1988) Induction of antibodies to the Epstein-Barr virus glycoprotein gp85 with a synthetic peptide corresponding to a sequence in the BXLF2 open reading frame. J Virol 62:1108–1114

Osato T, Imai S (1996) Epstein-Barr virus and gastric carcinoma. Semin Cancer Biol 7:175–182

Papworth MA, Van Dijk AA, Benyon GR, Allen TD, Arrand JR, Mackett M (1997) The processing, transport and heterologous expression of Epstein-Barr virus gp110. J Gen Virol 78:2179–2189

Penaranda ME, Lagenaur LA, Pierek LT, Berline JW, MacPhail LA, Greenspan D, Greenspan J, Palefsky JM (1997) Expression of Epstein-Barr virus BMRF-2 and BDLF-3 genes in hairy leukoplakia. J Gen Virol 78:3361–3370

Sinclair AJ, Farrell PJ (1995) Host cell requirements for efficient infection of quiescent primary B lymphocytes by Epstein-Barr virus. J Virol 69:5461–5468

Sixbey JW, Nedrud JG, Raab-Traub N, Hanes RA, Pagano JS (1984) Epstein-Barr virus replication in oropharyngeal epithelial cells. New Engl J Med 310:1225–1230

Spear PG (1993) Entry of alphaviruses into cells. Semin Virol 4:167–180

Spriggs MK, Armitage RJ, Comeau MR, Strockbine L, Farrah T, MacDuff B, Ulrich D, Alderson MR, Mullberg J, Cohen JI (1996) The extracellular domain of the Epstein-Barr virus BZLF2 protein binds the HLA-DR beta chain and inhibits antigen presentation. J Virol 70:5557–5563

Strnad BC, Schuster T, Klein R, Hopkins RFI, Witmer T, Neubauer R, Rabin H (1982) Production and characterization of monoclonal antibodies against the Epstein-Barr virus membrane antigen. J Virol 41:258–264

Sugano N, Chen W, Roberts ML, Cooper NR (1997) Epstein-Barr virus binding to CD21 activates the initial viral promoter via NF B induction. J Exp Med 186:731–737

Tanner J, Weis J, Fearon D, Whang Y, Kieff E (1987) Epstein-Barr virus gp350/220 binding to the B lymphocyte C3d receptor mediates adsorption, capping and endocytosis. Cell 50:203–213

Tanner J, Whang Y, Sample J, Sears A, Kieff E (1988) Soluble gp350/220 and deletion mutant glycoproteins block Epstein-Barr virus adsorption to lymphocytes. J Virol 62:4452–4464

Thorley-Lawson DA, Miyashita EM, Khan G (1996) Epstein-Barr virus and the B cell: that's all it takes. Trends Microbiol 4:204–208

Trkola A, Dragic T, Arthos J, Binley JM, Olson WC, Allaway C, Cheng-meyer C, Robinson J, Maddon PJ, Moore JP (1996) CD-4 dependent antibody sensitive interactions between HIV-1 and its co-receptor CCR-5. Nature 384:184–187

Wang X, Hutt-Fletcher LM (1998) Epstein-Barr virus lacking glycoprotein gp42 can bind to B cells but is not able to infect. J Virol 72:158–163

Wang X, Kenyon WJ, Li QX, Mullberg J, Hutt-Fletcher LM (1998) Epstein-Barr virus uses different complexes of glycoproteins gH and gL to infect B lymphocytes and epithelial cells. J Virol 72:5552–5558

Wu L, Gerard NP, Wyatt R, Choe H, Parolin C, Ruffing N, Borsetti A, Cardoso AA, Desjardin E, Newman W, Gerard C, Sodroski J (1996) CD-4-induced interaction of primary HIV-1 gp120 glycoproteins with the chemokine receptor CCR-5. Nature 384:179–183

Yaswen LR, Stephens EB, Davenport LC, Hutt-Fletcher LM (1993) Epstein-Barr virus glycoprotein gp85 associates with the BKRF2 gene product and is incompletely processed as a recombinant protein. Virology 195:387–396

Yoshiyama H, Imai S, Shimizu N, Takada K (1997) Epstein-Barr virus infection of human gastric carcinoma cells: implication of the existence of a new virus receptor different from CD21. J Virol 71:5688–5691

EBV Replication Enzymes

T. Tsurumi

1	Introduction	65
2	Life Cycle of the Epstein-Barr Virus	66
3	EBV Lytic Replication Origin, *oriLyt*	68
4	The EBV Gene Products Essential for *oriLyt*-Dependent DNA Replication	69
5	EBV Replication Enzymes	69
5.1	*oriLyt*-Binding Protein, BZLF1 Gene Product	69
5.2	Epstein-Barr Virus DNA Polymerase Holoenzyme	70
5.3	EBV *Pol* Catalytic Subunit (BALF5 Gene Product)	71
5.4	EBV *Pol* Accessory Subunit (BMRF1 Gene Product)	72
5.5	Functional Interactions Between the BALF5 *Pol* Catalytic Subunit and the BMRF1 *Pol* Accessory Subunit	72
5.5.1	Polymerase Activity	72
5.5.2	$3'$-to-$5'$ Exonuclease Activity	73
5.6	Comparison with the Prokaryotic and Eukaryotic Replicative DNA Polymerases	75
5.7	EBV Single-Stranded DNA-Binding Protein (BALF2 Gene Product)	76
5.7.1	General Properties	76
5.7.2	Helix Destabilizing Activity	77
5.8	Functional Interactions Between the EBV SSB and the EBV *Pol* Holoenzyme	78
5.9	EBV Putative Helicase–Primase Complex	79
5.10	Physical Interaction Between the EBV *Pol* Catalytic Subunit and the EBV BBLF4–BSLF1–BBLF2/3 Complex	80
5.11	Polymerase and Helicase–Primase Complex Interactions in Other Systems	80
6	Proposed Model for the Initiation Step of the Lytic Phase of EBV DNA Replication	81
7	Proposed Model for the EBV Replication Fork	82
	References	83

1 Introduction

The Epstein-Barr virus (EBV) can choose between two alternative life styles. It infects B lymphocytes, transforming them into lymphoblastoid lines and, in contrast to neurotropic herpesviruses such as herpes simplex virus type I that establish latency in nondividing neurons, must maintain its latent genomes in cells that have

---

Division of Virology, Aichi Cancer Center Research Institute, 1-1, Kanokoden, Chikusa-ku, Nagoya 464-8681, Japan

the potential to divide. In B lymphoblastoid cell lines established by EBV infection, the viral genome is maintained as covalently closed circular plasmids forming nucleosomal structures with histone proteins. The number of copies is maintained at 10–50 per cell to be duplicated once during each cell division cycle by the host cellular DNA replication machinery. When production of virus is induced, the circular genome becomes a ready template for amplification, generating thousands of copies per cell during lytic infection. Replication intermediates are head-to-tail concatamers, perhaps through a rolling-circle DNA replication, which are cleaved and packaged into infectious viral particles. The lytic phase of EBV DNA replication is dependent on seven viral replication proteins: BZLF1, BALF5, BMRF1, BALF2, BBLF4, BSLF1, and BBLF2/3 gene products. The BZLF1 protein is an *oriLyt*-binding protein and also acts as the lytic transactivator. The BALF5 gene encodes the DNA *Pol* catalytic subunit and the BMRF1 gene encodes the DNA *Pol* accessory subunit. They form a complex to act as the *Pol* holoenzyme. A single-stranded (ss)DNA-binding protein is encoded by the BALF2 gene. The enzymatic activities of the remaining three proteins, encoded by the BBLF4, BSLF1, and BBLF2/3 genes, have yet to be determined but they are predicted to act as helicase-, primase-, and helicase–primase-associated proteins, respectively, from sequence homology to the herpes simplex virus type 1 (HSV-1) UL5, UL52, and UL8 genes. Viral replication proteins other than the BZLF1 protein conceivably work together at replication forks to synthesize leading and lagging strands of the concatemeric EBV genome. Thus, EBV uses two distinct systems to replicate its DNA. In this review, *cis*- and *trans*-acting viral factors involved in the lytic phase of the viral DNA replication are described with an especial focus on viral replication enzymes. Reconstitution of the lytic phase of EBV replication remains a challenge for the future.

2 Life Cycle of the Epstein-Barr Virus

A scheme of Epstein-Barr virus genome replication is depicted in Fig. 1. The life cycle of the Epstein-Barr virus is bipartite. EBV infects B lymphocytes via the viral receptor, CD21. When the virus enters the nucleus, the linear viral double-stranded (ds)DNA circularizes to form an extrachromosomal plasmid that is then amplified to multiple copies by some unknown mechanism. The EBV genome appears to replicate once per cell cycle during the S phase, following the rules of chromosome replication (YATES and GUAN 1991; SHIRAKATA et al. 1999), and is a negatively supercoiled circular plasmid DNA covered with histone proteins. The plasmid DNA is synthesized via theta-like DNA replication by host-cell replication machinery, resulting in a stable number of genome copies per cell. The EBV latent replication origin, *oriP*, has been shown to be the only element in the viral genome capable of supporting efficient, stable replication of recombinant plasmids introduced into cells under selection conditions (YATES et al. 1984). The viral protein, EBNA1, binds to repeat sequences both in the family of repeated sequences (FR)

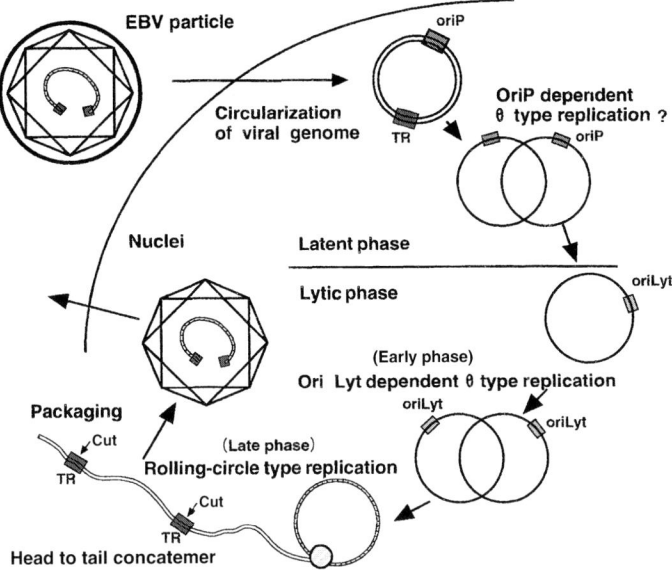

Fig. 1. Life cycle of the Epstein-Barr virus

and dyad symmetry (DS) regions of *oriP*. However, it now appears that its function as an initiation site for replication is not always needed. LITTLE and SCHILDKRAUT (1995) have revealed that initiation of replication on the EBV chromosome occurs not only at *oriP* but also at distant sites over a broad region. Both EBNA1 and FR are required for the EBV genome to be maintained in proliferating cells (AIYAR et al. 1998). At present EBNA1 is supposed to function in stable segregation of EBV episomes during cell division rather than in replication of the episome (AIYAR et al. 1998; LEE et al. 1999; SHIRE et al. 1999).

Full EBV replication appears to be highly dependent on the differentiated state of the epithelium. Also, replication of EBV, as well as other herpesviruses, can occur in cells treated with agents shown previously to arrest cell cycle progression. Thus, the lytic replication cycle can be induced in latently infected B cells experimentally by treatment with chemical agents such as phorbol esters, sodium *n*-butyrate, and calcium ionophores or by introduction of the BZLF1 gene encoded transactivator. Lytic replication differs from the latent amplification state in that multiple rounds of replication are initiated within *oriLyt* (HAMMERSCHMIDT and SUGDEN 1998), and the replication process has a greater dependence on EBV-encoded proteins (FIXMAN et al. 1995). The first genes expressed on induction are those for the immediate-early BZLF1 and BRLF1 transactivators. The initial transcription of these genes is further upregulated by the BZLF1 protein, which binds to specific sequences in both promoters. The BZLF1 and BRLF1 proteins then activate expression of the third transactivator BMLF1. The concerted action of these viral transactivators results in sequential activation of early gene expression followed by the lytic cascade of replication and late gene expression. The BZLF1 protein induces a G0/G1 cell-cycle block through the induction of $p53$ and the cyclin-dependent kinase inhibitors, $p21$ and $p27$, to facilitate the viral replicative program (CAYROL and FLEMINGTON 1996).

The *oriLyt*-mediated EBV DNA replication appears to be biphasic and a two-stage model has been proposed (PFULLER and HAMMERSCHMIDT 1996). Soon after induction of the lytic cycle, viral DNA is amplified to yield monomeric plasmid progeny DNA dependent on a functional *oriLyt* in *cis*. The BZLF1 protein binds to the *oriLyt*. The plasmid progeny DNAs have far fewer negative supercoils and fewer nucleosomes, which might be preferentially nicked by DNase I to provide the ideal template DNA for rolling-circle replication. In the late phase of the viral productive cycle, the EBV genome is amplified 100- to 1000-fold. Intermediates of viral DNA replication are found as large head-to-tail concatemeric molecules, probably resulting from rolling-circle DNA replication (HAMMERSCHMIDT and SUGDEN 1988), which are subsequently cleaved into unit length genomes and packaged into virions. At present, there is no information on how DNA replication switches from a presumed theta mode to a rolling-circle mode or what *trans*- and *cis*-acting factors are required for this transition.

3 EBV Lytic Replication Origin, *oriLyt*

The EBV genome contains two lytic-phase replication sequences, called *oriLyt*, which were determined by two essentially identical copies of a 695-bp gene, located in DR and DL, one of which is sufficient for lytic-cycle replication of EBV. *oriLyt* is located within the divergent promoter regions of the BHLF1 and BHRF1 genes and contains two essential core elements, separated by about 400bp, and several auxiliary components (HAMMERSCHMIDT and SUGDEN 1988; SCHEPERS et al. 1993). One core component, the upstream part of *oriLyt*, plays a dual role in transcription and replication, being a strong early promoter that is activated by the BZLF1 protein. Replication from *oriLyt* requires that the BZLF1 protein binds to these four sites (BZLF1-responsive elements; ZRE). Deletion of the TATA box, ZRE1, ZRE2, and the CAAT box abolishes replication, while mutation of the ZRE1 and ZRE2 sites reduces replication efficiency (SCHEPERS et al. 1996). The second essential domain, in the downstream part of *oriLyt*, is delineated by a central 225-bp region including two AT-rich palindromes and an adjacent polypurine-polypyrimidine tract. Such elements appear to facilitate the localized unwinding and helical destabilization essential for initiation of replication (GAHN and SCHILDKRAUT 1989; PORTES-SENTIS et al. 1997). Several cellular transcription factors have been found to bind to a functional subsequence of the downstream component, the TD element, which is extremely sensitive to mutations with regard to *oriLyt*-dependent replication (SCHEPERS et al. 1993; GRUFFAT et al. 1995). Two of these transcription factors, ZBP-89 and Sp1, stimulate replication and interact with the viral DNA polymerase-processivity factor complex (BAUMANN et al. 1999). The third component, a non-essential auxiliary element, is a powerful enhancer region that contains

DNA-binding sites for BRLF1 and BZLF1 proteins and responds synergistically to the presence of these transactivators. The transcriptional enhancer can be replaced by a heterologous enhancer (HAMMERSCHMIDT and SUGDEN 1988).

4 The EBV Gene Products Essential for *oriLyt*-Dependent DNA Replication

The essential EBV replication genes were identified by FIXMAN et al. (1992, 1995) with a transient DNA replication assay in which *oriLyt*-containing plasmids are replicated by transfected EBV sequences that supply *trans*-acting factors. This analysis allowed identification of seven essential core EBV replication genes (BZLF1, BALF5, BMRF1, BALF2, BBLF4, BSLF1, BBLF2/3) that are necessary and sufficient for *oriLyt*-specific DNA replication as well as the BRLF1 and BMLF1 lytic cycle transactivators. The BRLF1 protein is a transcriptional activator, which binds DNA as a dimer and acts in combination with the BZLF1 protein to produce a synergistic response in targets containing binding sites for both factors. BRLF1 augments DNA replication but is not required. BMLF1 primarily serves a post-transcriptional function and contributes to replication in an indirect manner. An absolute dependence was demonstrated for BZLF1 protein, which has an essential origin-binding function, mutation of all BZLF1 binding sites in the *oriLyt* abrogating replication function. The functions of the EBV replication gene products required for the lytic phase of viral replication are summarized in Table 1.

5 EBV Replication Enzymes

5.1 *oriLyt*-Binding Protein, BZLF1 Gene Product

The BZLF1 protein is the only transcription factor that is known to be essential for the replication function of the *oriLyt*. It composes 245 amino acids that contain two

Table 1. EBV DNA replication proteins

Protein	Gene	Size (kDa)	Function
Origin-binding protein	BZLF1	38	*oriLyt*-binding protein
DNA polymerase holoenzyme	BALF5	110	DNA polymerase, 3'-5' exonuclease
	BMRF1	50	dsDNA binding activity; DNA polymerase processivity factor
Single-stranded DNA-binding protein	BALF2	130	Single-stranded DNA-binding protein; Helix-destabilizing activity
DNA helicase–primase complex	BBLF4	98	Putative DNA helicase?
	BSLF1	89	Putative primase?
	BBLF2/3	80	Helicase–primase associated protein?

specific domains: a transactivation domain, which binds to TFIID, and a DNA-binding/dimerization domain. The protein activates its target genes by binding to BZLF1-responsive elements (ZRE) which are similar to AP1 sites and are present in many of the EBV early gene promoters as well as in the promoters of the two immediate-early genes, BZLF1 itself and BRLF1 (PACKHAM et al. 1990). The BZLF1 protein is a member of the bZIP family and shares homology within the DNA-binding domain to the c-Fos and c-Jun proteins (FARRELL et al. 1989). It binds as a homodimer to ZRE sites and its binding to DNA is modified by phosphorylation. A region within the N terminus genetically separates transcriptional activity from replication function within the polypeptide (SARISKY et al. 1996). The mechanisms by which the BZLF1 protein functions to fulfill its essential role in the replication process are not currently understood, but it has recently reported that it interacts with the CREB-binding protein (CBP) in a functional manner to activate EBV early gene expression (ADAMSON and KENNY 1999). CBP possesses histone acetylase activity and functions as a coactivator for several cellular transcriptional activators. In addition to the role in inducing early transcription, this interaction could potentially be required for histone acetylation near *oriLyt* ZRE sites, thereby weakening the association of histones and *oriLyt* allowing access of the replication machinery.

5.2 Epstein-Barr Virus DNA Polymerase Holoenzyme

The EBV DNA polymerase has been purified to varying degrees from cultured EBV-infected cells and characterized (KALLIN et al. 1985; LI et al. 1987; TSURUMI 1991a). The purified EBV DNA polymerase contains the 110-kDa catalytic polypeptide encoded by the BALF5 open reading frame with the 48- to 55-kDa nuclear phosphoproteins encoded by the BMRF1 open reading frame identified as a part of the EA-D (early antigen diffuse component). Neutralization of the EBV DNA polymerase activity by monoclonal antibody to the BMRF1 protein (CHIOU et al. 1985) and low activity in the DNA polymerase fraction lacking BMRF1 (KIEHL and DORSKY 1991) strongly suggests that the EBV DNA polymerase catalytic subunit forms a complex with the BMRF1 protein in EBV-infected cells to function as the EBV *Pol* holoenzyme.

DNA polymerase activity of the purified EBV DNA polymerase is stimulated by ammonium sulfate when activated DNA is used as template-primer and is inhibited by aphidicolin or phosphonoacetic acid (TSURUMI 1991b). Poly(dC) oligo(dG) is a good template-primer for the EBV DNA polymerase. Furthermore, the EBV *Pol* holoenzyme can efficiently extend both RNA and DNA primers on the template DNA without ATP hydrolysis and exhibits strikingly high processivity, which is a desirable feature for the synthesis of multiple copies of the EBV genome in rolling circle DNA replication (TSURUMI 1991b).

The 3'-to-5' exonuclease activity has been also demonstrated to be associated with the purified EBV DNA *Pol* holoenzyme, which liberates 5'-deoxynucleoside monophosphates from primer termini (TSURUMI 1991a). The exonuclease activity is stimulated by ammonium sulfate and is proposed to have a proofreading function as it preferentially excises terminally mismatched nucleotides incorporated at the primer terminus. The 3'-to-5' exonuclease activity can be selectively inhibited by purine ribonucleoside 5-monophosphates, while no inhibition of the DNA polymerase activity is observed (TSURUMI 1992). Kinetic studies have shown that 5-GMP, the most potent inhibitor of the exonuclease, inhibits its activity competitively with respect to DNA template-primer. The exonuclease domain appears to be separated from the polymerase domain in the EBV DNA polymerase molecule as is the case with the Klenow fragment of *Escherichia coli Pol* I (JOYCE 1989) and 5-GMP binds to the catalytic site of the 3'-to-5' exonuclease domain, rather than the polymerase domain, blocking entry of a primer terminus into the catalytic site of the exonuclease. The EBV DNA polymerase catalyzes DNA-dependent conversion of complementary or noncomplementary deoxynucleoside triphosphates to the monophosphate form with poly(dT).oligo(rA) as a template primer, suggesting functional association of exonuclease with polymerase activity (TSURUMI 1992).

5.3 EBV *Pol* Catalytic Subunit (BALF5 Gene Product)

It has been demonstrated in an in vitro transcription-translation system that the BALF5 open reading frame encodes an EBV DNA polymerase catalytic subunit (KIEHL and DORSKY 1991; LIN et al. 1991). To characterize enzymatic activity and functional interactions between the subunits of EBV DNA polymerase holoenzyme in detail, the EBV DNA *Pol* catalytic subunit (BALF5 protein) and its accessory subunit (BMRF1 protein) have been independently overexpressed in insect cells and purified (TSURUMI 1993a,c). The expressed EBV *Pol* catalytic polypeptide (BALF5 gene product) purified from recombinant virus AcBALF5-infected insect cells, with a molecular mass of 110kDa, exhibited both DNA polymerase and 3'-to-5' exonuclease activities (TSURUMI et al. 1993c). Thus, the 3'-to-5' exonuclease activity associated with the EBV DNA polymerase (TSURUMI 1991a) is an inherent feature of the BALF5 polymerase catalytic polypeptide. The DNA polymerase and the exonuclease activities associated with the EBV DNA polymerase catalytic subunit are sensitive to ammonium sulfate (TSURUMI et al. 1993b,c, 1994) in contrast to those of the polymerase complex purified from EBV producing lymphoblastoid cells, which are stimulated by the salt (TSURUMI 1991b). Thus, it might be speculated that the binding affinity of the *Pol* catalytic subunit for the primer terminus is weak and that high ionic strength destabilizes the catalytic protein-primer terminus interaction. The BMRF1 *Pol* accessory protein may increase the affinity of the polymerase for the primer terminus by its double stranded (ds)DNA binding activity and decrease the dissociation of the polymerase from the template DNA.

The template-primer preference for the polymerase catalytic subunit was found to be different from that for the polymerase complex (TSURUMI et al. 1993c). The EBV DNA polymerase purified from EBV-producing cells utilized poly(dC).oligo(dG) 29-fold more efficiently than activated DNA as template-primer, while the single subunit of the EBV DNA polymerase utilized it poorly. The latter, however, is sensitive to PAA or aphidicolin as well as the EBV *Pol* holoenzyme.

5.4 EBV *Pol* Accessory Subunit (BMRF1 Gene Product)

The BMRF1 protein is a major component of the EBV EA-D complex and is the major early phosphoprotein induced during EBV infection (PEARSON et al. 1983; EPSTEIN 1984). In Burkitt's lymphoma-cell lytic EBV replication, the BMRF1 and BALF5 proteins co-localize to intranuclear replication compartments where DNA synthesis occurs (KIEHL and DORSKY 1991). The 48- to 55-kDa phosphoprotein BMRF1 gene products appear to function as EBV DNA *Pol* accessory proteins since the EBV DNA polymerase activity is neutralized by specific monoclonal antibodies (CHIOU et al. 1985). Overexpressed in the recombinant baculovirus AcBMRF1-infected insect cells, the BMRF1 gene products have been purified and characterized (TSURUMI 1993a) as phosphorylated forms of 52 and 50kDa and an unphosphorylated form of 48kDa. The functional significance of the phosphorylation has yet to be elucidated. The BMRF1 DNA *Pol* accessory subunits have neither DNA polymerase nor exonuclease activity, but exhibit higher binding affinity for double stranded than for single-stranded DNA without ATP hydrolysis. The protein–DNA interaction does not require a primer terminus.

5.5 Functional Interactions Between the BALF5 *Pol* Catalytic Subunit and the BMRF1 *Pol* Accessory Subunit

5.5.1 Polymerase Activity

The DNA polymerase activity catalyzed by the BALF5 protein in the presence or absence of the BMRF1 *Pol* accessory subunit has been compared in vitro using short or long single-stranded (ss)DNA templates in order to facilitate the study of the role of each of these two components in the EBV DNA polymerase reaction (TSURUMI et al. 1993b).

The BALF5 catalytic subunit alone was sensitive to 100mM ammonium sulfate with activated DNA template (90% inhibition), but addition of the *Pol* accessory subunit greatly enhanced the DNA polymerase activity under these conditions (tenfold stimulation). Optimal stimulation was obtained with a molar ratio of BMRF1 protein/BALF5 protein of two or more. The DNA polymerase activity of the combined subunits was neutralized by the monoclonal antibody to the BMRF1 protein, whereas that of BALF5 protein alone was not (TSURUMI et al. 1993b). Bipartite DNA-binding region of the BMRF1 protein is essential for the DNA polymerase accessory function (KIEHL and DORSKY 1995).

The BALF5 *Pol* catalytic subunit alone extended the primer slightly (~50 nucleotides) and no full-length product was observed (TSURUMI et al. 1993b). Analyses of the replication products in an alkaline agarose gel showed no increase in size of the products throughout the time course, indicating a distributive action of the *Pol* catalytic subunit in the absence of the BMRF1 *Pol* accessory protein. Addition of BMRF1 protein, however, resulted in accumulation of full-length replicative form II. Some of the products demonstrated specific bands for pause sites on the template, which presumably represent sites of substantial helical regions on the ssDNA template. In the presence of the BMRF1 protein the BALF5 *Pol* catalytic protein moved through these barriers and completed synthesis of the 7.2kb M13mp18 template within 20min. Most of the primed ssDNA remained unchanged as detected by UV-induced ethidium bromide fluorescence, during the time in which full-length products (RF II) were formed (data not shown). These observations support a highly processive mode of nucleotide polymerization (>7200 nucleotides) by the EBV DNA *Pol* holoenzyme reconstituted in vitro, consistent with previous observations of the EBV DNA *Pol* holoenzyme purified from EBV-producing lymphoblastoid cells (TSURUMI 1991b). The 20min required for the complete replication of a M13 ssDNA circle (7.2kb) yields an average nucleotide turnover of 6 nucleotides/s/polymerase molecule. In the absence of the accessory protein, the template is replicated at a rate of about 1.5 nucleotides/s, quantified by measuring the kinetics of deoxyribonucleotide incorporation. Thus, addition of BMRF1 protein results in a highly processive mode of polymerization with at least fourfold stimulation of the rate of incorporation to a value of 6 nucleotides/s. These observations suggest that the BMRF1 protein acts at growing primer terminus to increase the processivity of the DNA polymerase by decreasing the dissociation of the polymerase from the 3'-primer terminus of the growing chain during polymerization.

5.5.2 3'-to-5' Exonuclease Activity

The EBV DNA polymerase catalytic subunit, BALF5 gene product, possesses an intrinsic 3'-to-5' proofreading exonuclease activity in addition to 5'-to-3' DNA polymerase activity (TSURUMI et al. 1993c). The exonuclease hydrolyses both double- and ssDNA substrates with 3'-to-5' directionality, releasing deoxyribonucleoside 5'-monophosphates, which for double strands are very sensitive to high ionic strength, whereas the single-strand exonucleolytic activity is moderately resistant (TSURUMI et al. 1994). Addition of the BMRF1 polymerase accessory subunit to the reaction enhances the double-strand exonucleolytic activity in the presence of high concentrations of ammonium sulfate (fourfold stimulation at 75mM ammonium sulfate). Optimal stimulation was obtained again when the molar ratio of BMRF1 protein/BALF5 protein was two or more, identical to the case for reconstituting optimum DNA polymerizing activity (TSURUMI et al. 1993b). Furthermore, product size analyses revealed that the polymerase catalytic subunit alone excised a few nucleotides from the 3' termini of the primer hybridized to template DNA and that the addition of the BMRF1 polymerase accessory subunit stimulated the nucleotide excision several fold. In contrast, the hydrolysis of single stranded DNA by the BALF5 protein was not affected by the addition of the BMRF1 polymerase accessory subunit. These observations suggest that the BMRF1 polymerase accessory subunit forms a complex with the BALF5 polymerase catalytic subunit to stabilize the interaction of the holoenzyme complex with the 3'-OH end of the primer on template DNA during exonucleolysis.

Our current concepts regarding the role of the BMRF1 *Pol* accessory subunit in the 3'-to-5' exonuclease activity of the EBV *Pol* catalytic subunit acting on single stranded or duplex DNA ends are described below and depicted in Fig. 2 (TSURUMI et al. 1994). (1) When the *Pol* catalytic subunit and partially dsDNA substrate are present in the reaction, the *Pol* catalytic subunit excises a few nucleotides from the 3' end. The exonucleolytic movement of the enzyme on the duplex DNA is very slow. If the physical model for the editing function of the Klenow fragment can be applied for the EBV DNA polymerase, the last several nucleotides of the primer end must become single-stranded to reach the 3'-to-5' exonuclease active site. (2) The addition of the *Pol* accessory subunit stimulates the rate of hydrolysis of the dsDNA substrates by the *Pol* catalytic subunit strongly. The *Pol* accessory subunit and the *Pol* catalytic subunit appear to assemble at the primer terminus to hold the polymerase holoenzyme on the primer end for a prolonged period during exonucleolysis. (3) When the *Pol* catalytic subunit and ssDNA substrate are present in the reaction, the degradation of the single-stranded template appears to be processive. The *Pol* catalytic subunit may preferentially bind to the 3'-OH primer end of the single stranded DNA substrate and hydrolyze rapidly. (4) Even when the *Pol* catalytic subunit interacts with the *Pol* accessory subunit to form the *Pol* holoenzyme, the catalytic subunit of the holoenzyme appears to interact with the 3' hydroxyl terminus of the single-stranded oligonucleotide and degrade the single-strand substrates at the same rate as the *Pol* catalytic subunit alone.

Thus, the BMRF1 *Pol* accessory subunit does not affect the single strand exonuclease activity at all. Therefore, the 3'-to-5' exonuclease active site of the EBV

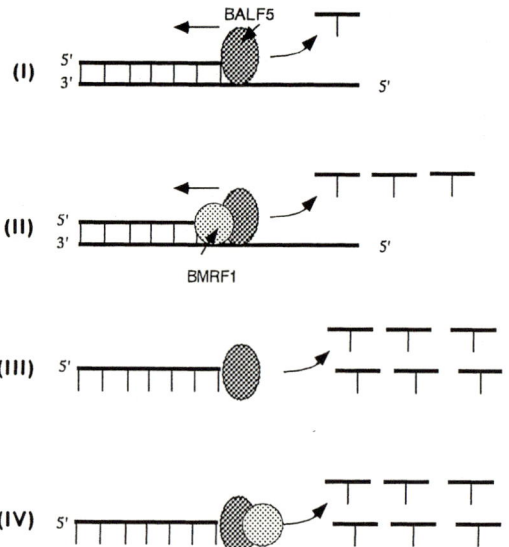

Fig. 2. The action of the 3'-5' exonuclease activities of the BALF5 *Pol* catalytic subunit in the presence and absence of the BMRF1 *Pol* accessory subunit on single-stranded and partially duplex DNA substrates

Pol holoenzyme appears to bind and act at the 3' end of the ssDNA substrate. Alternatively, it is possible that the *Pol* accessory subunit does not make a complex with *Pol* catalytic subunit in the absence of a primer template junction.

5.6 Comparison with the Prokaryotic and Eukaryotic Replicative DNA Polymerases

Most replicative DNA polymerases consist of a catalytic subunit and accessory proteins and display highly processive polymerization (Table 2). The BMRF1 protein plays an analogous role to the accessory subunits of phage T4 DNA polymerase, DNA polymerase δ, and *E. coli* DNA *Pol* III holoenzyme to increase the processivity and thereby the macroscopic rate of DNA synthesis by the enzymatically active core polymerases. In the case of phage T4, the gene 43 protein itself is a DNA polymerase, and a 3'-to-5' exonuclease (YOUNG et al. 1992). The three accessory proteins encoded by genes 44/62 and 45 of phage T4 increase the processivity and hence the rate of polymerase activity (JAVIS et al. 1989; YOUNG et al. 1992). The T4 gene 44/62 complex has both DNA-dependent ATPase activity and primer terminus binding activity that are stimulated by gene 45 protein. The ATPase activity is thought to reflect an energy requirement to maintain a stable polymerase accessory protein complex at the primer terminus. The gene 45 protein by itself has little DNA binding activity, but forms a sliding clamp to increase the processivity. As a second example, processive DNA synthesis by DNA polymerase δ requires eukaryotic replication factors, RF-C and PCNA (HURWITZ et al. 1990; TURIMOTO and STILLMAN 1991). RF-C exhibits DNA-dependent ATPase activity and primer terminus binding activity as well as T4 gene 44/62 complex (TSURIMOTO and STILLMAN 1990). PCNA cooperates with RF-C to stimulate processive DNA synthesis by DNA polymerase δ, like the T4 gene 45 protein. PCNA by itself has little DNA binding activity. The DNA *Pol* III holoenzyme of *E. coli* shows a similar functional homology (HURWITZ et al. 1990), accessory subunits γ/δ, and β cooperating to increase processivity (FAY et al. 1982). The γ/δ complex has a DNA-dependent ATPase and ATP hydrolysis is required for these subunits to form a preinitiation complex on a primed DNA template. The β subunit of *E. coli Pol* III has no intrinsic affinity for DNA but dimerizes to form a torus that is sterically retained on the nucleic acid strand (O'DONNELL and STUDWELL 1990). In all three

Table 2. Replicative DNA polymerase holoenzymes

	E. coli *Pol* III	Bacteriophage T4	Eukaryotic Pol δ	EBV
Pol catalytic subunit	Core (α, ε, θ)	Gene43	δ *Pol*	BALF5
Pol accessory subunits	β (Clamp) γ/δ (Clamp loader; DNA-dependent ATPase)	Gene45 (Clamp) Gene44/62	PCNA (Clamp) RF-C	BMRF1 (dsDNA binding)

systems the action of the accessory proteins can be visualized as clamping the core polymerase to the primer template, enabling processive DNA synthesis to occur. Processivity of the bacteriophage T7 DNA polymerase, the product of gene 5, is increased by *E. coli* thioredoxin (HUBER et al. 1987). The gene 5 protein has low processivity, dissociating from the primer template after catalyzing the incorporation of 1–50 nucleotides. *E. coli* thioredoxin binds tightly to the gene-5 protein with a 1:1 stoichiometry and increases the processivity of polymerization by 1000-fold. Thioredoxin by itself possesses neither DNA-binding activity nor ATPase activity. It has been speculated that thioredoxin binds gene-5 protein at the edge of the DNA-binding crevice such that the two proteins together clamp the primer template into position.

Although the BMRF1 protein does not exhibit DNA dependent ATPase activity, it possesses dsDNA-binding activity for primer recognition and acts to increase processivity. However, the DNA-binding properties of EBV BMRF1 protein differ from those of the accessory proteins in these systems. For example, the BMRF1 protein binds to dsDNA without ATP hydrolysis and does not require a primer terminus, whereas RF-C, for example, binds specifically to primer terminus junctions but not to ssDNA or dsDNA (TSURIMOTO and STILLMAN 1990). The HSV-1 Pol accessory subunit increases the processivity of polymerization catalyzed by the HSV-1 Pol catalytic subunit (GOTTLIEB et al. 1990). GOTTLIEB and CHALLBERG (1994) demonstrated that HSV-1 *Pol* catalytic subunit protects 14bp of the 3' duplex region and an adjacent 18 bases of the single-stranded template. The HSV-1 *Pol* accessory subunit results in additional protection of a contiguous 5–14bp in the duplex region but does not affect the 5' position of the *Pol* subunit. From these observations, it can be speculated that the increase in processivity in the presence of the BMRF1 protein is related to the dsDNA-binding activity of free BMRF1 protein and that the role of the latter in the EBV DNA polymerase complex is to act as a clamp stabilizing the BALF5 protein at the primer terminus without energy requirements. However, it remains to be clarified how the BMRF1 *Pol* accessory protein can slide to follow the elongating primer without acting as a brake.

5.7 EBV Single-Stranded DNA-Binding Protein (BALF2 Gene Product)

5.7.1 General Properties

The EBV ssDNA-binding protein (EBV SSB) is the product of the BALF2 gene (TSURUMI et al. 1996, 1998) and consists of 1128 amino acids with a calculated molecular mass of 123,122Da. Although Raji cell is a virus-nonproducer line, the lytic cycle can be rescued by the expression of the BALF2 protein (DECAUSSIN et al. 1995). Studies on a number of SSB proteins have already revealed an distinctive structural motif (PRASAD and CHIU 1987; WANG and HALL 1990; GUTIERREZ et al. 1991): (K/R)-X3–12-(K/R)-X4–15-(Y/F/W)-X3–36-(Y/F)-X4–13-(Y/F)-X0–27-(K/R)-X6–11-(K/R). This is based on the best alignment of aromatic and positively charged residues in a number of SSB proteins, which otherwise lack a strong amino acid homology. There is considerable evidence that the aromatic and positively charged residues in the ssDNA binding motif are directly involved in the binding of T4 gene 32 protein (SHAMOO et al. 1989), *fd* gene 5 protein (KING and COLEMAN 1988), *E. coli* SSB (KHAMIS et al. 1987), and Adenovirus SSB (NEAL and KITCHINGMAN 1990) to ssDNA. Since the EBV SSB possesses the consensus motif (741–785aa) as with other SSB proteins, we speculate that a similar binding mechanism also occurs in this case. EBV SSB lacks any zinc finger (GAO et al. 1988).

The BALF2 protein has been purified to near homogeneity from nuclear extracts of B95-8 cells in the virus-productive cycle (TSURUMI et al. 1996). Sodium dodecyl sulfate (SDS)-polyacrylamide gel electrophoresis showed the presence of a single polypeptide with a molecular mass of 130kDa, which was identified as BALF2 protein by Western immunoblot analysis. On Superose 6 HR 10/30 gel filtration the BALF2 protein eluted at a position corresponding to an apparent molecular mass of approximately 128kDa, indicating that the BALF2 protein behaves as a monomer in solution. The native forms of SSB proteins differ in different species (KORNBERG and BAKER 1992): T4 gene 32 protein exists as monomer of 33.5kDa, M13 gp5 is a dimer of 9.69kDa subunits and *E. coli* SSB is a tetramer of 18.9kDa subunits. However, the binding mode of these SSB proteins to ssDNA appears to be similar in that it is always cooperative. The purified BALF2 protein binds to ssDNA preferentially over dsDNA or ssRNA. The curve for ssDNA binding to nitrocellulose filters obtained with increasing concentrations of EBV SSB provides information on the cooperativity of binding (TSURUMI et al. 1996). Saturation of the labeled ssDNA by the EBV SSB occurs at a protein/nucleic acid weight ratio of 13:1, indicating that the EBV SSB binds to ssDNA at a molar ratio of approximately one EBV SSB monomer/30 nucleotides, assuming that the native protein has a molecular mass of 128kDa.

5.7.2 Helix Destabilizing Activity

The BALF2 protein can displace short DNA strands from their complementary sequences in the single stranded form of M13 (TSURUMI et al. 1998) and the EBV SSB can displace labeled oligonucleotide in a concentration-dependent fashion. However, incubation of the substrate in the absence of the EBV SSB does not result in the release of labeled 20-mer oligonucleotide, indicating that the DNA duplex is stable under the conditions employed for the reaction. DNA unwinding activity is not stimulated by the addition of ATP, CTP, UTP, dATP, or dCTP and the activity is not influenced by the addition of the ATP analogue, ATPγS, indicating that the EBV SSB does not function as a helicase. Maximum displacement occurs at concentrations saturating the DNA substrate. The helix destabilization reaction mediated by the EBV SSB is highly cooperative and extremely rapid, with no directionality or any requirement for either ATP or $MgCl_2$, two essential cofactors for DNA helicase activity. Displacement of the 59-mer from complementary sequence in M13 ssDNA requires higher levels of the EBV SSB compared with the 20-mer displacement. Moreover, the EBV SSB fails to displace oligonucleotides more than 100 bases long. In contrast, human SSB, RP-A (GEORGAKI et al. 1992) and adenovirus DNA-binding protein (DBP) (MONAGHAN et al. 1994; ZIJDERVELD and VAN DER VLIET 1994) are found to possess significant unwinding activity up to 200–350 nucleotides. The helix-destabilizing property of the EBV SSB may function to melt out the secondary structures on an ssDNA template, thereby facilitating the movement of the EBV DNA polymerase.

5.8 Functional Interactions Between the EBV SSB and the EBV *Pol* Holoenzyme

Replication of singly primed M13 ssDNA by the EBV DNA *Pol* holoenzyme in the absence of the BALF2 protein exhibits a highly processive mode of replication with generation of full-length products in addition to some bands of pausing sites. Regions that block DNA polymerase passage are known to form particularly stable duplex hairpin structures (VILLANI et al. 1981). Although addition of the BALF2 protein has been shown not to affect the replication rate, the average chain length of the replication products is slightly increased with elimination of bands of pausing sites. Similar effects are observed with the in vitro reconstituted polymerase complex composed of the BALF5 and BMRF1 *Pol* subunits. On the other hand, in the absence of the EBV SSB, most of the BALF5 *Pol* catalytic subunit pauses at secondary structures on the M13 ssDNA template, incapable of synthesizing efficiently through duplex regions. However, addition of the BALF2 protein stimulates DNA synthesis and yields a distribution of replication products with long lengths in addition to full-length products. Although the BALF2 protein behaves as if it converts a low processive EBV *Pol* catalytic subunit to a highly processive form, like the BMRF1 *Pol* accessory subunit, challenger DNA experiments have revealed that the BALF5 *Pol* catalytic subunit is frequently transferred to challenger DNAs in the presence of the BALF2 protein while the *Pol* complex is not transferred during polymerization (TSURUMI et al. 1996). From these observations the function of the EBV SSB on the EBV DNA replication appears not to be as processivity factor of the EBV DNA *Pol* catalytic subunit but rather to remove secondary structures in the DNA, thus reducing and eliminating pausing of the EBV DNA polymerase at specific sites. The ability of the EBV SSB to transiently destabilize dsDNA duplexes could be utilized in the progress of the replication fork movement. The helix-destabilizing activity would disrupt loops, hairpins, or other secondary structures formed by complementary sequences in the ssDNA template. It is therefore likely that the EBV BALF2 protein functions to melt out the regions of secondary structure on the ssDNA template, thereby reducing and eliminating pausing of the EBV DNA polymerase at specific sites and keeping the DNA template in the optimal conformation for DNA elongation. It was demonstrated that the BALF2 protein physically interacts with the EBV *Pol* holoenzyme (ZENG et al. 1997).

5.9 EBV Putative Helicase–Primase Complex

The enzymatic activities of the BBLF4, BSLF1, and BBLF2/3 proteins have yet to be demonstrated. The BBLF4 gene product shares 34% amino acid sequence identity with the HSV-1 UL5 gene product and, further, possesses the six conserved motifs found in all members of helicase superfamily I, as well as the HSV-1 UL5 gene product (GRAVES-WOODWARD and WELLER 1996; SPECTOR et al. 1998). Substitutions of the conserved residues in each of the six helicase motifs of the HSV-1 UL5 protein abolished the ability of UL5 to support viral DNA replication in vivo (ZHU and WELLER 1992) and the helicase activity of the purified UL5–UL52 subcomplex in vitro, but not the primase activity (GRAVES-WOODWARD and WELLER 1996). The BSLF1 gene product has 23% amino acid sequence identity with the HSV-1 UL52 gene product and contains five regions conserved among other identified herpesvirus UL52 homologs, including a DXD motif, which resembles the putative metal-binding site found in other primases. Changing either of the two aspartate residues in the primase DXD motif of the HSV-1 UL52 gene product was found to abolish the primase activity of the purified UL5–UL52–UL8 complex, but not the ATPase and helicase activities (KLINEDINST and CHALLBERG 1994; DRACHEVA et al. 1995). The BBLF 2/3 gene product has no significant similarity with the HSV-1 UL8 gene product in overall identity but has a stretch of 55 amino acids with considerable homology. Unlike the HSV-1 UL8 protein, the BBLF2/3 protein has a potential ATP binding motif, whose function is unclear (FIXMAN et al. 1992). Although the UL8 gene product is not absolutely required for the helicase and primase activities in vitro, it interacts with the UL5–UL52 subcomplex and is essential for viral DNA replication in vivo.

Relatively few studies of the EBV putative helicase-primase have been performed. GAO et al. (1998) provided evidence of BSLF1–BBLF4–BBLF2/3 complexes through immunofluorescence assays with nuclear re-translocation, expressing three BBLF4, BSLF1, and BBLF2/3 proteins fused to the myc epitope in Vero cells by transfecting their expression vectors. When individually transfected, Myc-BBLF2/3 showed mixed nuclear and cytoplasmic staining, Myc-BSLF1 was perinuclear, and Myc-BBLF4 localized to the cytoplasm. The concurrent presence of all three members resulted in nuclear localization of the BBLF4, BBLF2/3, and BSLF1 proteins, suggesting the existence of a BSLF1–BBLF4–BBLF2/3 complex. When expressed in B95-8 cells after induction of virus productive cycle, these proteins demonstrated apparent molecular masses of 89kDa, 90kDa, and 80kDa, respectively (YOKOYAMA et al. 1999). Anti-BSLF1 or anti-BBLF2/3 protein-specific antibodies immunoprecipitate all of the BSLF1, BBLF4, and BBLF2/3 proteins from extracts of cells in the virus productive cycle, indicating that these viral proteins are assembled in vivo. Assembly has been reproduced in insect cells triply infected with the three recombinant baculoviruses, indicating that the complex formation does not require other EBV replication proteins (YOKOYAMA et al. 1999). Glycerol density gradient centrifugation analysis revealed that these proteins form a heteromeric complex (N. Yokoyama et al., unpublished results). Furthermore, experiments performed with double infection of pairs of recombinant viruses

at within the BBLF4–BSLF1–BBLF2/3 complex each component
...tly with the other two (YOKOYAMA et al. 1999).
...ducts of HSV-1 UL5, UL8, and UL52 genes form a heterotrimeric
...h helicase and primase activities (CRUTE et al. 1988, 1989; DODSON et al.
...former presumably acting to unwind duplex DNA ahead of the pro-
...eplication fork, thereby producing the open configuration needed for both
...us and discontinuous strand synthesis. By analogy with other primases, the
...of HSV-1 can be concluded to initiate discontinuous DNA synthesis on the
... strand by providing oligonucleotide primers that are elongated by the
... DNA polymerase. Although the enzymatic activities of the EBV BBLF4–
...1–BBLF2/3 heterotrimeric complex have yet to be demonstrated, it may act as
...licase and primase like the HSV-1 UL5–UL52–UL8 complex.

10 Physical Interaction Between the EBV *Pol* Catalytic Subunit and the EBV BBLF4–BSLF1–BBLF2/3 Complex

The multiple steps essential for DNA replication are catalyzed by a number of proteins whose enzymatic reactions must be closely coordinated. This is most apparent at the replication fork where both leading and lagging strand synthesis must occur simultaneously in order to achieve movement. It is, therefore, not surprising that replication proteins are frequently isolated as complexes or interact physically one another, so that their individual enzymatic reactions are coordinated. The specific interactions that occur among the six viral replication proteins appear to be essential for EBV DNA replication. A physical interaction between the EBV DNA *Pol* holoenzyme and the EBV putative helicase primase complex via the BALF5 DNA *Pol* catalytic subunit was demonstrated by immunoprecipitation analyses using anti-BSLF1 or anti-BBLF2/3 protein-specific antibody with clarified lysates of B95-8 cells in a viral productive cycle (FUJII et al. 2000). Although the *Pol* holoenzyme-the BBLF4–BSLF1–BBLF2/3 complex was stable in 500mM NaCl and 1% NP-40, the BALF5 protein became dissociated in the presence of 0.1% SDS. By experiments using lysates from insect cells superinfected with combinations of recombinant baculoviruses capable of expressing each viral replication protein, it was shown that not the BMRF1 *Pol* accessory subunit but rather the BALF5 *Pol* catalytic subunit directly interacts with the BBLF4–BSLF1–BBLF2/3 complex. Furthermore, double infection with pairs of recombinant viruses revealed that each component of the BBLF4–BSLF1–BBLF2/3 complex makes contact with the BALF5 *Pol* catalytic subunit.

5.11 Polymerase and Helicase–Primase Complex Interactions in Other Systems

Polymerase and helicase–primase complex interactions have also been observed in other replication systems. In the case of bacteriophage T7, gene 5 DNA polymerase interacts with the gene 4 helicase–primase via its carboxyl terminus (NAKAI and

RICHARDSON 1986; NOTARNICOLA et al. 1997), playing an important role in its coordination of leading and lagging strand DNA synthesis at the replication fork (DEBYSER et al. 1994). Also, the catalytic subunit of the HSV-1 DNA polymerase interacts with the carboxyl terminus of the UL8 protein of the HSV-1 helicase–primase heterotrimeric complex (MARSDEN et al. 1997). Considering these observations, the interaction of the EBV DNA *Pol* holoenzyme with the helicase–primase complex might be central to the coordination of the leading and lagging strand synthesis at the replication fork of EBV.

The interaction of the helicase–primase and DNA polymerase is required not only for strand displacement DNA synthesis but also for priming DNA synthesis on the lagging strand of the replication fork. Little is known about DNA primase-DNA polymerase interactions involved in the transition from RNA to DNA synthesis. The bacteriophage T7 gene 4A protein catalyzes the synthesis of tetraribonucleotides at specific sequences on ssDNA in a template-mediated reaction (TABER and RICHARDSON 1981), these then being stabilized on the template by gene 4A protein until T7 DNA polymerase can use them as primers to initiate DNA synthesis (NAKAI and RICHARDSON 1986). Such short oligonucleotides prime T7 DNA polymerase extremely poorly in the absence of the gene 4 protein. The effective use in its presence implies that a specific protein–protein interaction is required. In the case of bacteriophage T4, a complex of two proteins encoded by genes 41 and 61 is necessary to catalyze synthesis of the pentaribonucleotide pppACN3 efficiently, which in turn primes DNA synthesis by T4 DNA polymerase on ssDNA (LIU and ALBERTS 1980). In a variety of eukaryotic systems, DNA polymerase α has been purified in a tight complex with DNA primase (FRY and LOEB 1986). Interaction between DNA primases and polymerases may in general play an important role in promoting efficient transition from RNA primer to DNA synthesis on lagging strands. Thus, the interaction of the BBLF4–BSLF1–BBLF2/3 complex and the EBV *Pol* holoenzyme at the replication fork may be an important aspect of the replication process and a possible new target for antiviral agents.

6 Proposed Model for the Initiation Step of the Lytic Phase of EBV DNA Replication

Considering other DNA replication systems, it is likely that the initiation of the lytic phase EBV DNA replication involves the formation of an initiation complex at *oriLyt*, which consists of two essential domains, upstream and downstream components. Whereas the upstream component contains several BZLF1 binding sites, the downstream includes binding sites of several cellular proteins. The first step in this process would be the binding of the BZLF1 protein and two of transcription factors, ZBP-89 and Sp1, to recognition sequences within *oriLyt* to form an initial complex. ZBP-89 and Sp1 stimulate replication (BAUMANN et al. 1999). The interaction of the BZLF1 protein with the BBLF4–BSLF1–BBLF2/3 complex reported by GAO et al.

(1998) supposes that the BZLF1 protein recruits the viral helicase–primase complex to *oriLyt*. Also, the BZLF1 protein can interact with both BALF5 and BMRF1 proteins (ZHANG et al. 1996; BAUMANN et al. 1999). The BALF2 ssDNA-binding protein then appears to interact with the BZLF1-BBLF4–BSLF1–BBLF2/3 prepriming complex (GAO et al. 1998). These proteins together, therefore, would have the potential to open up the duplex DNA in the origin region and synthesize RNA primers. The interaction between the EBV *Pol* holoenzyme and the BBLF4–BSLF1–BBLF2/3 complex (FUJII et al. 2000) may play an important role in bringing the viral polymerase into the prepriming complex to initiate DNA synthesis. Furthermore, Sp1 and ZBP-89 transcription factors binding to the downstream component of *oriLyt* are able to tether EBV *Pol* holoenzyme (BAUMANN et al. 1999). It is possible, for example, that binding of the *Pol* holoenzyme to the BBLF4–BSLF1–BBLF2/3 complex reduces its affinity for BZLF1, allowing the polymerase-helicase-primase complex to migrate away from *oriLyt* to the replication forks.

7 Proposed Model for the EBV Replication Fork

The proposed model for EBV replication fork is depicted in Fig. 3. BALF5 *Pol* catalytic and BMRF1 *Pol* accessory subunits form a heterodimer to function as

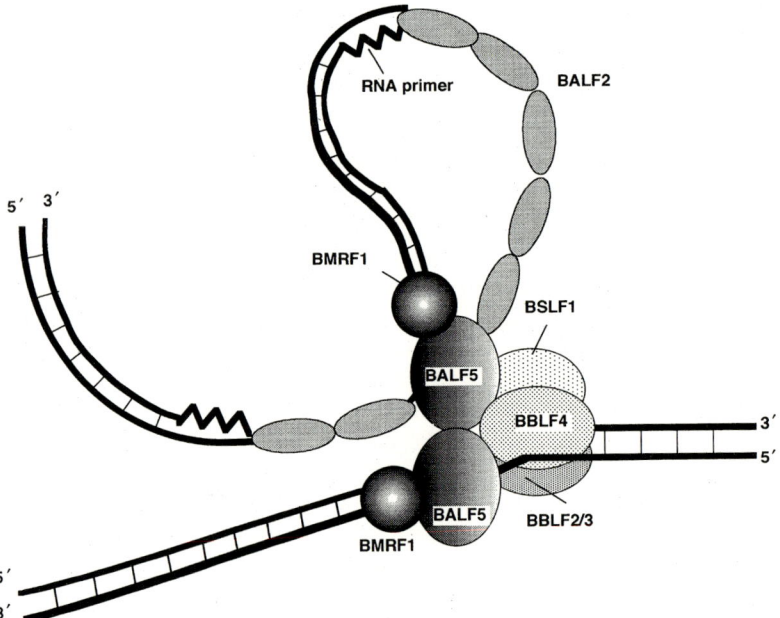

Fig. 3. Proposed model for the EBV replication fork

the *Pol* holoenzyme, with high polymerase processivity, presumably synthesizing both leading and lagging strands. The BALF2 protein, EBV SSB, binds ssDNA templates and may function to melt out secondary structures in the viral displaced ssDNA of the replication fork, thereby facilitating movement of the EBV DNA *Pol* holoenzyme on ssDNA template. As expected with the HSV-1 helicase/primase complex, the proposed EBV helicase and primase complex consisting of the BBLF4–BSLF1–BBLF2/3 proteins may bind to the lagging strand at the fork, translocate in the 5'-to-3' direction and synthesize the RNA primer. The BALF5 *Pol* catalytic subunit physically interacts with each component of the BBLF4–BSLF1–BBLF2/3 complex (Fujii et al. 2000). Thus, the six viral replication proteins appear to all work at the replication fork as the replication machinery.

In the HSV case, aphidicolin-resistant mutants have been isolated (Nishiyama et al. 1984). The drug inhibits host-cell DNA *Pols* α, δ, and ϵ and blocks cellular DNA synthesis. However, even in the drug's presence, the mutant viruses can synthesize viral DNA well. Therefore, the possibility that cellular polymerases are involved in viral replicative DNA synthesis is very low. By analogy with bacteriophage T4, or T7, the EBV *Pol* holoenzyme might synthesize both leading and lagging strands. However, we cannot be precluded that a dimer consisting of BALF5 and BMRF1 proteins catalyzes leading strand synthesis and the BALF5 protein alone with its relatively low processivity catalyzes lagging strand synthesis.

References

Adamson AL, Kenney S (1999) The Epstein-Barr virus BZLF1 protein interacts physically and functionally with the histone acetylase CREB-binding protein. J Virol 73:6551–6558

Aiyar A, Tyree C, Sugden B (1998) The plasmid replication of EBV consists of multiple *cis*-acting elements that facilitate DNA synthesis by the cell and a viral maintenance element. EMBO J 17:6394–6403

Baumann M, Feederle R, Kremmer E, Hammerschmidt W (1999) Cellular transcription factors recruit viral replication proteins to activate the Epstein-Barr virus origin of lytic DNA replication, *oriLyt*. EMBO J 18:6095–6105

Cayrol C, Flemington EK (1996) The Epstein-Barr virus bZIP transcription factor Zta causes G0/G1 cell cycle arrest through induction of cyclin-dependent kinase inhibitors. EMBO J 15:2748–2759

Chiou JF, Li JKK, Cheng YC (1985) Demonstration of a stimulatory protein for virus-specified DNA polymerase in phorbol-ester treated Epstein-Barr virus carrying cells. Proc Natl Acad Sci USA 82:5728–5731

Crute JJ, Mocarski ES, Lehman IR (1988) A DNA helicase induced by herpes simplex virus type 1. Nucleic Acids Res 16:6585–6596

Crute JJ, Tsurumi T, Zhu L, Weller SK, Olivo PD, Challberg MD, Mocarski ES, Lehman IR (1989) Herpes simplex virus 1 helicase-primase: a complex of three herpes-encoded gene products. Proc Natl Acad Sci USA 86:2186–2189

Debyser Z, Tabor S, Richardson CC (1994) Coordination of leading and lagging strand DNA synthesis at the replication fork of bacteriophage T7. Cell 77:157–166

Decaussin G, Leclerc V, Ooka T (1995) The lytic cycle of Epstein-Barr virus in the nonproducer Raji line can be rescued by the expression of a 135-kilodalton protein encoded by the BALF2 open reading frame. J Virol 69:7309–7314

Dodson MS, Crute JJ, Bruckner RC, Lehman IR (1989) Overexpression and assembly of the herpes simplex virus type 1 helicase-primase in insect cells. J Biol Chem 264:20853–20838

Dracheva S, Koonin EV, Crute JJ (1995) Identification of the primase active site of the herpes simplex virus type 1 helicase-primase. J Biol Chem 270:14148–14153

Epstein AL (1984) Immunochemical characterization with monoclonal antibodies of Epstein-Barr virus-associated early antigens in chemically induced cells. J Virol 50:372–379

Farrell P, Rowe D, Rooney C, Kouzarides T (1989) Epstein-Barr virus BZLF1 trans-activator specifically binds to consensus Ap1 site and is related to c-fos. EMBO J 8:127–132

Fay PJ, Johanson KO, McHenry CS, Bambara RA (1982) Size classes of products synthesized processively by two subassemblies of Escherichia coli DNA polymerase III holoenzyme. J Biol Chem 257:5692–5699

Fixman ED, Hayward GS, Hayward SD (1992) trans-acting requirements for replication of Epstein-Barr virus ori-Lyt. J Virol 66:5030–5039

Fixman ED, Hayward GS, Hayward SD (1995) Replication of Epstein-Barr virus ori Lyt: Lack of a dedicated virally encoded origin-binding protien and dependence on Zta in cotransfection assays. J Virol 69:2998–3006

Fry M, Loeb LA (1986) Animal cell DNA polymerases. CRC Press

Fujii K, Yokoyama N, Kiyono T, Kuzushima K, Homma M, Nishiyama Y, Fujita M, Tsurumi T (2000) The Epstein-Barr virus BALF5 *Pol* catalytic subunit physically interacts with the BBLF4/BSLF1/BBLF2/3 complex. J Virol 74:2550–2557

Gahn TA, Schildkraut CL (1989) The Epstein-Barr virus origin of plasmid replication, oriP, contains both the initiation and termination sites of DNA replication. Cell 58:527–535

Gao M, Bouchey J, Curtin K, Knipe DM (1988) Genetic identification of a portion of the herpes simplex virus ICP8 protein required for DNA-binding. Virology 163:319–329

Gao Z, Krithivas A, Finan JE, Semmes OJ, Zhou S, Wang Y, Hayward SD (1998) The Epstein-Barr virus lytic transactivator Zta interacts with the helicase-primase replication proteins. J Virol 72:8559–8567

Georgaki A, Strack B, Podust V, Hubscher U (1992) DNA unwinding activity of replication protein A. FEBS lett 308:240–244

Gottlieb J, Challberg MD (1994) Interaction of herpes simplex virus type 1 DNA polymerase and the UL42 accessory protein with a model primer template. J Virol 68:4937–4945

Gottlieb J, Marcy AI, Coen DM, Challberg MD (1990) The herpes simplex virus type 1 UL42 gene product: a subunit of DNA polymerase that functions to increase processivity. J Virol 64:5976–5987

Graves-Woodward KL, Weller SK (1996) Replacement of Gly815 in helicase motif V alter the single-stranded DNA-dependent ATPase activity of the herpes simplex virus type 1 helicase-primase. J Biol Chem 271:13629–13635

Gruffat H, Renner O, Pich D, Hammerschmidt W (1995) Cellular proteins bind to the downstream component of the lytic origin of DNA replication of Epstein-Barr virus. J Virol 69:1878–1886

Gutierrez C, Martin G, Sogo JM, Salas M (1991) Mechanism of stimulation of DNA replication by bacteriophage f29 single stranded DNA-binding protein p5. J Biol Chem 266:2104–2111

Hammerschmidt W, Sugden B (1988) Identification and characterization of *oriLyt*, a lytic origin of DNA replication of Epstein-Barr virus. Cell 55:427–433

Huber HE, Tabor S, Richardson CC (1987) Escherichia coli thioredoxin stabilizes complexes of bacteriophage T7 DNA polymerase and primed templates. J Biol Chem 262:16224–16232

Hurwitz J, Dean FB, Kwng AD, Lee S-H (1990) The in vitro replication of DNA containing the SV40 origin. J Biol Chem 262:16224–16232

Javis TC, Paul LS, Hockensmith JW, von Hippel PH (1989) Structure and enzymatic studies of the T4 DNA replication system. II ATPase properties of the polymerase accessory protein complex. J Biol Chem 264:12717–12729

Joyce CM (1989) How DNA travels between the separate polymerase and 3′-5′-exonuclease sites of DNA polymerase I (Klenow fragment). J Biol Chem 264:10858–10866

Kallin B, Stern SL, Saemundssen AK, Luka J, J rnvall H, Eriksson B, Tao P-Z, Nilsson MT, Klein G (1985) Purification of Epstein-Barr virus DNA polymerase from P3HR-1 cells. J Virol 54:561–568

Khamis MI, Casas-Finet JR, Maki AH, Murphy JB, Chase JW (1987) Investigation of the role of individual tryptophan residues in the binding of Escherichia coli single-stranded DNA binding protein to single stranded polynucleotides. J Biol Chem 262:10938–10945

Kiehl A, Dorsky DL (1991) Coorperation of EBV DNA polymerase and EA-D (BMRF1) in vitro and colocalization in nuclei of infected cells. Virology 184:330–340

Kiehl A, Dorsky DL (1995) Bipartite DNA-binding region of the Epstein-Barr virus BMRF1 product essential for DNA polymerase accessory function. J Virol 69:1669–1677

King GC, Coleman J 3E (1988) The fd gene 5 protein-d(pA)40–60 complex: ^{1}H NMR supports a localized base-binding model. Biochemistry 27:6947–6953

Klinedinst DK, Challberg MD (1994) Helicase-primase complex of herpes simplex virus type 1: a mutation in the UL52 subunit abolishes primase activity. J Virol 68:3693–3701

Kornberg A, Baker T (1992) DNA replication, 2nd ed., WH Freeman & Co., New York

Lee M-A, Diamond ME, Yates JL (1999) Genetic evidence that EBNA-1 is needed for efficient, stable latent infection by Epstein-Barr virus. J Virol 73:2974–2982

Li J-S, Zhou B-S, Dutschman GE, Grill SP, Tan R-S, Cheng Y-C (1987) Association of Epstein-Barr virus early antigen diffuse component and virus-specified DNA polymerase activity. J Virol 61:2947–2949

Lin J-C, Sista ND, Besencon F, Kamine J, Pagano JS (1991) Identification and functional characterization of Epstein-Barr virus DNA polymerase by in vitro transcription-translation of a cloned gene. J Virol 65:2728–2731

Little R, Schildkraut C (1995) Initiation of latent DNA replication in the Epstein-Barr virus genome can occur at sites other than the genetically defined origin. Mol Biol Cell 5:2893–2903

Liu C-C, Alberts BM (1980) Pentaribonucleotides of mixed sequence are synthesized and efficiently prime de novo DNA chain starts in the T4 bacteriophage DNA replecation system. Proc Natl Acad Sci USA 77:5698–5703

Marsden HS, McLean GW, Barnard EC, Francis GJ, MacEachran K, Murphy M, McVey G, Cross A, Abbotts AP, Stow ND (1997) The catalytic subunit of the DNA polymerase of herpes simplex virus type 1 interacts specifically with the C terminus of the UL8 component of the viral helicase-primase complex. J Virol 71:6390–6397

Monaghan A, Weber A, Hay RT (1994) Adenovirus DNA binding protein: helix destabilizing properties. Nucl Acid Res 22:742–748

Nakai H, Richardson CC (1986) Interactions of the DNA polymerase and gene 4 protein of bacteriophage T7. J Biol Chem 261:15208–15216

Neal GAM, Kitchingman GR (1990) Conserved region 3 of the adenovirus type 5 DNA-binding protein is important for interaction with single stranded DNA. J Virol 64:630–638

Nishiyama Y, Suzuki S, Yamauchi M, Maeno K, Yoshida S (1984) Characterization of an aphidicolin-resistant mutant of herpes simplex virus type 2 which induces an altered viral DNA polymerase. Virology 135:87–96

Notarnicola SM, Mulcahy HL, Lee J, Richardson CC (1997) The acidic carboxyl terminus of the bacteriophage T7 gene 4 helicase/primase interacts with T7 DNA polymerase. J Biol Chem 272:18425–18433

O'Donnell M, Studwell PS (1990) Total reconstitution of DNA polymerase III holoenzyme reveals dual accessory protein clamps. J Biol Chem 265:1179–1187

Packham G, Economou A, Rooney CM, Rowe DT, Farrel PJ (1990) Structure and function of the Epstein-Barr virus BZLF1 protein. J Virol 64:2110–2116

Pearson GR, Vroman B, Chase B, Sculley T, Hummel H, Kieff E (1983) Identification of polypeptide components of the Epstein-Barr virus early antigen complex with monoclonal antibodies. J Virol 47:193–201

Pfuller R, Hammerschmidt W (1996) Plasmid-like replicative intermediates of the Epstein-Barr virus lytic origin of DNA replication. J Virol 70:3423–3431

Portes-Sentis S, Sergeant A, Gruffat H (1997) A particular DNA structure is required for the function of a *cis*-acting component of the Epstein-Barr virus *oriLyt* origin of replication. Nucl Acid Res 25:1347–1354

Prasad BVV, Chiu W (1987) Sequence comparison of single stranded DNA binding proteins and its structural implications. J Mol Biol 193:579–584

Sarisky RT, Gao Z, Lieberman PM, Fixman ED, Hayward GS, Hayward D (1996) A replication function associated with the activation domain of the Epstein-Barr virus Zta transactivator. J Virol 70:8340–8347

Schepers A, Pich D, Hammerschmidt W (1996) Activation of *oriLyt*, the lytic origin of DNA replication of Epstien-Barr virus, by BZLF1. Virology 220:367–376

Schepers A, Pich D, Mankertz J, Hammerschmidt W (1993) *cis*-acting elements in the lytic origin of DNA replication of Epstein-Barr virus. J Virol 67:4237–4245

ShamooY, Ghosaini LR, Keating KM, Williams KR, Sturtevant JM, KonigsbergWH (1989) Site-specific mutagenesis of T4 gene 32: The role of tyrosine residues in protein nucleic acid interactions. Biochemistry 28:7409–7417

Shirakata M, Imadome K, Hirai K (1999) Requirement of replication licensing for the Dyad Symmetry element-dependent replication of the Epstein-Barr virus oriP minichromosome. Virology 263:42–54

Shire K, Ceccarelli DFJ, Avolio-Hunter TM, Frappier L (1999) EBP2, a human protein that interacts with sequences of the Epstein-Barr virus nuclear antigen 1 important for plasmid maintenance. J Virol 73:2587–2595

Spector FC, Giordano LH, Sivaraja M, Peterson MG (1998) Inhibition of herpes simplex virus replication by a 2-amino thiazole via interactions with the helicase component of the UL5-UL8-UL52 complex. J Virol 72:6979–6987

Tabor S, Richardson CC (1981) Template recognition sequence for RNA primer synthesis by gene 4 protein of bacteriophage T7. Proc Natl Acad Sci USA 78:205–209

Tsurimoto T, Stillman B (1990) Functions of replication factor C and proliferating cell nuclear antigen: functional similarity of DNA polymerase accessory proteins from human cells and bacteriophage T4. Proc Natl Acad Sci USA 87:1023–1027

Tsurimoto T, Stillman T (1991) Replication factors required for SV40 DNA replication in vitro. I. DNA structure-specific recognition of a primer-template junction by eukaryotic DNA polymerase and their accessory proteins. J Biol Chem 266:1950–1960

Tsurumi T (1991a) Characterization of 3′-to-5′ exonuclease activity associated with Epstein-Barr virus DNA polymerase. Virology 182:376–381

Tsurumi T (1991b) Primer terminus recognition and highly processive replication by Epstein-Barr virus DNA polymerase. Biochem J 280:703–708

Tsurumi T (1992) Selective inhibition of the 3′-to-5′ exonuclease activity associated with Epstein-Barr virus DNA polymerase by ribonucleoside monophosphates. Virology 189:803–807

Tsurumi T (1993a) Purification and characterization of the DNA-binding activity of the Epstein-Barr virus DNA polymerase accessory protein BMRF1 gene products, as expressed in insect cells by using the baculovirus system. J Virol 67:1681–1687

Tsurumi T, Daikoku T, Kurachi R, Nishiyama Y (1993b) Functional interaction between Epstein-Barr virus DNA polymerase catalytic subunit and its accessory subunit in vitro. J Virol 67:7648–7765

Tsurumi T, Kobayashi A, Tamai K, Daikoku T, Kurachi R, Nishiyama Y (1993c) Functional expression and characterization of the Epstein-Barr virus DNA polymerase catalytic subunit. J Virol 67:4651–4658

Tsurumi T, Daikoku T, Nishiyama Y (1994) Further characterization of the interaction between the Epstein-Barr virus DNA polymerase catalytic sub unit and its accessory subunit with regard to the 3′-to-5′ exonucleolytic activity and stability of initiation complex at primer terminus. J Virol 68:3354–3363

Tsurumi T, Kobayashi A, Tamai K, Yamada H, Daikoku T, Yamashita Y, Nishiyama Y (1996) Epstein-Barr virus single-stranded DNA-binding protein: purification, characterization, and action on DNA synthesis by the viral DNA polymerase. Virology 222:352–364

Tsurumi T, Kishore J, Yokoyama N, Fujita M, Daikoku T, Yamada H, Yamashita Y, Nishiyama Y (1998) Overexpression, purification and helix-destabilizing properties of Epstein-Barr virus ssDNA-binding protein. J Gen Virol 79:1257–1264

Villani G, Fay PJ, Bambara RA, Lehman IR (1981) Elongation of RNA-primed DNA templates by DNA polymerase a from Drosophila melanogaster embryos. J Biol Chem 256:8202–8207

Wang Y, Hall JD (1990) Characterization of a major DNA-binding domain in the herpes simplex type 1 DNA-binding protein (ICP8). J Virol 64:2082–2089

Yates JL, Guan N (1991) Epstein-Barr virus-derived plasmids replication only per cell cycle and not amplified after entry into cells. J Virol 65:483–488

Yates JL, Warren N, Reisman D, Sugden B (1984) Stable replication of plasmids derived from Epstein-Barr virus in various mammalian cells. Nature 313:812–815

Yokoyama N, Fujii K, Hirata M, Tamai K, Kiyono T, Kuzushima K, Nishiyama Y, Fujita M, Tsurumi T (1999) Assembly of the Epstein-Barr virus BBLF4, BSLF1, BBLF2/3 proteins and their interactive properties. J Gen Virol 80:2879–2888

Young MC, Reddy MK, von Hippel PH (1992) Structure and function of the bacteriophage T4 DNA polymerase holoenzyme. Biochemistry 31:8675–8690

Zeng Y, Middeldorp J, Madjar J, Ooka T (1997) A major DNA binding protein encoded by BALF2 open reading frame of Epstein-Barr virus (EBV) forms a complex with other EBV DNA-binding proteins: DNAase, EA-D, and DNA polymerase. Virology 239:285–295

Zhang Q, Hong Y, Dorsky D, Holley-Guthrie E, Zalani S, Elshiekh NA, Kiehl A, Le T, Kenney S (1996) Functional and physical interactions between the Epstein-Barr virus (EBV) proteins BZLF1 and BMRF1: Effects on EBV transcription and lytic replication. J Virol 70:5131–5142

Zhu L, Weller SK (1992) The six conserved helicase motifs of the UL5 gene product, a component of the herpes simplex virus type 1 helicase-primase, are essential for its function. J Virol 66:469–479

Zijderveld DC, van der Vliet PC (1994) Helix-destabilizing properties of the adenovirus DNA-binding protein. J Virol 68:1158–1164

II
EBV-Associated Malignancies

Pathology and Molecular Pathology of Epstein-Barr Virus-Associated Gastric Carcinoma

M. Fukayama, J.-M. Chong, and H. Uozaki

1	Introduction	91
2	EBVaGC and Early Gastric Carcinoma	92
2.1	EBV in Early Gastric Carcinoma	92
2.2	Pathological Features of Early Cases of EBVaGC	95
3	EBV in Nonneoplastic Gastric Mucosa	95
3.1	Controversy Over Target Epithelial Cells for EBV Infection	95
3.2	EBVaGC and Gastritis	97
4	Molecular and Cellular Abnormalities in EBVaGC	98
5	Model Systems for EBVaGC	99
6	Concluding Remarks	100
	References	100

1 Introduction

Epstein-Barr virus (EBV) is the first virus that was identified in a human neoplastic cell in 1963 (Epstein 1994). More than 90% of the world population is infected with EBV before adolescence, and it is thought that a small population develops EBV-associated malignancy in an endemic manner, such as Burkitt's lymphoma in equatorial Africa (Osato 1998) and nasopharyngeal carcinoma (NPC) in Southern China (Raab-Traub 1992). However, recent advances in molecular biological techniques have demonstrated that an unexpectedly wide variety of neoplasms in the general population is associated with EBV infection (Anagnostopoulos and Hummel 1996), among which EBV-associated gastric carcinoma (EBVaGC) is the most common with a worldwide distribution. In Japan, for example, more than 5,000 patients are estimated to develop gastric carcinoma annually in association with EBV (less than 10% of total gastric cancer) (Fukayama et al. 1998).

EBVaGC occurs in two forms, lymphoepithelioma-like carcinoma and ordinary gastric carcinoma (Fig. 1). The relative frequency of the two types is roughly

---

Department of Pathology, Faculty of Medicine, the Tokyo University, 7-3-1 Hongo, Bunkyo-ku, Tokyo 113-0033, Japan

1:4. Both types of carcinomas exhibit predominance in males, occur primarily in the proximal stomach, show a moderately or poorly differentiated type of histology, and exhibit various degrees of lymphocytic infiltration, which at one extreme corresponds to lymphoepithelioma-like carcinoma. Since EBVaGC seems to be a distinct entity among gastric carcinomas, its molecular pathology should be compared with that of EBV-negative gastric carcinomas (CHONG et al. 1994, 1996).

2 EBVaGC and Early Gastric Carcinoma

2.1 EBV in Early Gastric Carcinoma

Few studies have focused on the early stages of EBVaGC (ARIKAWA et al. 1997b), and controversy still remains as to whether EBV infects the gastric epithelial cells before or after the development of invasive carcinoma (TAKANO et al. 1999). According to the definition of the JAPANESE RESEARCH SOCIETY FOR GASTRIC CANCER (1995), gastric carcinoma can be classified into two groups, early and advanced. Early gastric carcinoma is confined to the mucosa or the submucosa, irrespective of the presence of lymph node metastasis, while advanced carcinoma invades the muscular or deeper layers. In most cases, early gastric carcinomas represent the preinvasive and early invasive stages of advanced gastric carcinoma, but in some it could represent a specific type of gastric carcinoma, which spreads superficially rather than vertically. Nevertheless, the macroscopic appearances of early and advanced carcinoma are quite characteristic, enabling us to take the fresh tissues selectively from cases with early gastric carcinoma by gross inspection of the resected stomachs.

We evaluated the collected cases by in situ hybridization (ISH) using EBV-encoded small RNA (EBER), and thereafter performed Southern blot analysis on the DNA extracted from the early cases of EBVaGC (M. Fukayama et al., manuscript submitted), using *Bam*H I-digestion and hybridization with unique *Xho*-I or *Eco*R-1I probes (clonal analysis of EBV; RAUB-TRAUB and FLYNN 1986; GULLEY et al. 1992). With these probes, single or double fragments were observed in the intramucosal carcinomas and single fragments in all of the submucosal carcinomas. All the fragments were larger than 6kb in size, and each of the fragments was of the same size with both probes. These findings indicate that the EBV is monoclonal or biclonal in the early carcinoma cases and that EBV is present in an episomal form without being integrated to the host genome. Furthermore, the infection is latent, with no viral replication, since the length of the linear infectious virus is less than 6kb.

According to our observations (FUKAYAMA et al. 1994), as well as the reports of others (TOKUNAGA et al. 1994; ARIKAWA et al. 1997a), nearly all of the carcinoma cells in all cases of EBVaGC, whether intramucosal (Fig. 2) or early invasive stage, showed a positive signal in EBER-ISH. This finding, together with the presence of clonal EBV in Southern blot analysis, indicates that all of the carci-

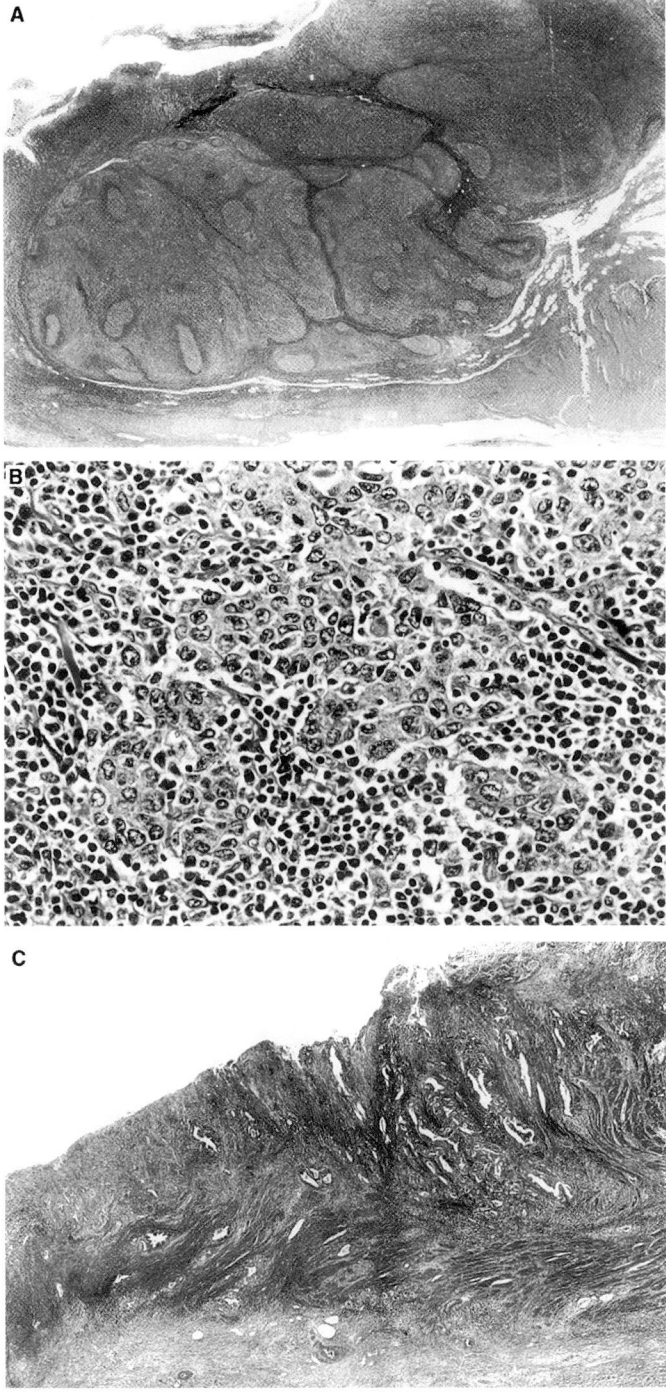

Fig. 1A–F. Histological features of EBV-associated gastric carcinoma (EBVaGC)

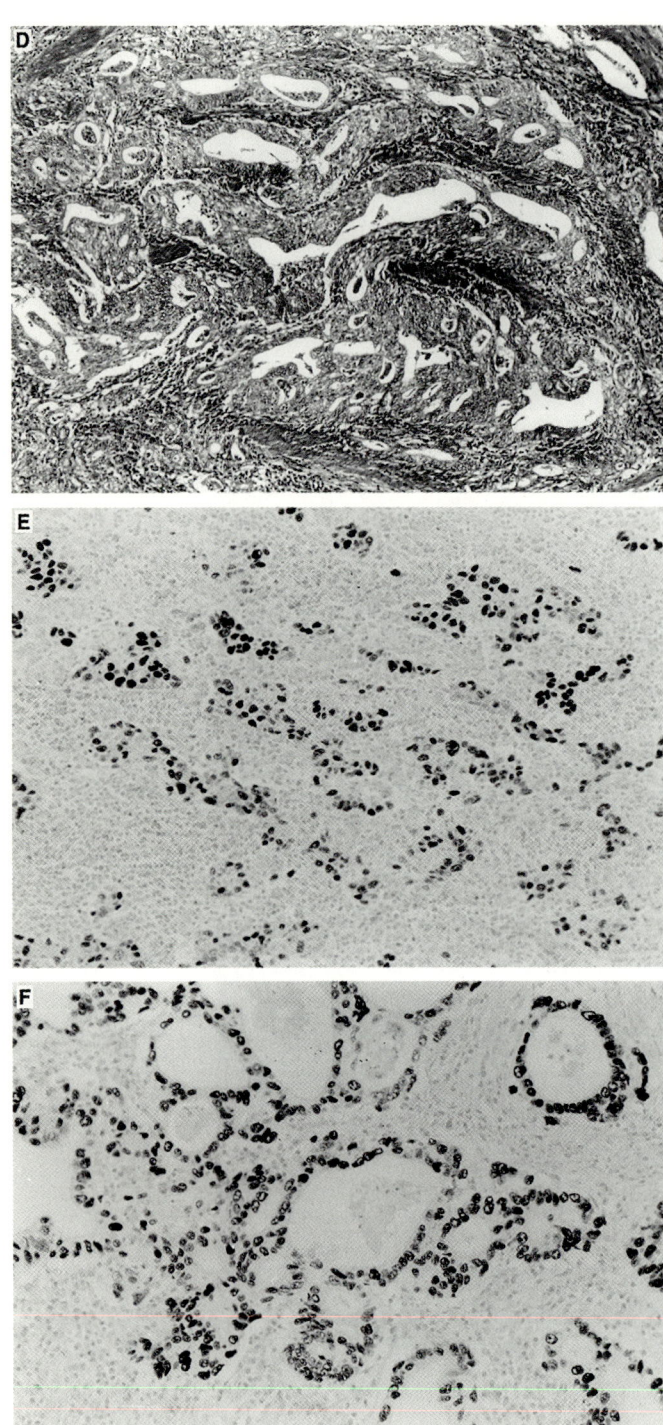

Fig. 1. A lymphoepithelioma-like gastric carcinoma (**A, B**) is accompanied by diffuse infiltration by lymphocytes and lymphoid follicles. Gastric carcinoma showing ordinary histology (**C, D**) consists of tubular structures with infiltration by lymphocytes to varying degrees. In EBV-encoded small RNA in situ hybridization (EBER-ISH), a positive signal is demonstrated in the nuclei of carcinoma cells in both types of gastric carcinoma (**E, F**). (With permission from FUKAYAMA et al. 1998)

noma cells of EBVaGC in the mucosa are clonal. Thus, EBVaGC develops from clonal expansion of EBV-infected cells within the gastric mucosa.

2.2 Pathological Features of Early Cases of EBVaGC

EBV-negative carcinomas can be divided histologically into two subtypes, intestinal and diffuse, according to LAUREN (1965). The histological features of intramucosal EBVaGC appear to be different from both types, i.e., irregular budding and fusion of neoplastic tubules at the neck zone of the gastric glands (Fig. 2). When the carcinomas invade beyond the muscularis mucosae, lymphocytic infiltration becomes marked and prominent enough in some of early invasive carcinomas to classify these as lymphoepithelioma-like carcinoma. According to our observation, EBVaGC also showed differences from both of these types in clinicopathological features: the patients with EBVaGC are significantly younger than those with intestinal-type of EBV-negative gastric carcinoma. EBVaGC occurs predominantly in males and is most often located at the gastric cardia, compared to the diffuse-type of EBV-negative gastric carcinoma. Remarkably, the incidence of multiple carcinomas is higher in EBVaGC than in both types of EBV-negative carcinomas, as suggested by ARIKAWA et al. (1997a) and MATSUNOU et al. (1996). Furthermore, most carcinomas associated with EBVaGC also showed a positive signal in EBER-ISH. On the other hand, almost all of the carcinomas associated with EBV-negative gastric carcinomas were negative in EBER-ISH. Interestingly, in the clonal analysis of EBV as mentioned above, the fragments of the EBV termini showed different lengths in three different samples derived from the same patient, indicating that EBVaGC develops at multiple sites independently in the same stomach. Thus, EBVaGC is likely to occur multiply, in other words, the stomachs bearing EBVaGC may have been conditioned to develop gastric carcinomas by EBV infection (field cancerization; GUOREN et al. 1996).

3 EBV in Nonneoplastic Gastric Mucosa

3.1 Controversy Over Target Epithelial Cells for EBV Infection

According to our previous study using EBER-ISH on nonneoplastic stomach mucosa (FUKAYAMA et al. 1994), shedding epithelial cells of the fundic gland mu-

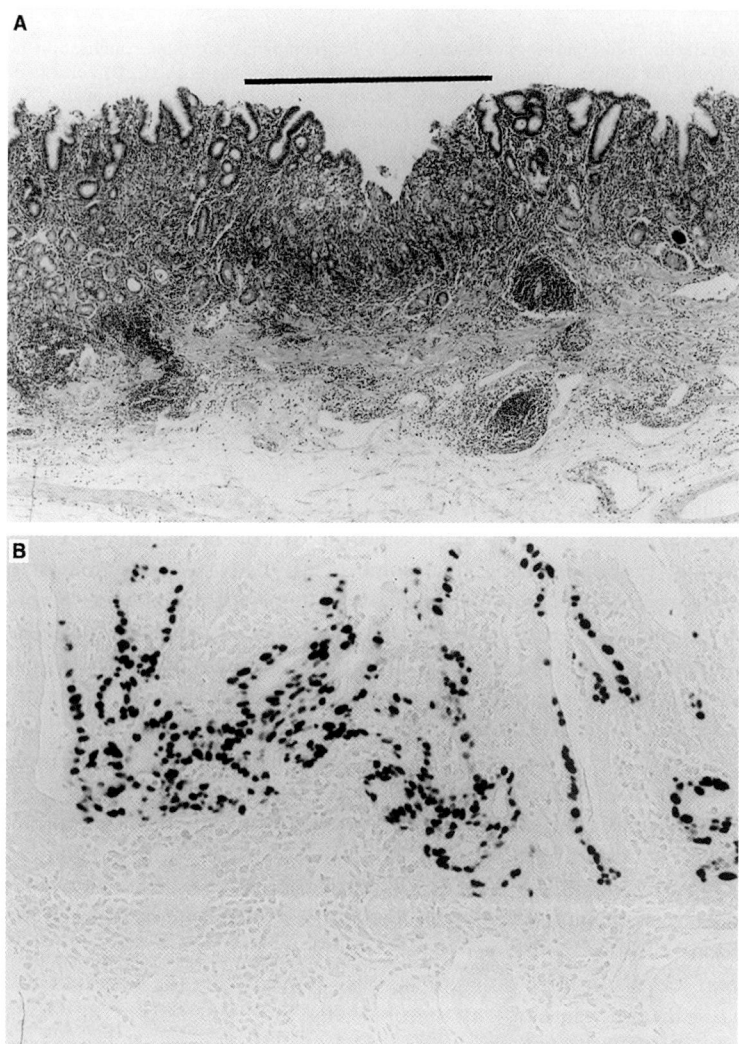

Fig. 2A,B. Histologic features of intramucosal EBVaGC and its surrounding nonneoplastic mucosa. Low-power view of intramucosal EBVaGC (**A**), which is present over a limited area of the mucosa (*bar*). Note the paucity of glands in the surrounding nonneoplastic mucosa. There is marked lymphocytic infiltration, but no intestinal metaplasia around the carcinoma. In EBER-ISH, nearly all of the carcinoma cells show a positive signal in the cellular nuclei of intramucosal EBVaGC (**B**). The carcinoma, classified as moderately differentiated tubular adenocarcinoma, consists of branching of abortive glands at the neck zone of the gastric glands. Nonneoplastic pyloric glands, negative for EBER-ISH, are observed below the carcinomatous region. (With permission from KAIZAKI et al. 1999)

cosa were rarely positive in solitary or cluster form only in patients with a high titer of anti-EBV antibodies. We speculate that EBV may infect some proliferating cells or surface epithelium-committed cells, possibly through EBV-carrying lymphocytes, and that the infected cells are shed when EBER is expressed in the infected cells. On the other hand, YANAI et al. (1997) and JING et al. (1997), using a commercially available DNA-ISH kit, recently reported that EBV-infection could occasionally be observed in the epithelial cells in intestinal metaplasia. However, compared to the theoretically expected level, the signals were too strong in their studies. The copy number of EBV is generally considered to reach a few hundred at the most even in EBVaGC (FUKAYAMA et al. 1994; IMAI et al. 1994), which is the lower limit of ISH using DNA probes without any enhancing procedure. Since the methods in both studies lacked sufficient control experiments, the findings are open to question.

As we could not confirm their findings of DNA-ISH using the same method, we applied PCR for the *Bam*HI W region of EBV DNA to microdissected tissues of nonneoplastic gastric mucosa (KAIZAKI et al. 1999). Two of 118 microdissected samples from stomachs with EBVaGC and 5 of 62 samples from those with EBV-negative gastric carcinoma showed amplification of EBV DNA. The positive samples consisted of three from the pyloric glands and four from the fundic glands, while none of the metaplastic gland samples showed such amplification. These findings suggest that infection of intestinal metaplastic cells is unlikely. Thus, at present, we believe that EBV infection is a rare event in the stomach, and that the primary target of EBV infection is not the epithelial cells of intestinal metaplasia.

3.2 EBVaGC and Gastritis

We mentioned the possibility that stomachs harboring EBVaGC may be conditioned to develop the virus-associated carcinoma. To clarify the existence of a local predisposing factor, we histologically evaluated gastritis in nonneoplastic gastric mucosa that surrounded early carcinoma with or without associated EBV infection (Figs. 2 and 3) (KAIZAKI et al. 1999). For the evaluation, the tissue samples were taken from early carcinoma as well as nonneoplastic mucosa, encompassing a length of at least 1cm either from the proximal or distal edge of the carcinoma. The whole carcinomatous lesion was subjected to routine histologic examination by the step-section method. The factors that were evaluated were the localization of the carcinoma relative to the mucosal element, atrophy of the mucosa, topographical relationship of intestinal metaplasia, and lymphocytic infiltration. The grades of atrophy and lymphocytic infiltration were determined using the visual analogue scales of the updated Sydney System (DIXON et al. 1996) to avoid bias of estimation. Infection by *Helicobacter pylori* in nonneoplastic mucosa was determined by immunohistochemistry.

As for the location of the carcinoma, EBVaGC was most frequently present at the intermediate zone between the fundic and pyloric gland mucosa, where gastric ulcers occur frequently. Atrophy and lymphocytic infiltration were more marked compared to that in the mucosa of both the intestinal and diffuse types of EBV-negative carcinomas. Only 13% of EBVaGC were surrounded by intestinal metaplasia, in contrast to 41% of intestinal-type EBV-negative gastric carcinomas.

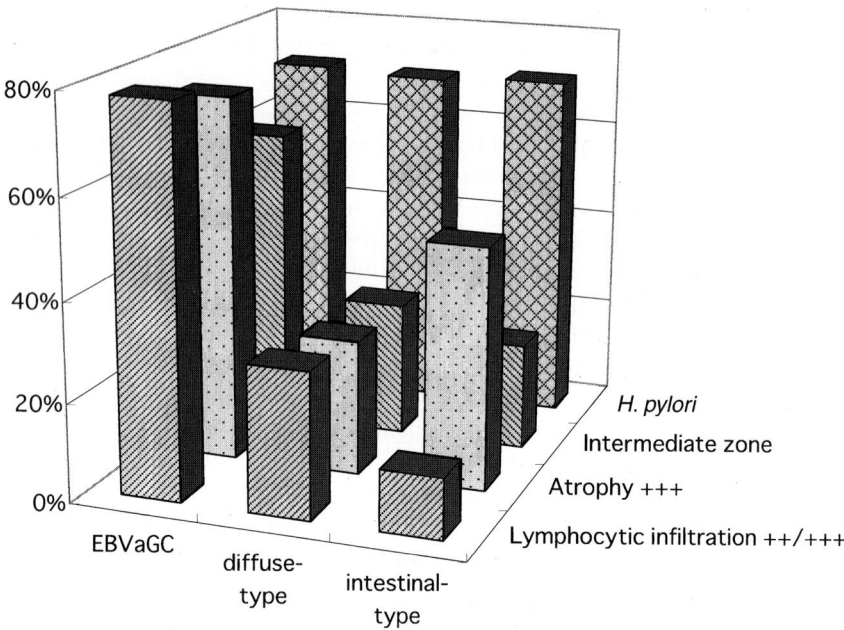

Fig. 3. Histological features of nonneoplastic mucosa of EBVaGC and EBV-negative gastric carcinoma. The nonneoplastic mucosa of early carcinomas (EBVaGC, $n=23$; EBV-negative carcinomas, intestinal-type, $n=139$, and diffuse-type, $n=44$) were histologically evaluated. The presence of *Helicobacter pylori* was immunohistochemically evaluated in the representative cases (EBVaGC, $n=14$; carcinomas, intestinal-type, $n=14$, and diffuse-type, $n=10$)

The frequency of *H. pylori* infection was not different in EBVaGC and EBV-negative carcinomas. Thus, the stomachs bearing EBVaGC is characterized by severe atrophic gastritis, which does not show intimate relationship with intestinal metaplasia. EBVaGC may develop from EBV-infected epithelial cells in a specific type of severe atrophic gastritis.

4 Molecular and Cellular Abnormalities in EBVaGC

Studies on the molecular mechanism underlying EBVaGC have only begun, compared to those on NPC. We investigated the deletion of 5q and/or 17p and microsatellite instability using polymerase chain reaction-restriction fragment length polymorphism (PCR-RFLP) and microsatellite markers respectively, in EBVaGC and EBV-negative gastric carcinomas (Chong et al. 1994). These abnormalities were extremely rare in EBVaGC, in contrast to their high frequency in EBV-negative carcinoma, especially the intestinal type. These findings indicate that the genetic pathways underlying the development of EBVaGC and EBV-negative carcinoma may be different. An immunohistochemical study revealed that EBVaGC did not show any specific pattern of *bcl*-2 expression or p53 accumulation (GULLEY et al. 1996).

Studies on the cellular characteristics of EBVaGC have similarly been scarce. Both the frequency of apoptosis and the proportion of proliferative cells were significantly lower in EBV-associated lymphoepithelioma-like carcinoma than in conventional EBV-negative gastric carcinoma (OHFUJI et al. 1996). Some isoforms of CD44, an adhesion molecule expressed on the cell surface, have been associated with the metastatic potential of carcinomas, such as colon and breast carcinomas. When CD44 variants 3–5 and 6 were immunohistochemically determined (CHONG et al. 1997), EBV-associated gastric carcinomas are strongly positive for both variants. By multivariate analysis, EBV-infection and lymph node metastasis contribute independently to CD44 variant expression. Thus, the mechanism and significance of CD44 variant expression are different in gastric carcinoma with and without EBV infection. EBV infection may influence CD44 expression by interacting with cytokine genes, such as those for TNFα, γ, and interleukin-10, which are known to modulate CD44 expression. Infiltration by lymphocytes in EBVaGC, most of which are CD8-positive, may be induced by such a mechanism, rather than as a reaction to carcinoma cells (KUZUSHIMA et al. 1999).

5 Model Systems for EBVaGC

Several experimental systems are now being applied for the study of EBVaGC, such as an in vitro infection system using the virus-producing cell line Akata and genetically engineered EBV (YOSHIYAMA et al. 1997; IMAI et al. 1998; NISHIKAWA et al. 1999) to investigate the mechanism of EBV infection in epithelial cells. On the other hand, in vitro cell culture and in vivo transplantation of neoplastic cells, which retain the characteristics of the original tumor, are also useful for studying the cell biology and molecular mechanism underlying EBVaGC. Since a stable cell line of NPC that carries the EBV genome in the nucleus is difficult to establish, we attempted to transplant a human EBVaGC in mice with severe combined immunodeficiency (SCID). We established a transplantable strain designated as KT after the patient from whom the tumor was removed (IWASAKI et al. 1998). Mucin and cytokeratin expression and the Alu sequence in tumor DNA confirmed that the KT tumor was derived from human epithelial tissue. The identity of clonal EBV in the original and KT tumors was demonstrated by clonal analysis of EBV DNA. The pattern of the latency-gene expression of EBV was the same in both tumors: EBER1 was also found in tumor cell nuclei by ISH. Reverse transcription-PCR (RT-PCR) analysis also demonstrated Qp-driven EBNA1 expression, but not EBNA2- or LMP1-expression (IMAI et al. 1994; SUGIURA 1997). Thus, the transplantable human EBVaGC retains the

Table 1. Infectious agents closely associated with gastric carcinoma. Comparison between *Helicobacter pylori* and Epstein-Barr virus

	Helicobacter pylori	Epstein-Barr virus
General population	75% of adults older than 40	More than 95% after adolescence
Detection of organism in the stomach	Nonneoplastic mucosa in most of gastric cancer	10% or less in the carcinoma tissues
Relationship with gastric carcinomas	Indirect, through chronic inflammation	Direct?
Relationship with lymphoma	MALToma	Various types of lymphomas, but rare in the stomach
Preventive or therapeutic strategy	Eradication of *H. pylori* in high-risk group	Immunotherapy? Gene therapy?

original EBV with the same latency-gene expression pattern, and serves as a model system.

6 Concluding Remarks

EBVaGC is a unique type of gastric carcinoma that is tagged by clonal EBV, and is expected to become a relatively more important gastric carcinoma as the frequency of other risk factors seems to be declining. Although *H. pylori* is regarded as a causative agent for most gastric carcinomas, its effect might be indirect and mediated through sustained injury to the mucosa, resulting in atrophic gastritis, intestinal metaplasia, and precancerous lesion (Table 1). On the other hand, while EBVaGC accounts for 10% or less of gastric carcinomas, EBV seems to play a direct role in the development of this carcinoma. As for the therapeutic strategy, while prevention of the infection or eradication of the organism in high-risk groups is the primary strategy for *H. pylori*, other types of therapy should be considered in the case of EBV. Gene therapy specific for EBV-associated neoplasms (GUTIERREZ et al. 1996), if established, should establish EBVaGC as a distinct clinical entity in gastric cancer. Our in vivo model will also be useful in this regard.

Acknowledgements. The authors thank Dr. M. Koike and Ms. Y. Hayashi (Tokyo Metropolitan Komagome Hospital, Tokyo, Japan) for their encouragement. This work was supported by a Grant in Aid for Scientific Research on Priority Areas (09253103) from the Ministry of Education, Science, Sports, and Culture of Japan.

References

Anagnostopoulos I, Hummel M (1996) Epstein-Barr virus in tumours. Histopathology 29:297–315
Arikawa J, Tokunaga M, Tashiro Y, Tanaka S, Sato E, Harabuchi K, Yamamoto A, Toyohira O, Tuchimochi A (1997a) Epstein-Barr virus-positive multiple early gastric cancers and dysplastic lesions: a case report. Pathol Int 47:730–734

Arikawa J, Tokunaga M, Satoh E, Tanaka S, Land CE (1997b) Morphological characteristics of Epstein-Barr virus-related early gastric carcinoma: a case control study. Pathol Int 47:360–367

Chong J, Fukayama M, Hayashi Y, Takizawa T, Koike M, Konishi M, Kikuchi-Yanoshita R, Miyaki M (1994) Microsatellite instability in the progression of gastric carcinoma. Cancer Res 54:4595–4597

Chong J, Fukayama M, Hayashi Y, Funata N, Takizawa T, Koike M, Muraoka M, Kikuchi-Yanoshita R, Miyaki M, Mizuno S (1997) Expression of CD44 variants in gastric carcinoma with or without Epstein-Barr virus. Int J Cancer 74:450–454

Dixon MF, Genta RM, Yardley JH, Correa P, and participants in the International Workshop on the Histopathology of Gastritis, Houston 1994 (1996) Classification and grading of gastritis. The updated Sydney System. Am J Surg Pathol 20:1161–1181

Epstein A (1994) Thirty years of Epstein-Barr virus. Epstein-Barr virus Report 1:3–4

Fukayama M, Hayashi Y, Iwasaki Y, Chong J, Ooba T, Takizawa T, Koike M, Miyaki M, Mizutani S, Hirai K (1994) Epstein-Barr virus-associated gastric carcinoma and Epstein-Barr virus infection of the stomach. Lab Invest 71:73–81

Fukayama M, Chong J, Kaizaki Y (1998) Epstein-Barr virus and gastric carcinoma. Gastric Cancer 1:104–114

Gulley ML, Raphael M, Lutz CT, Ross DW, Raab-Traub N (1992) Epstein-Barr virus integration in human lymphomas and lymphoid cell lines. Cancer 70:185–191

Gulley ML, Pulitzer DR, Eagan PA, Schneider BG (1996) Epstein-Barr virus infection is an early event in gastric carcinogenesis and is independent of bcl-2 expression and p53 accumulation. Hum Pathol 27:20–27

Gutierrez MI, Judde J, Magrath IT, Bhatia KG (1996) Switching viral latency to viral lysis: a novel therapeutic approach for Epstein-Barr virus-associated neoplasia. Cancer Res 56:969–972

Imai S, Koizumi S, Sugiura M, Tokunaga M, Uemura Y, Yamamoto N, Tanaka S, Sato E, Osato T (1994) Gastric carcinoma: monoclonal epithelial malignant cells expressing Epstein-Barr virus latent infection protein. Proc Natl Acad Sci USA 91:9131–9135

Imai S, Nishikawa J, Takada K (1998) Cell-to-cell contact as an efficient mode of Epstein-Barr virus infection of diverse human epithelial cells. J Virol 72:4371–4378

Iwasaki Y, Chong J, Hayashi Y, Ikeno R, Arai K, Kitamura M, Koike M, Hirai K, Fukayama M (1998) Establishment and characterization of a human EBV-associated gastric carcinoma in SCID mice. J Virol 72:8321–8326

Japanese Research Society for Gastric Cancer (1995) Japanese classification of gastric carcinoma. Kanahara, Tokyo

Jing X, Nakamura Y, Nakamura M, Yokoi T, Shan L, Taniguchi E, Kakudo K (1997) Detection of Epstein-Barr virus DNA in gastric carcinoma with lymphoid stroma. Viral Immunol 10:49–58

Kaizaki Y, Sakurai S, Chong J, Fukayama M (1999) Atrophic gastritis, Epstein-Barr virus-associated gastric carcinoma. Gastric Cancer 2:101–108

Kuzushima K, Nakamura S, Nakamura T, Yamamura Y, Yokoyama N, Fujita M, Kiyono T, Tsurumi T (1999) Increased frequency of antigen-specific CD8 + cytotoxic T lymphocytes infiltrating an Epstein-Barr virus-associated gastric carcinoma. J Clin Invest 104:163–171

Lauren P (1965) The two histological main types of gastric carcinoma: diffuse and so-called intestinal type carcinoma. Acta Pathol Microbiol Scand (A) 64:31–49

Matsunou H, Konishi F, Hori H, Ikeda T, Sasaki K, Hirose Y, Yamamichi N (1996) Characteristics of Epstein-Barr virus-associated gastric carcinoma with lymphoid stroma in Japan. Cancer 77:1998–2004

Ohfuji S, Osaki M, Tsujitani S, Ikeguchi M, Sairenji T, Ito H (1996) Low frequency of apoptosis in Epstein-Barr virus-associated gastric carcinoma with lymphoid stroma. Int J Cancer 68:710–715

Nishikawa J, Imai S, Oda T, Kojima T, Okita K, Takada K (1999) Epstein-Barr virus promote epithelial cell growth in the absence of EBNA2 and LMP1 expression. J Virol 73:1286–1292

Osato T (1998) Epstein-Barr virus infection and oncogenesis. In: Osato T, Takada K, Tokunaga M (eds) Epstein-Barr virus and human cancer. Japan Scientific Societies Press, Tokyo, pp 3–16

Raab-Traub N, Flynn K (1986) The structure of the termini of the Epstein-Barr virus as a marker of clonal cellular proliferation. Cell 47:833–839

Raab-Traub N (1992) Epstein-Barr virus and nasopharyngeal carcinoma. Cancer Biol 3:297–307

Sugiura M, Imai S, Tokunaga M, Koizumi S, Uchizawa M, Okamoto K, Osato T (1996) Transcriptional analysis of Epstein-Barr virus gene expression in EBV-positive gastric carcinoma: unique viral latency in the tumor cells. Br J Cancer 74:625–631

Takano Y, Kato Y, Saegusa M, Mori S, Shiota M, Masuda M, Mikami T, Okayasu I (1999) The role of Epstein-Barr virus in the oncogenesis of EBV (+) gastric carcinoma. Virchows Arch 434:17–22

Tokunaga M, Land CE, Uemura Y, Tokudome T, Tanaka S, Sato E (1993) Epstein-Barr virus in gastric carcinoma. Am J Pathol 143:1250–1254
Yanai H, Takada K, Shimizu N, Mizugaki Y, Tada M, Okita K (1997) Epstein-Barr virus infection in non-carcinomatous gastric epithelium. J Pathol 183:293–298
Yoshiyama H, Imai S, Shimizu N, Takada K (1997) Epstein-Barr virus infection of human gastric carcinoma cells: implication of the existence of a new virus receptor different from CD21. J Virol 71:5688–5691

EBV and Malignant Lymphoma with Special Emphasis on Pyothorax-Associated Lymphoma

K. Aozasa, H. Kanno, H. Miwa, and Y. Tomita

1	Introduction.	103
2	Nasal Lymphoma.	104
2.1	General Aspects.	104
2.2	EBV Association.	104
2.3	HLA Allele Frequency.	106
2.4	Mutations of p53 Gene.	107
3	Hodgkin's Disease.	107
4	Pyothorax-Associated Lymphoma.	108
4.1	General Aspects.	108
4.2	Clinical Findings.	109
4.3	Pathologic Findings.	110
4.4	EBV Association.	111
4.5	Characterization of Cell Lines from PAL.	111
4.6	Escape Mechanism of PAL Cells from Cytotoxic T-Lymphocytes.	113
4.6.1	HLA Allele Frequency.	113
4.6.2	Immunosuppressive Cytokine.	114
4.6.3	Mutations of the EBNA4 Epitope of CTL.	114
4.7	Mutations of p53 Gene.	117
4.8	Kaposi's Sarcoma-Associated Herpesvirus in PAL.	117
	References.	117

1 Introduction

The etiologic role of Epstein-Barr virus (EBV) for B lymphomagenesis has been postulated from epidemiologic and in vitro studies. There is now a substantial body of evidence linking EBV to African Burkitt's lymphoma, which is a B-cell lymphoma (KLEIN 1985). An association between EBV and human malignancies, including Hodgkin's disease (WEISS et al. 1987) and non-Hodgkin's lymphomas of either B- or T-cell immunophenotypes (HOCHBERG et al. 1983; HAMILTON-DUTOIT et al. 1992), has been reported. EBV-positive rate in immunocompetent patients with nodal lymphomas is less than 10% in B-cell and approximately 50% in T-cell lymphoma (AOZASA et al. 1998). The criteria for defining cases as EBV-positive

Department of Pathology (C3), Osaka University Medical School, 2-2 Yamadaoka, Suita, Osaka 565-0871, Japan

includes (1) presence of EBV genome by polymerase chain reaction (PCR), and (2) positive signals for EBV in large tumor cells by DNA or RNA in situ hybridization (ISH) using EBV-encoded small nuclear early region-1 (EBER1) probe. Among extranodal lymphomas, nasal natural killer (NK)-cell lymphoma, pyothorax-associated lymphoma (PAL), and adrenal lymphoma are EBV-associated; the EBV-positive rate is over 90% in the nasal lymphoma and PAL and approximately 50% in the adrenal lymphoma.

2 Nasal Lymphoma

2.1 General Aspects

Nasal NK-cell lymphoma is one of the constituents of lethal midline granuloma (LMG), which is a clinical term for progressive, destructive lesions in the midline of the face. Histologically, LMG comprises three different diseases, i.e., Wegener's granulomatosis (WG), NK-cell lymphoma (Fig. 1), and ordinary malignant lymphoma (KASSEL et al. 1969). WG is an inflammatory disease of an autoimmune nature. Nasal NK-cell lymphoma was formerly known as polymorphic reticulosis (PR) because of its polymorphic character of infiltrates (Fig. 1). Because the proliferating cells in PR showed a positive reaction for anti-T-lymphocyte antibody, the term nasal T-cell lymphoma was once used. Recent studies indicated the NK-cell nature of proliferating cells in PR (EMILE et al. 1996; OHSAWA et al. 1999).

Our previous epidemiologic study revealed a much higher frequency of nasal NK-cell lymphoma in Asian countries, including Japan, Korea, and China than in Western countries (AOZASA et al. 1989; AOZASA et al. 1992), with the frequency being highest in Korea (Table 1). The frequency of nasal NK-cell lymphoma in Japan was approximately three times higher than that in the UK. Even in Japan, the disease is much more frequent in Okinawa, islands situated in the southwestern part of the country with a subtropical climate, than in other Japanese areas.

2.2 EBV Association

In the sino-nasal regions, an etiologic role of EBV in the development of nasopharyngeal carcinoma (zur HAUSEN et al. 1990) and nasal NK-cell lymphoma (HARABUCHI et al. 1990) was suggested by the combined study of PCR and ISH for EBV DNA or its transcripts and immunohistochemistry. To examine whether the higher frequency of nasal NK-cell lymphoma in Okinawa than in Osaka is linked to the higher rate of EBV infection, we studied the presence of EBV genomes in cases with nasal NK-cell lymphoma from both locations (TOMITA et al. 1995). All but one case in Osaka and all of the Okinawa cases were EBV positive, showing no difference in frequency by district.

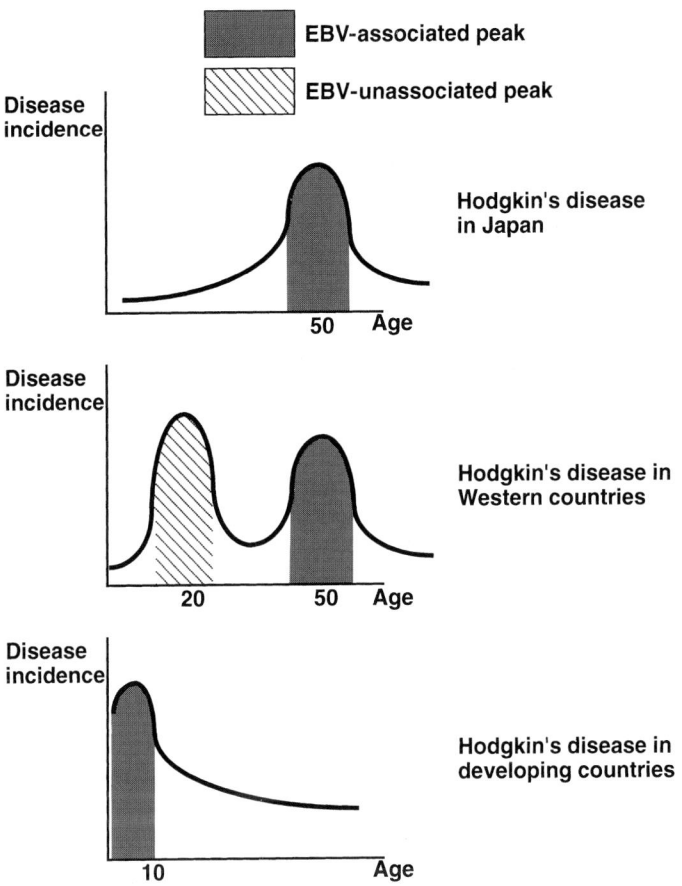

Fig. 1. Peak of age in Hodgkin's disease. EBV association is found in younger age in developing countries and older age in Japan and Western countries

Table 1. Frequency of lethal midline granuloma (AOZASA et al. 1989)

Diseases	Number of patients (frequency per 100,000 ENT patients)		
	Korea (1979–1989)	China (1979–1990)	Japan (1965–1986)
WG	0 (0)	1 (0.1)*	64 (4)*
NNKL	56 (40.8)**	73 (9.8)**	114 (8)**
ML	15 (10.9)	54 (7.2)	82 (6)
CI	6 (4.4)	0 (0)	42 (3)
Total	77 (56)	128 (17)	302 (21)

CI, chronic inflammation; ML, malignant lymphoma; NNKL, nasal NK-cell lymphoma; WG, Wegener's granulomatosis.
* $p < 0.03$; ** $p < 0.0001$.

We recently examined the chronological changes in incidence of nasal NK-cell lymphoma in Korea and Japan (OHSAWA et al. 1999). A total of 102 cases and 655 cases of LMG admitted to Yonsei University, Korea from 1977 to 1966 and 59 university hospitals in Japan from 1965 to 1996 were examined. The frequency rate of nasal NK-cell lymphoma per 100,000 outpatients of ears, nose, and throat clinics in Korea decreased from 40 to 20 between the periods of 1977–1989 and 1990–1996. However, there were no significant changes in Japan during the period studied. Socioeconomic factors together with EBV might be important in the development of nasal NK-cell lymphoma.

Two types of EBV, which differ at the EBV nuclear Ag (EBNA) gene loci, have been described (ROWE et al. 1989). Difference of subtype frequency was reported between geographic areas (ZIMBER et al. 1986) and immunologic conditions, i.e., immunocompromised or immunocompetent (BOYLE et al. 1991). Majority of nasal NK-cell lymphoma in East Asian countries had type B EBV (TOMITA et al. 1995).

2.3 HLA Allele Frequency

Tumor cells in nasal NK-cell lymphoma are known to express EBV-nuclear antigen (EBNA)-1, and frequently latent membrane protein (LMP)-1 and LMPs, but not EBNA 2, 3, 4, or 6 (MINAROVITS et al. 1994). This pattern of EBV latent infection is Lat II. Then how the lymphomas cells expressing EBV LMPs evade host immune surveillance by cytotoxic T lymphocyte (CTL) is an important issue. The cells expressing viral antigens are eliminated primarily by CTL in a MHC class I-restricted manner (RICKINSON and MOSS 1997). Several CTL-defined epitopes have been mapped in LMPs restricted with human leukocyte antigen (HLA)-A2, -A11, or -A24. To examine the possibility that the HLA-A allele may affect the development of nasal NK-cell lymphoma, HLA-A alleles of 25 patients were determined with low-resolution PCR-based typing using HLA-A locus sequence-specific primer combinations (Table 2; Kanno et al. 2000). The frequency of HLA-A alleles including HLA-A2 and -A24 antigens in the patients was lower than that in the

Table 2. Antigen frequencies at the HLA-A locus in PR patients and the normal Japanese population

Allele	PR ($n=25$)		Japanese ($n=303$)	
	n	Frequency (%)	Frequency (%)	P
A2	7	28.0	45.5	0.0665
A11	8	32.0	18.1	0.0820
A24	13	52.0	61.0	0.8644
A26	9	36.0	20.1	0.0594
A31	1	4.0	13.5	0.1413
A33	2	8.0	20.5	0.0994
A*0201	1[a]	4.0	20.1	0.0314
A*0206	3	12.0	15.2	0.4693
A*0207	4	16.0	8.6	0.1813

[a] Not exactly identified, and the possibility of A*0222 and A*0224 could not be ruled out.

Japanese population, but the difference was not significant. Since HLA-A2-restricted CTL responses are well delineated at A2-subtype level, the A2-subtype of the cases with HLA-A2 antigen was further determined by high-resolution genetic typing. The frequency of the HLA-A*0201 in PR was significantly lower than in the normal population ($p = 0.0314$). The HLA-A*0201-restricted CTL responses might work in vivo to suppress the development of overt lymphoma.

2.4 Mutations of p53 Gene

p53 is a well-known tumor suppressor gene which causes cells with damaged DNA to arrest at the G_1 phase of the cell cycle or stimulating expression of the *bax* gene, the protein of which promotes apoptosis (LEVINE et al. 1991). In a wide variety of human cancers, p53 gene mutations have been detected mainly in the exon 5-8 (HOLLSTEIN et al. 1991). A high incidence of malignant lymphoma in p53 knockout mice has been reported (DONEHOWER et al. 1992), suggesting an important role of p53 gene mutations in lymphomagenesis. Mutated p53 gene encodes mutant p53 protein, which has a much longer half-life than that of wild-type p53 protein, thus accumulating in the cytoplasm in a sufficient amount to be detected immunohistochemically. A previous immunohistochemical study showed that p53 is overexpressed in a high percentage of nasal NK/T-cell lymphoma cases (QUINTANILLA-MARTINEZ et al. 1998).

We then examined the expression and mutations of p53 gene in the paraffin-embedded specimens of nasal lesions from 42 Chinese (Beijing and Chengdu) and Japanese (Okinawa and Osaka) patients with nasal NK-cell lymphoma by immunohistochemistry and single-strand conformation polymorphism (SSCP) analysis of PCR-amplified products followed by direct sequencing. Thirty single-nucleotide substitution mutations were observed in 20 of 42 cases (47.6%; LI et al. 2000). Among the 30 mutations, 18 were missense (mainly G:C to A:T transitions), 9 were silent, and 1 was nonsense. The remaining two mutations involved intron 5 and exon 5 terminal points. Abnormal expression of p53 protein was also observed in 19 of 42 (45.2%) cases. The incidence was significantly (fourfold) higher in the cases of Osaka than those in other areas, although the incidence of p53 mutations in the cases of Osaka was 1/2 to 1/3 of those in other three areas (Table 3). The results may suggest some racial, environmental, and/or lifestyle differences in the cause of human nasal tumorigenesis.

3 Hodgkin's Disease

The association between EBV and Hodgkin's disease (HD) is of current interest. A combined study using the PCR and ISH methods demonstrated the EBV genome in Reed-Sternberg (R-S) cells (WEISS et al. 1989) and pathognomonic cells in HD. Several studies showed clonal integration of the EBV genome in R-S cells, sug-

Table 3. Overexpresssion and mutations of p53 gene in the cases with nasal NK-cell lymphoma of four different areas

Regions	No. of cases	p53 overexpression (%)	p53 mutation (%)
China			
Beijing	14	6 (42.9)	8 (57.1)
Chengdu	5	2 (40.0)	3 (60.0)
Japan			
Okinawa	14	3 (21.4)	7 (50.0)
Osaka	9	8 (88.9)*	2 (22.2)
Total	42	19 (45.2)	20 (47.6)

*$p < 0.01$ vs Okinawa by Fisher's exact text.

gesting prior infection with the virus before development of HD (WEISS et al. 1989). Serologic study has demonstrated that elevation of the antibody titers to EBV antigens occurred in HD patients before their diagnosis (MUELLER et al. 1989). Epidemiologic study also indicated a causal association between EBV and HD (SERRAINO et al. 1991). The EBV positivity ratio in HD is reported to be much higher in developing countries than in the United States and Europe (AMBINDER et al. 1993; CHANG et al. 1993). The initial peak incidence of HD is found in childhood in developing countries but in young adulthood in industrialized countries (CORREA et al. 1971). Therefore, correlation of primary EBV infection at a younger age in the developing countries and a peak incidence of HD in childhood in these countries has been suggested.

We investigated the EBV association with HD in Japan (TOMITA et al. 1996), where a distinct peak incidence is found in older adults. EBV genomes were detected in the R-S cells of 32 of the 50 patients examined (64%). This association was independently affected by histologic subtype (84% in mixed cellularity and 44% in others), sex (76% in males and 31% in females), and age (76% in patients aged 40 years and older and 38% in patients below age 40: $p < 0.01$). High EBV association is found at the peak in older adults predominantly with mixed cellularity type. Previous studies revealed high EBV association in the older peak of the bimodal peaks in Western HD, and a unimodal peak in childhood in developing countries. Recently MACK et al. (1995) suggested that genetic susceptibility underlies HD in young adults in Western countries (Fig. 1).

4 Pyothorax-Associated Lymphoma

4.1 General Aspects

In 1987, we reported three patients who developed pleural lymphoma after a 22–30-year history of pyothorax resulting from artificial pneumothorax for the treatment of pulmonary tuberculosis or tuberculous pleuritis (IUCHI et al. 1987). These cases were found among 134 patients with chronic pyothorax (CP) at one of the hospitals specializing in chest diseases in Osaka, Japan, during the period 1971–1985 (Fig. 2). We regarded the CP to be etiologically important in the development of pleural lymphoma because (1) no occurrences of pleural lymphoma were found in over 2,000 cases of malignant lymphoma seen in the general hospitals in Osaka; (2) all six cases of pleural lymphoma registered in the Annual of the Pathological Autopsy Cases in Japan (1974–85) were associated with CP; and (3) all cases of pleural lymphoma reported in the Japanese journals on chest diseases were associated with CP.

Other malignancies, such as squamous-cell carcinoma, malignant mesothelioma, and soft tissue sarcoma, have been reported to develop in the pleural cavity of patients with CP, in both Japan and Western countries (AOZASA et al. 1997). We also have emphasized the etiological role of CP in the development of angiosarcoma (AOZASA et al. 1994). Malignant lymphoma, however, has only rarely been reported as a complication of CP in Western countries but relatively frequent in Japan. (IUCHI et al. 1987). This is most interesting, although it remains unclear why this difference occurs between CP patients in Japan and Western countries (IUCHI et al. 1989).

4.2 Clinical Findings

The clinical findings in 37 patients with pleural lymphoma are summarized in Table 4 (IUCHI et al. 1989). The age at first diagnosis of lymphoma was 46–81 years (mean, 63); the male:female ratio was 5.2:1. All patients were admitted to the

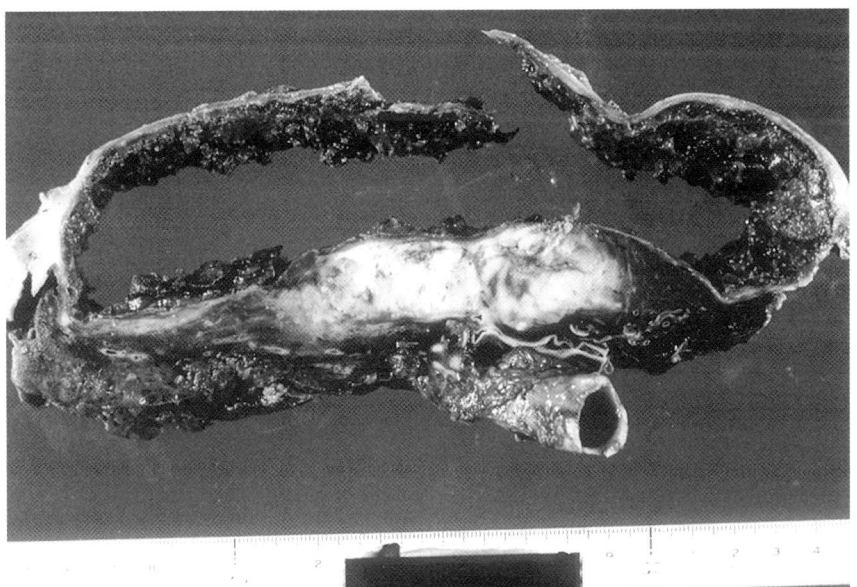

Fig. 2. Lymphoma develops adjoining the irregularly thickened pleural wall

Table 4. Summary of clinical findings in pleural lymphoma patients

Age at diagnosis	46–81 (mean 63 years)
Male:female ratio	5.2:1
Tuberculosis history	
Lung tuberculosis	81%
Tuberculous pleuritis	16%
Interval between pyothorax history and onset	22–55 (mean 33 years)
Presenting symptoms	
Chest pain	51%
Productive cough and dyspnea	54%
Tumor of chest wall	14%
Diagnosis at admission	
Chronic pyothorax	49%
Chronic pyothorax complicated with malignancy	38%
Lung tumor	5%
Detection:	
Chest X-ray	35%
Computed tomographic scan	77%
Definitive diagnosis	
Biopsy	84%
Autopsy	16%

hospital with a >20-year history of CP resulting from artificial pneumothorax for the treatment of pulmonary tuberculosis or tuberculous pleuritis. Common presenting symptoms were pain in the chest, back, or shoulder, and respiratory symptoms such as productive cough, often with hemoptysis, fever, or dyspnea. Five patients presented with a tumor of the chest wall. Chest X-ray films and computed tomographic (CT) scans revealed masses in 13 (35%) of 37 patients and 24 (77%) of 31 patients, respectively. The main masses detected by these investigations were situated in the pleura (28 patients), the lung near the pleura (5 patients), and the pleura and lung (4 patients). Combined findings from the physical examination, X-ray films, CT, and echogram revealed direct invasions to adjacent structures, such as the chest wall, lung, pericardium, and diaphragm, in 21 of 37 patients. Six patients had regional and nine had distant nodal enlargement. No patient had a leukemic blood picture.

4.3 Pathologic Findings

The pathologic findings are summarized in Table 5. The pleural tissue adjoining the tumors generally showed marked fibrous thickening, with sparse nonneoplastic inflammatory cells mainly comprising small lymphocytes and plasma cells. All cases were non-Hodgkin's lymphomas with approximately 80% being of the diffuse large cell type. Immunohistochemical studies revealed that all cases but one were of B-cell lineage. Autopsies showed that the malignant lymphomas were localized in the thoracic cavity in approximately half of the cases. Extrathoracic dissemination was found in the liver, intra-abdominal lymph nodes, adrenal gland, stomach, kidney, central nervous system, spleen, superficial lymph nodes, small intestine, and pancreas.

Table 5. Summary of pathologic findings

Histologic classification of tumors[a]	
Diffuse large cell types	30 (81%)
Immunoblastic	22 (59%)
Noncleaved cell	6 (16%)
Cleaved	2 (5%)
Diffuse lymphoplasmacytic type	5 (14%)
Immunologic and immunohistologic marker[b]	
B-cell type	32
T-cell type	1
Extent of tumor at autopsy:	
Intrathoracic disease	48%
Extrathoracic extension	52%

[a] Two cases (5%) could not be classified because of inadequate materials.
[b] Undetermined in four cases.

From these findings, we proposed the term pyothorax-associated lymphoma (PAL). PAL is a non-Hodgkin's lymphoma of exclusively B-cell phenotype developing in the pleural cavity of patients with a >20-year history of chronic pyothorax.

4.4 EBV Association

Because of the B-cell nature and development of the tumor under particular circumstances of chronic pyothorax (CP), we proposed an investigation of the etiological role of EBV in the development of PAL (AOZASA and MISHIMA 1992). In 1993 Japanese investigators suggested an association of EBV with PAL (FUKAYAMA et al. 1993; SASAJIMA et al. 1993); detection of the EBV genome was accomplished by PCR and ISH, clonal form of the structure of the EBV-fused termini, immunohistochemical detection of EBV-encoded latent gene products, EBV nuclear antigen-2 (EBNA2) and LMP1. High serum titers against EBV were also observed. Because the number of PAL cases examined in these reports was small and the presence of EBV in cases with CP alone was not recorded, we determined the presence of EBV genome in 26 PAL cases and 16 of CP alone (OHSAWA et al. 1995). Median duration of CP in patients with this condition alone and with CP complicated with PAL was 33 and 37 years, respectively. Combined PCR and ISH showed that the EBV genome was detected in lymphoma cells in all cases with PAL, but in only one of those with CP alone (Fig. 3). These findings confirmed the role of EBV in pleural lymphomagenesis. Immunohistochemistry on the paraffin-embedded specimens showed that PAL cells expressed EBNA-2 and LMP1.

4.5 Characterization of Cell Lines from PAL

We have established two lymphoma cell lines from biopsy specimens of PAL cases, OPL-1 and OPL-2, and examined their growth characteristics and the expression of

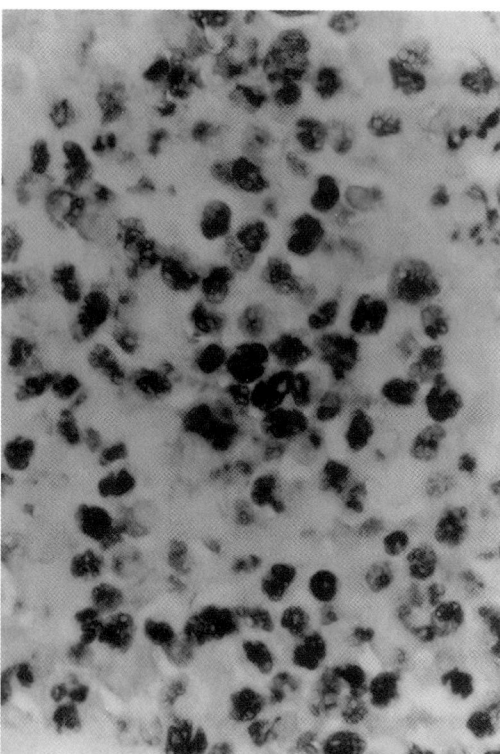

Fig. 3. In situ hybridization using EBER-1 probe reveals positive signals in the nucleus of large cells

EBV latent infection genes and oncogenes (KANNO et al. 1996a). OPL-2 exhibited a more rapid growth and higher saturation density than OPL-1, and only OPL-2 exhibited colony-forming activity in soft agar. A patient from whom OPL-1 was derived entered into complete remission with chemotherapy and has remained alive for more than 3 years since biopsy. Aggressive chemotherapy was not effective in the patient from whom OPL-2 was derived, and this patient died approximately 2 months after biopsy (Table 6). We regarded OPL-2 (case 2) to be at an advanced stage and OPL-1 (case 1) at an early stage of disease. OPL-1 and -2 had B-differentiation markers and clonal surface immunoglobulins. Both lines contained a single predominant form of episomal EBV DNA, indicating clonal cellular proliferation of an EBV-infected progenitor cell. OPL-1 and -2 contained type B and A EBV genomes, respectively.

Since PALs develop in the sites of chronic inflammation, inflammatory cytokines might be involved in the lymphomagenesis. To address this point, we examined the regulation of growth of OPLs by human interleukin-6 (IL-6) (KANNO et al. 1996b). Human rIL-6 enhanced the growth rate of OPLs. OPL-1 responded to rIL-6 to grow faster, whereas OPL-2 required a higher concentration of rIL-6. OPLs expressed IL-6 receptor messenger (m)RNA detectable by RT-PCR analysis and IL-6 receptor on the cell surface by flow cytometric analysis using anti-IL-6

Table 6. Clinical and biological characteristics of the PAL cell lines and of the patients from which they were derived

	Case 1 (OPL-1)	Case 2 (OPL-2)
Clinicopathological features of original cases		
Age (years)	76	67
Sex	M	M
Duration of pyothorax (years)	46	40
Artificial pneumothorax	Done	Done
Histologic diagnosis	DIB	DIB
Prognosis	CR	Died
Growth characteristics		
Doubling time (h)	48	24
Saturation density (cells/ml)	5×10^5	1×10^6
Colony formation on soft agar	(−)	(+)

CR, complete remission; DIB, diffuse lymphoma of immunoblastic cell type.

receptor antibodies. On the other hand, only OPL-1 showed the expression of IL-6 mRNA detectable solely by RT-PCR and secreted IL-6 protein into the culture media. The culture supernatant of OPL-1 exhibited growth-enhancing effects on OPL-1 and OPL-2. Addition of anti-IL-6 antibodies to the cultures inhibited the growth of OPL-1, but not OPL-2. OPL-2 did not secrete IL-6 protein into the media, and the culture supernatant from OPL-2 did not enhance the growth of OPL-2. These findings suggested that IL-6 enhanced the proliferation of PAL-derived cell lines, and was involved in the lymphomagenesis of PAL.

4.6 Escape Mechanism of PAL Cells from Cytotoxic T-Lymphocytes

PAL is a diffuse large-cell lymphoma of B-cell type, contain EBV DNA and express some EBV latent infection genes, EBNA 2 and LMP 1. In immunocompromised hosts, B lymphocytes expressing these molecules can escape from cytotoxic T-lymphocyte (CTL) immune surveillance. However, systemic immunodeficient conditions have not been noted in PAL patients. How can PAL cells evade host immune surveillance? To clarify this point, we examined three hypotheses: (1) HLA allele frequency, (2) immunosuppressive cytokine, (3) mutations of EBNA 4 epitope of CTL.

4.6.1 HLA Allele Frequency

Through literature review, it is evident that PAL patients are clustered in Japan. Therefore, we carried out HLA-A typing under the same hypothesis as nasal NK-cell lymphoma (2–3): that frequencies of HLA-A2 and A-11 might be lower than those of the normal population. HLA alleles of 16 patients with PAL were determined with low-resolution polymerase chain reaction-based typing using HLA-A locus sequence-specific primer combinations (Kanno et al. 1999). The antigen frequencies of HLA-A2 and -A11 in PAL patients were not significantly different from those in the normal Japanese population, suggesting that no HLA-A alleles influence the development of overt PAL.

4.6.2 Immunosuppressive Cytokine

It has been postulated that immunoregulatory cytokines and other factors affecting CTL induction and proliferation play an important role in the host immune reaction to EBV latent infection gene-positive cells (RICKINSON et al. 1992). IL-10 and transforming growth factor (TGF)-β exert immunosuppressive effects by suppressing antigen-specific CTL induction (RANGES et al. 1987; MOORE et al. 1993) and by inhibiting cytokine production by helper T cells and macrophages (MOORE et al. 1993). Intracellular binding of TGF-β protein with latent TGF-β-binding protein (LTBP) is required for its efficient secretion and activation (MIYAZONO et al. 1991). Therefore, examination of LTBP expression is necessary for the estimation of the function of TGF-β protein. In addition, there is another EBV gene, BCRF-1, that bears partial homology to the human IL-10 gene and shows some of the biological activities of IL-10 (HSU et al. 1990). We examined the expression of these immunosuppressive factors as well as BCRF-1 in the PAL cell lines (KANNO et al. 1997). Both OPL-1 and OPL-2 expressed TGF-β1 mRNA. However, neither expressed LTBP. The expression of IL-10 mRNA and protein was observed only in OPL-1, an EBNA-2- expressing PAL cell line (Fig. 4). IL-10 might contribute to the development of overt lymphoma by inducing locally immunosuppressive circumstances in the early stage of PAL development (OPL-1).

4.6.3 Mutations of the EBNA4 Epitope of CTL

EBV latently infects B cells in healthy individuals. Latent infection genes of EBV including EBNAs and latent membrane proteins (LMPs) are expressed in latently infected and immortalized B cells (RICKINSON et al. 1992). The infected B cells expressing all of the latent infection genes are, however, assumed to be killed by host CTLs, and those expressing only EBNA1, which is not recognized as a target by CTLs, are assumed to evade host immune surveillance in vivo (RICKINSON et al. 1992). EBNA3, -4, and -6 are immunodominant antigens for CTL responses (RICKINSON and MOSS 1997), and among them, the CTL-epitopes in EBNA4 are well-characterized (DE CAMPOS-LIMA et al. 1994). There are several HLA-A11-restricted CTL-epitopes in EBNA4 and mutated sequences in the two immuno-dominant epitopes compared with the sequence of the prototype type A B95-8 are reported in endemic strains in Southeast Asia and Papua New Guinea (DE CAMPOS-LIMA et al. 1993, 1994). These mutated sequences reduce HLA-A11-restricted CTL responses (DE CAMPOS-LIMA et al. 1994). Since the population with HLA-A11 allele in Southeast Asia and Papua New Guinea is more than half of the total population, the genetic pressure of the population with HLA-A11 is likely to make these mutated EBV endemic (DE CAMPOS-LIMA et al. 1993). The population with HLA-A11 in Europe and Africa is approximately 15% and 0%, respectively, thus the strains with prototype sequence could be endemic (DE CAMPOS-LIMA et al. 1994). The

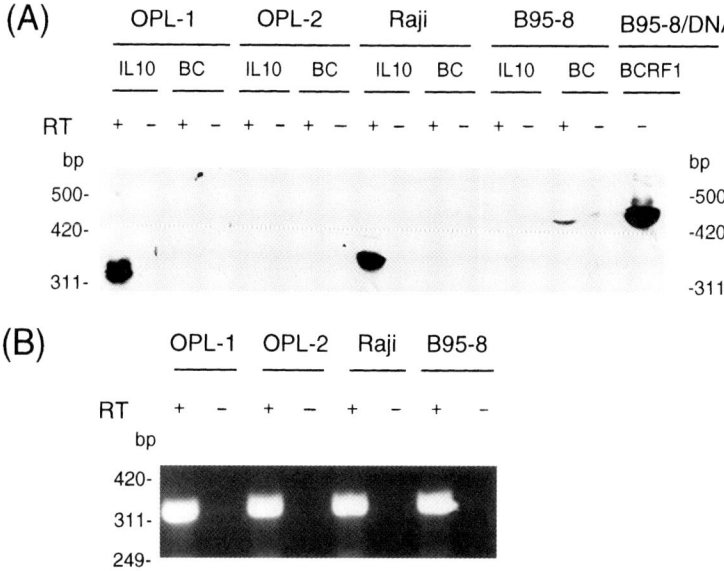

Fig. 4A,B. RT-PCR analyses of human IL-10 and BCRF-1. The 0.5-μg samples of DNAse-treated mRNA prepared from the indicated cell lines were reverse transcribed and amplified by PCR for IL-10 (IL10) or BCRF-1 (BC; **A**) or for actin mRNA (**B**). Amplified products were electrophoresed in 2% agarose gels, Southern blotted, and hybridized with the IL-10/BCRF-1 oligonucleotide probe (**A**) or visualized with an ultraviolet transilluminator (**B**). mRNA samples without reverse transcription (RT) were also subjected to the same PCR amplification, and 0.5-μg of DNA from EBV-producing B95-8 cells was used as a positive control for PCR amplification of the BCRF-1 sequence. DNA size markers are indicated on both sides. EBV-producing B95-8 cells shows the expression of BCRF-1 mRNA. Raji and OPL-1, but not OPL-2, show the expression of human IL-10 mRNA

sequence of the CTL-epitope region of EBNA4 in normal individuals, nasal NK-cell lymphoma, and PAL in Japan, where individuals with HLA-A11 allele reach approximately 20% of the total population (DE CAMPOS-LIMA et al. 1994), was examined (Kanno et al. 2000). The EBNA4 CTL-epitope region in the normal Japanese population and in two lymphoid neoplasias, pyothorax-associated lymphoma (PAL) and nasal NK-cell lymphoma, was directly sequenced by PCR. Most EBV in peripheral blood leukocytes (PBLs) from healthy Japanese donor exhibited prototype type A sequence, with mutations in approximately 20% (3/16). EBNA4 sequence in lymphoma tissue was obtained in six PAL cases, and five exhibited mutations compared with the prototype type A sequence. Furthermore, the EBNA4 sequence in PAL tissue was different from those in PBLs of the same patients or one of the sequences found in PBLs. On the other hand, the EBNA4 gene in nasal lymphoma tissues exhibited predominantly prototype type A sequence. Because PAL cells expressed EBNA4 mRNA detected with reverse transcriptase-PCR, but nasal lymphoma cells did not, mutations of CTL-epitopes in EBNA4 sequences were specific findings to EBNA4-positive lymphoma. Anti-tumor and virus immune response might work in vivo and affect the development of EBV antigen-positive lymphomas.

Table 7. p53 mutations in lymphoma and other tumors in humans

Tumors	Histology	EBV + total cases (%)	Cases with mutations/total cases (%)	G:C to A:T transitions/total substitution (%)	G:C to A:T transitions at CpG sites/total substitutions (%)	G:C to A:T transitions at dipyrimidine sites/total substitutions (%)
Malignant lymphoma						
PAL	Diffuse immunoblastic type	21/21 (100)	14/21 (67)[a],*	12/13 (92)	2/13 (15)	10/13 (77)**
Burkitt's lymphoma	Burkitt's type	2/4 (50)	3/4 (75)	1/3 (33)	1/3 (33)	1/3 (33)
	Burkitt's type	ND	9/27 (33)	5/11 (45)	4/11 (36)	2/11 (18)
AIDS-related lymphoma	Diffuse large cell type	22/34 (65)	1/34 (3)	1/1 (100)	0/1 (0)	1/1 (100)
MALT	Small noncleaved cell type	10/24 (42)	10/27 (37)	8/10 (80)	5/10 (50)	5/10 (50)
	B-cell type	ND	21/75 (28)	6/21 (29)	4/21 (19)	3/21 (14)
Carcinoma						
Nasopharyngeal SCC		23/25 (92)	1/25 (4)	1/1 (100)	1/1 (100)	1/1 (100)
Skin cancer	BCC	ND	20/59 (34)	15/20 (75)	8/20 (40)	13/20 (65)
	SCC	ND	22/47 (47)	10/22 (45)	0/22 (0)	10/22 (45)
Radon-associated lung cancer	Large cell type or SCC	ND	36/72 (50)	6/32 (19)	2/32 (6)	2/32 (6)
A-bomb-associated lung cancer	Adeno-Ca or SCC	ND	4/9 (44)	2/4 (50)	0/4 (0)	1/4 (25)

Adeno-Ca, adenocarcinoma; BCC, basal-cell carcinoma; MALT, mucosa-associated lymphoid tissue lymphoma. ND, not done; SCC, squamous-cell carcinoma.
* $p < 0.05$ by χ^2 test with Yate's correction and/or Fisher's exact tests versus other kinds of lymphomas or except squamous cell carcinoma of the skin and radon-associated lung cancer.
** $p < 0.05$ expect basal cell carcinoma and squamous cell carcinoma of the skin and AIDS-related lymphoma. Statistical analyses were not done for the samples with less than four cases.
[a] One deletion.

4.7 Mutations of p53 Gene

We analyzed p53 mutations on paraffin-embedded specimens from 21 patients with PAL by PCR-SSCP followed by direct sequencing (HONGYO et al. 1998). An unusually high frequency of p53 mutations (14 of 21 cases; 67%) was detected in the PAL specimens (Table 7), and mutations consisted of 13 nucleotide substitutions and 1 deletion. Furthermore 10 of 13 substitutions (77%) occurred at the dipyrimidine site (CC:GG to CT:GA substitution). Dipyrimidine sites are known to be susceptible to radiation. Therefore, these findings suggest that long-term radiation during the artificial pneumothorax or specific drug exposure may have caused specific mutations in the p53 gene.

4.8 Kaposi's Sarcoma-Associated Herpesvirus in PAL

DNA sequences belonging to the recently discovered Kaposi's sarcoma-associated herpesvirus (KSHV) (CHANG et al. 1994), now called human herpes virus 8 (HHV-8), have been previously identified in an uncommonly occurring subset of AIDS-related lymphomas, body-cavity-based lymphomas (BCBL), which present as lymphomatous effusions (CESARMAN et al. 1995). Although PALs present as solid tumor masses, they are otherwise similar to BCBL in that they also are B-cell lymphomas, usually exhibit immunoblastic morphology, and contain EBV. KSHV sequences were present in two BCBL in patients without AIDS, but not in 12 Japanese or 2 French PAL patients (CESARMAN et al. 1996). From these findings, the BCBL could be divided into two types; one is pyothorax-associated lymphoma (PAL) with formation of a mass in the pleural cavities and without KSHV, and the other is primary effusion lymphoma (PEL) with pleural effusion without mass formation but with KSHV.

References

Ambinder RF, Browning PJ, Lozenzana I (1993) Epstein-Barr virus and childhood Hodgkin's disease in Honduras and the United States. Blood 81:462–467
Aozasa K, Mishima K (1992) Etiological role of chronic inflammation in the development of sarcoma. Biotherapy 6:1351–1357 (in Japanese)
Aozasa K, Naka N, Tomita Y, et al. (1994) Angiosarcoma developing from chronic pyothorax. Mod Pathol 7:906–911
Aozasa K, Ohsawa M, Kanno H (1997) Pyothorax-associated lymphoma: A distinctive type of lymphoma strongly associated with Epstein-Barr virus. Adv Anat Pathol 4:58–63
Boyle MJ, Sewell WA, Sculley TB, Apolloni A, Turner JJ, Swanson CE, et al. (1991) Subtypes of Epstein-Barr virus in human immunodeficiency virus- associated non-Hodgkin's lymphoma. Blood 78:3004–3011
de Campos-Lima PO, Levitsky V, Brooks J, Lee SP, Hu LF, Rickinson AB, Masucci MG (1994) T cell responses and virus evolution: Loss of HLA A11 restricted CTL epitopes in Epstein-Barr virus isolates from highly A-11 positive populations by selective mutation of anchor residues. J Exp Med 179:1297–1305

de Campos-Lima PO, Gavioli R, Zhang Q-J, Wallace LE, Dolcetti R, Rowe M, Rickinson AB (1993) HLA-A11 epitope loss isolates of Epstein-Barr virus from a highly A11+ population. Science 260: 98–100

Cesarman E, Chang Y, Moore PS, Said JW, Knowles DM (1995) Kaposi's sarcoma-associated herpesvirus-like DNA sequences are present in AIDS-related body cavity based lymphomas. N Engl J Med 332:1186–1191

Cesarman E, Nador RG, Aozasa K, Delsol G, Said JW, Knowles DM (1996) Kaposi's sarcoma-associated herpesvirus in non-AIDS-related lymphomas occurring in body cavities. Am J Pathol 149:53–57

Chang KL, Albugar PF, Chen YY, Johnson RM, Weiss LM (1993) High prevalence of Epstein-Barr virus in the Reed-Sternberg cells of Hodgkin's disease occurring in Peru. Blood 81:496–501

Chang Y, Cesarman E, Pessin MS, Lee F, Culpepper J, Knowles DM, Moore PS (1994) Identification of herpesvirus-like DNA sequences in AIDS-related Kaposi's sarcoma. Science 266:1865–1869

Correa P, O'Conor GT (1971) Epidemiologic patterns of Hodgkin's disease. Int J Cancer 8:192–201

Donehower LA, Harrey M, Slagle BL, McArthur MJ, Montgomery CA Jr, Butel J, Bradley A (1992) Mice deficient for p53 are developmentally normal but susceptible to spontaneous tumors. Nature 356:215–221

Emile JF, Boulland ML, Haioun C, Kanavaros P, Petrella T, Delfau-Larue MH, Besussan A, Farcet JP, Gaulard P (1996) CD5- CD56+ T-cell receptor silent peripheral T-cell lymphomas are natural killer cell lymphomas. Blood 87:1466–1473

Fukayama M, Ibuka T, Hayashi Y, Ooba T, Koike M, Mizutani S (1993) Epstein-Barr virus in pyothorax-associated pleural lymphoma. Am J Pathol 143:1044–1049

Hamilton-Dutoit SJ, Pallesen G (1992) Survey of Epstein-Barr virus gene expression in sporadic non-Hodgkin's lymphoma. Detection of Epstein-Barr virus in a subset of peripheral T-cell lymphomas. Am J Pathol 140:1315–1325

Harabuchi Y, Yamanaka N, Kataura A, Imai S, Kinoshita T, Mizuno F (1990) Epstein-Barr virus in nasal T-cell lymphomas with lethal midline granuloma. Lancet 335:128–130

zur Hausen H, Schulte-Holthausen H, Klein G, Henle W, Henle G, Glifford P, Santesson L (1970) EBV DNA in biopsies of Burkitt tumors and anaplastic carcinomas of the nasopharynx. Nature 228:1056–1058

Hochberg FH, Miller G, Schooley RT, Hirsch MS, Feorino P, Henle W (1983) Central nervous system lymphoma related to Epstein-Barr virus. N Engl J Med 309:745–748

Hollstein M, Sidransky D, Vogelstein B, Harris SR (1991) p53 mutations in human cancers. Science 253:49–53

Hongyo T, Kurooka M, Taniguchi E, Iuchi K, Nakajima Y, Aozasa K, Nomura T (1998) Frequent p53 mutations at dipyrimidine sites in patients with pyothorax-associated lymphoma. Cancer Res 58:1105–1107

Hsu D-H, de Waal Malefyt R, Fiorentino DF, Dang M-N, Vieira P, DeVries S, Spits H, Mosmann TR, Moore KM (1990) Expression of interleukin-10 activity by Epstein-Barr virus protein BCRF1. Science 250:830–832

Iuchi K, Ichimiya A, Akashi A, Mizuta T, Lee Y-E, Tada H, Mori T, Sawamura K, Lee Y-S, Furuse K, Yamamoto S, Aozasa K (1987) Non-Hodgkin's lymphoma of the pleural cavity developing from long-standing pyothorax. Cancer 60:1771–1775

Iuchi K, Aozasa K, Yamamoto S, Mori T, Tajima K, Minato K, Mukai K, Komatsu H, Takagi T, Kobashi Y, Yamabe H, Shimoyama M (1989) Non-Hodgkin's lymphoma of the pleural cavity developing from long-standing pyothorax. Summary of clinical and pathological findings in thirty-seven cases. Jpn J Clin Oncol 19:249–257

Kanno H, Yasunaga Y, Ohsawa M, Taniwaki M, Iuchi K, Naka N, Torikai K, Shimoyama M, Aozasa K (1996a) Expression of Epstein-Barr virus latent infection genes and oncogenes in lymphoma cell lines derived from pyothorax-associated lymphoma. Int J Cancer 67:86–94

Kanno H, Yasunaga Y, Iuchi K, Yamauchi S, Tatekawa T, Sugiyama H, Aozasa K (1996b) Interleukin-6-mediated growth enhancement of cell lines derived from pyothorax-associated lymphoma. Lab Invest 75:167–173

Kanno H, Naka N, Yasunaga Y, Iuchi K, Yamauchi S, Hashimoto M, Aozasa K (1997) Production of the immunosuppressive cytokine interleukin-10 by Epstein-Barr virus-expressing pyothorax-associated lymphoma. Possible role in the development of overt lymphoma in immunocompetent hosts. Am J Pathol 150:349–357

Kanno H, Ohsawa M, Hashimoto M, Iuchi K, Nakajima Y, Aozasa K (1999) HLA-A alleles of patients with pyothorax-associated lymphoma: delineation of anti-Epstein-Barr virus (EBV) host immune response during the development of EBV latent antigen-positive lymphomas. Int J Cancer 82:630–634

Kanno H, Nakatsuka S, Iuchi I, Aozasa K (2000) Sequences of cytotoxic T-lymphocytes epitopes in Epstein-Barr virus (EBV) nuclear antigen-3B gene in Japanese population with or without EBV-positive lymphoid malignancies. Int J Cancer 88:626–632

Kanno H, Kojya S, Li T, Ohsawa M, Nakatsuka S, Miyaguchi M, Harabuchi Y, Aozasa K (2000) Low frequency of HLA-A*0201 allele in patients with Epstein-Barr virus -positive nasal lymphomas with polymorphic reticulosis morphology. Int J Cancer 87:195–199

Kassel SH, Echevaria RA, Guzzo FP (1969) Midline malignant reticulosis (so-called lethal midline granuloma). Cancer 23:920–925

Klein G (1985) Lymphoma development in mice and humans: diversity of initiation is followed by convergent cytogenetic evolution. Proc Natl Acad Sci USA 76:2442–2446

Levine AJ, Momand J, Finley CA (1991) The p53 tumor suppressor gene. Nature 351:453–456

Li T, Hongyo T, Syaifudin M, Nomura T, Dong Z, Shingu N, Kojya S, Nakatsuka S, Aozasa K (2000) Mutations of the p53 gene in nasal NK/T-cell lymphoma. Lab Invest 80:493–499

Mack TM, Cozen W, Shibata D, Weiss LM, Nathwani BN, Hernandes AM, Taylor CR, Hamilton AS, Deapen DM, Rappaport EB (1995) Concordance for Hodgkin's disease in identical twins suggesting genetic susceptibility to the young-adult form of the disease. N Engl J Med 332:413–418

Minarovits J, Hu L-F, Imai S, Harabuchi Y, Kataura A, Minarovits-Kormuta S, Osato T, Klein G (1994) Clonality, expression and methylation patterns of the Epstein-Barr virus genomes in lethal midline granulomas classified as peripheral angiocentric T cell lymphomas. J Gen Virol 75:77–84

Miyazono K, Olofsson A, Colosetti P, Heldin CH (1991) A role of the latent TGF-β1-binding protein in the assembly and secretion of TGF-β1. EMBO J 10:1091–1101

Moore KW, O'Garra A, de Waal Malefyt R, Veira P, Mosmann TR (1993) Interleukin 10. Ann Rev Immunol 11:165–190

Mueller N, Evans A, Harris N, et al. (1989) Hodgkin's disease and Epstein-Barr virus: altered antibody pattern before diagnosis. N Engl J Med 320:689–695

Ohsawa M, Tomita Y, Kanno H, Iuchi K, Kawabata Y, Nakajima Y, Komatsu H, Mukai K, Shimoyama M, Aozasa K (1995) Role of Epstein-Barr virus in pleural lymphomagenesis. Mod Pathol 8:848–853

Ohsawa M, Nakatsuka S, Kanno H, Miwa H, Kojya S, Harabuchi Y, Yang W-I, Aozasa K (1999) Immunophenotypic and genotypic characterization of nasal lymphoma with polymorphic reticulosis morphology. Int J Cancer 81:865–870

Ohsawa M, Shingu N, Inohara H, Kubo T, Yang W-I, Yoo J-H, Aozasa K (1999) Chronological changes in incidences of polymorphic reticulosis in Korea and Japan. Oncol 56:202–207

Quintanilla-Martinez L, Guerro I, Franklin JL, Naresh KN, Krenacs L, Rama-Rao C, Bhatia K, Magrath IT, Raffeld M (1998) Nasal NK/T-cell lymphoma from Peru: High prevalence of p53 overexpression. Mod Pathol 11:139

Ranges GE, Figari IS, Espevik T, Palladino MA Jr (1987) Inhibition of cytotoxic T cell development by transforming growth factor-β and reversal by recombinant tumor necrosis factor-α. J Exp Med 166:991–998

Rickinson AB, Murray RJ, Brooks J, Griffin H, Moss DJ, Masucci MG (1992) T cell recognition of Epstein-Barr virus associated lymphomas. Cancer Surv 13:53–80

Rickinson AB, Moss DJ (1997) Human cytotoxic T lymphocyte responses to Epstein-Barr virus infection. Ann Rev Immunol 15:405–431

Rowe M, Young LS, Cadwallader K, Petti L, Kieff E, Rickinson AB (1989) Distinction between Epstein-Barr virus type A (EBNA 2A) and type B (EBNA 2B) isolates extends to EBNA 3 family of nuclear proteins. J Virol 63:1031–1039

Sasajima Y, Yamabe H, Kobashi Y, Hirai K, Mori S (1993) High expression of the Epstein-Barr virus latent protein EB nuclear antigen on pyothorax-associated lymphoma. Am J Pathol 143:1280–1285

Serraino D, Franceschi S, Takamini R (1991) Socio-economic indicators, infectious disease and Hodgkin's disease. Int J Cancer 47:352–357

Tomita Y, Ohsawa M, Kanno H, Mishiro Y, Kojya S, Noda Y, Aozasa K (1995) The presence and subtype of Epstein-Barr virus in B and T cell lymphomas of the sino-nasal region from Osaka and Okinawa districts of Japan. Lab Invest 73:190–195

Tomita Y, Ohsawa M, Kanno H, Hashimoto M, Ohnishi A, Nakanishi H, Aozasa K (1996) Epstein-Barr virus in Hodgkin's disease in Japan. Cancer 77:186–192

Weiss LM, Strickler JG, Warnke RA, Purtilo DT, Sklar J (1987) Epstein-Barr viral DNA in tissue of Hodgkin's disease. Am J Pathol 129:86–91

Weiss LM, Movahed LA, Warnke RA, Sklar J (1989) Detection of Epstein-Barr viral genomes in Reed-Sternberg cells of Hodgkin's disease. N Engl J Med 302:502–506

Zimber U, Adldinger HK, Lenoir GM, Vuillaume M, Kaebel-Doeritz MV, Vaux G, et al. (1986) Geographical prevalence of two types of Epstein-Barr virus. Virol 154:56–66

AIDS Lymphoma: Its Virological Aspects

H. KATANO[1], T. SATA[1], and S. MORI[2]

1	Introduction	121
2	Epidemiology	122
3	Clinical Findings	122
4	Pathological Findings	123
5	Etiology	124
5.1	Genetic Pathogenesis of AIDS Lymphoma	124
5.2	EBV and AIDS Lymphoma	126
5.2.1	Detection of EBV in AIDS Lymphoma	126
5.2.2	Pathogenesis of EBV Lymphoma	126
5.2.3	Animal Model of EBV Lymphoma	127
5.3	HHV-8 and AIDS Lymphoma	127
5.3.1	HHV-8 and Kaposi's Sarcoma	127
5.3.2	HHV-8 and Primary Effusion Lymphoma	129
5.3.3	HHV-8 and AIDS-ALCL	130
5.3.4	HHV-8 and MCD	130
5.3.5	HHV-8 Serology	131
6	Conclusion	132
References		133

1 Introduction

Malignant lymphoma is one of the most common malignancies that associate with AIDS (acquired immunodeficiency syndrome) (HERNDIER et al. 1994; KNOWLES 1999). The revised criteria for the diagnosis of AIDS by the CENTERS FOR DISEASE CONTROL AND PREVENTION (CDC) in 1987 includes diffuse aggressive intermediate- or high-grade non-Hodgkin's lymphoma (NHL) of B-cell or indeterminate phenotype as one of the major complications in human immunodeficiency virus (HIV)-infected individuals (CDC 1987). Recently, the introduction of highly active

---

[1] Department of Pathology, National Institute of Infectious Diseases, 1-23-1 Toyama, Shinjuku-ku, Tokyo 162-8640, Japan
[2] Department of Pathology, Institute of Medical Science, University of Tokyo, 4-6-1 Shirokanedai, Minato-ku, Tokyo 108-8639, Japan

anti-retroviral therapy (HAART) has dramatically decreased various complications in AIDS patients. In contrast, however, the incidence of AIDS lymphoma did not decrease as did the other complications (FRANCESCHI et al. 1999; GRULICH 1999).

Ninety-five percent of AIDS lymphomas are the B-cell NHL. EBV has been thought to be the most causative agent for large portion of AIDS lymphomas (HERNDIER et al. 1994; GAIDANO et al. 1998; KNOWLES 1999). However, it was shown recently that a new human herpes virus, human herpesvirus 8 (HHV-8, Kaposi's sarcoma-associated herpesvirus, KSHV) associates with a part of AIDS lymphomas. This new virus was also shown to associate with Kaposi's sarcoma, and multicentric Castleman's disease (MCD). Thus, HHV-8 is now becoming one of the most important causative agents of AIDS-associated illnesses.

In this article we will review the recent knowledge regarding AIDS-associated NHLs and Hodgkin's disease.

2 Epidemiology

NHL is the most common malignancy found in Japanese AIDS patients. This is in contrast to Western countries in which Kaposi's sarcoma is the most common. In Western countries the clinical incidence of NHL in AIDS was reported to be 2.9% (BERAL et al. 1991), and NHLs have been observed in 9%–16% of autopsied AIDS patients (WELCH et al. 1984; FALK et al. 1987; MOHAR et al. 1992; KLATT et al. 1994). The incidence was shown to be much higher in Japanese AIDS patients (around 30%, MORI et al. 1991). We investigated 49 AIDS autopsy cases at the Institute of Medical Science Hospital, the University of Tokyo, from 1986 to 1997 and found that NHLs were detected in 17 cases (35%). While Kaposi's sarcoma usually occurs in homosexual AIDS patients, NHL complicates with various types of AIDS patients regardless of risk factors. Recently the incidence of AIDS-associated illnesses has changed dramatically because of the introduction of HAART. HAART decreased AIDS-associated illnesses such as opportunistic infections by *Pneumocystis carinii*, *mycobacterium avium-intracellulare complex*, cytomegalovirus, and other herpesviruses. However, the rate of NHL did not decrease (FRANCESCHI et al. 1999; GRULICH 1999).

3 Clinical Findings

Clinically, AIDS lymphomas are classified into systemic, primary central nervous system (PCNS), and primary effusion lymphoma (PEL; GAIDANO et al. 1998; KNOWLES 1999). Systemic AIDS lymphomas account for close to 80% of all AIDS lymphomas and are found much more commonly in the gastrointestinal tract, skin, lungs, liver, and adrenal glands. AIDS lymphomas usually occur when the CD4 cells decrease to below 200 cells/μl. The central nervous system is another major site of AIDS lymphomas. PCNS lymphoma occurs in AIDS patients with a much lower CD4 count (less than 50 cells/μl). In the terminal stage, malignant lymphomas spread into various organs. In patients with a low CD4 count, the progression of lymphoma is much more rapid.

PEL is a rare subtype of AIDS lymphomas (NADOR et al. 1996). PEL is defined as a lymphoma subtype causing effusion of the body cavity, involving the pleural, peritoneal, and pericardial cavity, without the formation of tumor masses. High serum human interleukin (hIL)-6 levels were reported in PEL patients.

4 Pathological Findings

Histologically, AIDS-NHL (B-cell type) can be categorized into Burkitt's lymphoma (AIDS-BL), diffuse large-cell lymphoma (AIDS-DLCL), and primary effusion lymphoma (AIDS-PEL) (GAIDANO et al. 1998; KNOWLES 1999), while Hodgkin's disease and T-cell NHL is rarely observed (HERNDIER et al. 1994). AIDS-DLCL can be further categorized into large noncleaved cell lymphoma (LNCCL) and immunoblastic lymphoma (IBL; GAIDANO et al. 1998). In Japan, almost all cases of AIDS-associated lymphoma are classified as DLCL (MORI et al. 1991). Microscopically, lymphoma occurs in various organs such as the gastrointestinal tract, brain, lungs, skin, liver, kidneys, and adrenal glands. In the brain, lymphoma cells tend to invade the cuff of blood vessels (Virshow Roban space; Fig. 1). Vast necrotic regions are often found in AIDS-DLCL. AIDS-DLCL is usually infected with EBV. Also, almost all AIDS-DLCLs express B cell markers such as CD19 and CD20. The clonality of AIDS lymphomas has been a matter of debate. Recent studies tend to conclude that most AIDS-associated NHLs are monoclonal, or oligoclonal at least (KNOWLES 1999; MEEKER et al. 1991; SHIRAMIZU et al. 1992).

PEL is a very rare disease, which usually occurs in homosexuals (CESARMAN et al. 1995; NADOR et al. 1996). Neoplastic cells in PEL exhibit an immunoblastic morphology, i.e., large and irregular-shaped cells with abundant amphotrophic cytoplasms and atypical nuclei (Fig. 2). Whereas many AIDS-BLs and AIDS-DLCLs express some B-cell markers, most PEL cases lack the expression of any B-cell or T-cell markers, but express CD45 (leukocyte common antigen), and CD138 (Syndecan-1, a plasma cell marker) (GAIDANO et al. 1997b). Because the Ig gene rearrangement is detected, the origin of PEL is speculated to be from B-cell lineages.

Some anaplastic large-cell lymphomas (ALCL) were also reported as rare complications of AIDS (CHADBURN et al. 1993; HERNDIER et al. 1994; TIRELLI et al. 1995; NOSARI et al. 1996; BUSKE et al. 1997; DEPOND et al. 1997; NAKAMURA et al. 1999). However, the definition of 'ALCL' was ambiguous in these

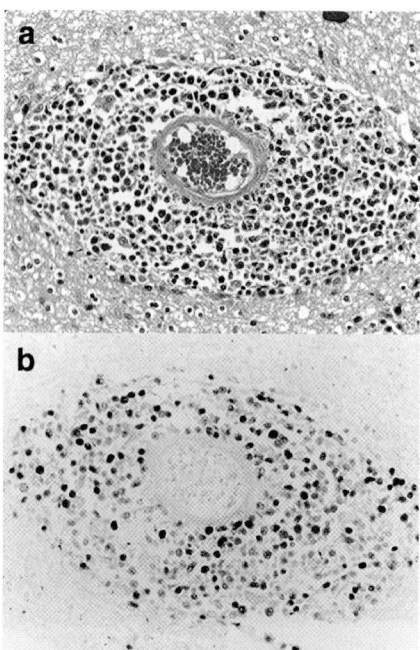

Fig. 1a,b. Primary central nervous system (PCNS) lymphoma. **a** Immunoblastic large-cell lymphoma cells invade the cuff of vessels (H&E staining). **b** In situ demonstration of EBER. Almost all lymphoma cells express EBER

reports. These cases expressed CD30, but its expression is varied among cases. The skin, lungs, and gastrointestinal tract are the main sites of involvement in AIDS-ALCL.

5 Etiology

5.1 Genetic Pathogenesis of AIDS Lymphoma

The etiology of AIDS lymphomas is variable even among the histological features.

In the lymph node of HIV-infected patients in the persistent generalized lymphadenopathy (PGL) phase, hyperplasia occurs in germinal centers, suggesting B-cell proliferation in the lymph node (PELICCI et al. 1986). Genetic alteration can occur in some populations of B cells at this phase. Reciprocal translocation including c-*myc* (8q24) is reported in almost all AIDS-BL (CHAGANTI et al. 1983; GROOPMAN et al. 1986; SUBAR et al. 1988; BALLERINI et al. 1993). Translocation induces the deregulation of the c-*myc* proto-oncogene, causing AIDS-BL. EBV is detected in about 30% of AIDS-BL cases, suggesting the association of EBV-infection with the translocation involving c-*myc* (GAIDANO et al. 1998).

In AIDS-DLCL, the translocation involving c-*myc* is rare. Chromosomal rearrangement of bcl-6 is reported in approximately 20% of AIDS-DLCL cases

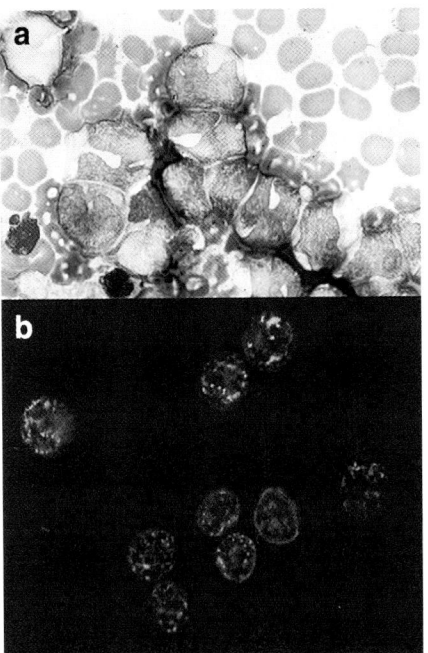

Fig. 2a,b. Primary effusion lymphoma (PEL). a Morphologically, immunoblastic large cells were observed in the pericardial effusion of a homosexual AIDS patient (Gimza staining). b Immunofluorescence assay of HHV-8. Serum derived from a Kaposi's sarcoma patient was used as the first antibody. Dot-like signals of LANA were found in the nucleus of HHV-8-infected PEL cell line

(GAIDANO et al. 1994). In addition, mutations of 5' regulatory sequences of bcl-6 were detected in 70% of AIDS-DLCL (MIGLIAZZA et al. 1995; GAIDANO et al. 1997a). BCL-6 is physiologically expressed in the germinal center B cells (ONIZUKA et al. 1995). Knockout-mice experiments revealed that BCL-6 is important for germinal center formation (DENT et al. 1997). In contrast to ordinary B-DLCL, only a minor portion of AIDS-DLCLs express BCL-6. Such BCL6-expressing AIDS-DLCLs include both bcl-6 gene-rearranged and non-rearranged cases. Peculiarly, they usually exhibit LNCCL morphology. In the case of EBV infection, LMP-1 is suppressed on such BCL-6-expressing AIDS-DLCLs (CARBONE et al. 1997b). BCL-6-non-expressing AIDS-DLCLs, which is the majority of AIDS-DLCLs, usually express LMP-1, suggesting that LMP-1 plays an important role in the pathogenesis of this type of lymphoma (GAIDANO et al. 1998).

Some AIDS-associated T cell-type NHLs were reported (HERNDIER et al. 1994). Usually, EBV is not detected in those lymphomas. The etiology of these T-cell lymphomas remains unclear; however, one hypothesis was proposed by Shiramizu et al. who reported a case of T-NHL in which HIV was integrated into the host genome (SHIRAMIZU et al. 1994). They showed that the c-fes/fps proto-oncogene was encoded near the integration site, and speculated that the integration of HIV caused abnormal expression of that proto-oncogene. Even though this case raised some interest, no further reports on this subject were published thereafter. In addition, T-cell lymphoma is exceptionally rare in AIDS patients.

5.2 EBV and AIDS Lymphoma

5.2.1 Detection of EBV in AIDS Lymphoma

EBV can be detected in various types of lymphomas regardless of HIV infection (Table 1). Meanwhile, the incidence of EBV infection is much higher in AIDS lymphomas than in non-HIV lymphomas (GUARNER et al. 1991; HAMILTON-DUTOIT et al. 1991, 1993a,b; HERNDIER et al. 1993).

In the nonneoplastic lymph nodes of patients in PGL, EBV-infected lymphocytes are occasionally observed by in situ hybridization detecting EBV-encoded small RNA (EBER). Those cells are predominantly large and blastic (ZHENG and MORI 1997). This finding indicates that cells with latent EBV gene expression proliferate in lymphoid organs of advanced stage HIV-infected individuals and that some of the cells are in a transformed state. It is possible to speculate that these cells are precursors of AIDS-NHL.

Among AIDS lymphomas, almost all cases of PCNS lymphoma harbor EBV (MACMAHON et al. 1991; CAMILLERI-BROET et al. 1995, 1997). EBER was detected in almost all lymphoma cells (Fig. 1b). Many PCNS lymphoma cells express LMP-1, which is one of the major transforming genes of EBV (CAMILLERI-BROET et al. 1995). Thus, it is quite clear that EBV plays an important role in the pathogenesis of AIDS-PCNS lymphoma.

While Hodgkin's disease is rare in AIDS (HERNDIER et al. 1994), EBV is detected in these cases at high incidence (HERNDIER et al. 1993), and EBER is usually detected in Reed-Sternberg cells, a diagnostic hallmark of HD. A recent study concluded that Reed-Sternberg cells are mostly derived from B cells (COSSMAN et al. 1999).

5.2.2 Pathogenesis of EBV Lymphoma

There are three pieces of evidence that indicate the association of EBV with AIDS lymphoma. The first is that EBV is detected at a high rate in AIDS lymphomas. The second is that EBV can immortalize B cells in vitro. The third is that EBV encodes some transforming genes, i.e., LMP-1 and EBNA-2. Particularly, transfection experiments showed that LMP-1 has a full transforming activity in rodent cell lines (WANG et al. 1985). LMP-1 signal transduction has been clarified recently. TNF-

Table 1. Positivity of EBV in AIDS-associated lymphoma

Reference	Lymphoma	Methods	Total no.	Positive no.	Positivity (%)
1 HERNDIER et al. 1993	AIDS-HD	ISH (EBER)	12	11	91.7
2 HAMILTON-DUTOIT et al. 1991	AIDS-NHL	ISH (EBV-DNA)	24	12	50.0
3 GUARNER et al. 1991	AIDS-NHL	ISH (EBER)	14	9	64.3
4 HAMILTON-DUTOIT et al. 1993a	AIDS-NHL	ISH (EBER)	128	85	66.4

HD, Hodgkin's disease; NHL, non-Hodgkin's lymphoma; ISH, in situ hybridization.

alpha receptor-associated factors (TRAFs) bind to the cytoplasmic domain of LMP-1 and the signals from TRAFs inducing nuclear localization of NFkB, resulting in cell proliferation (MOSIALOS et al. 1995). Simultaneously, LMP-1 induces BCL-2 expression, which at high levels inhibits apoptosis of host cells (HENDERSON et al. 1991). In fact, it was demonstrated that AIDS-PCNS lymphomas express BCL-2 highly (CAMILLERI-BROET et al. 1995).

Several EBV-encoded proteins, including LMP-1, -2A, -2B, and six EBNAs, are expressed in most AIDS-DLCLs. This pattern of EBV-gene expression is common among EBV-associated opportunistic lymphomas that occur in immunocompromised hosts, designated as latency type III proteins (ROWE et al. 1992). LMP-1, -2A and EBNA-2 are strong immunogens. They can be targets of cytotoxic T lymphocytes (CTLs) (KHANNA et al. 1998; MURRAY et al. 1992). Whereas LMP-1-, -2A- and EBNA-2-expressing cells are killed by specific CTL in immunocompetent host, they can escape attack and can proliferate in immunocompromised hosts as seen with AIDS. The escape from CTL attack is crucial in the development of AIDS lymphoma.

Recently, it was demonstrated that EBERs have transforming activity in BL cells (KOMANO et al. 1999). This suggests that EBERs play an important role in the development of AIDS lymphomas.

It is known that EBV-infected cells express various adhesive molecules, i.e., LFA1, LFA3, and ICAM1 (BILLAUD et al. 1990). Probably, these adhesion molecules on EBV-infected cells will work towards the settlement of those cells in extranodular organs, and the systemic insufficiency of T-cell function will allow those cells to grow at those organs, forming extranodal involvement.

5.2.3 Animal Model of EBV Lymphoma

Inoculation of human nonneoplastic B lymphocytes from EBV-infected individuals to severe combined immunodeficiency (SCID) mice produce B-cell lymphomas at the inoculation site (BASHIR et al. 1991; ITOH et al. 1993). Such B lymphocytes in SCID mice form tumors because these mice do not carry CTL and consequently cannot reject the EBV-immortalized B cells. Thus, such EBV-infected lymphomas in SCID mice are useful experimental models for NHL occurring in AIDS patients or immunocompromised hosts. These lymphomas usually express latency type III proteins including LMP-1 (Fig. 3). Flowcytometric analysis indicated that all lymphoma cells express LMP-1 at quantitatively different levels (KATANO et al. 1996), suggesting the possibility that LMP-1 can work as an immunotherapeutic target.

5.3 HHV-8 and AIDS Lymphoma

5.3.1 HHV-8 and Kaposi's Sarcoma

HHV-8 is a newly described human herpes virus identified from AIDS-associated Kaposi's sarcoma (CHANG et al. 1994). HHV-8 is a 170 kbp DNA virus, belonging

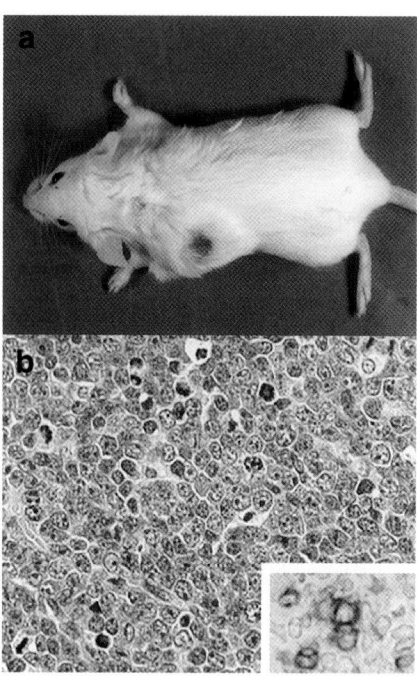

Fig. 3a,b. EBV-positive lymphoma in SCID mice. a Macroscopic view of SCID mice transplanted with EBV-infected human B lymphocytes. b Microscopic view of the lymphoma in SCID mice. Tumors in SCID mice exhibited DLCL morphologically. *Inset* an immunostaining for LMP-1

to the gammaherpesvirinae (RENNE et al. 1996a). HHV-8 has homologous genes to EBV and herpesvirus saimiri (HVS), which are known to be oncoviruses causing human lymphoma (EBV) or simian T-cell lymphomas (HVS), respectively. HHV-8 encodes several gene homologous to human cytokines or chemokines, i.e., IL-6, MIP-I, MIP-II, IRF-1 (RUSSO et al. 1996). IL-6 is known to be a growth factor for Kaposi's sarcoma (MILES et al. 1990).

Kaposi's sarcoma was first described in 1872 as a rare tumor in elderly men of Mediterranean descent (classic type; KAPOSI 1872). Recently, three additional clinical types, which are histologically indistinguishable, have been recognized: AIDS-associated, post-transplantational (iatrogenic or immunodeficient), and African (endemic) types (BUCHBINDER et al. 1996). DNA fragments of HHV-8 are detected in 95% of Kaposi's sarcoma tissues by polymerase chain reaction (PCR) regardless of their subtypes (DUPIN et al. 1995; MOORE and CHANG 1995). However, the copy number of this virus in spindle cells of Kaposi's sarcoma is low (approximately one copy per cell) (CESARMAN et al. 1995). Like other herpesviruses, HHV-8 has two phases of infection, the latent or lytic phase. Permissive cells have not been identified. It has been demonstrated that almost all spindle cells in Kaposi's sarcoma are in the latent phase (KATANO et al. 1999c). Therefore, it seems likely that the latent infection of HHV-8 is important for the pathogenesis of Kaposi's sarcoma.

5.3.2 HHV-8 and Primary Effusion Lymphoma

HHV-8 was first detected in 1995 in AIDS-associated body cavity-based lymphoma (PEL) (CESARMAN et al. 1995). Its immunophenotypes are undetermined; however, Ig gene rearrangement was detected (NADOR et al. 1996). PEL cells contain high copies (40–100 copies) of HHV-8 DNA (CESARMAN et al. 1995). Some PEL cases are co-infected with EBV, while others are infected with only HHV-8. Usually c-*myc* rearrangement is not detected in PEL (NADOR et al. 1996). In addition, expression of LMPs and EBNAs is suppressed in PEL cells. Several HHV-8-infected cell lines have been established from PEL (GAIDANO et al. 1996; RENNE et al. 1996b; CARBONE et al. 1997a, 1998; SAID et al. 1997; BOSHOFF et al. 1998; KATANO et al. 1999a). We established an HHV-8-positive, EBV-negative cell line, TY-1 from EBV-positive and HHV-8-positive PEL cases (KATANO et al. 1999a). These data suggest that HHV-8 plays a very important role in the pathogenesis of PEL.

HHV-8-positive PEL cell lines have given us many insights into the features of this virus. Chemical stimulation using 12-O-tetradecanoyl phorbol-13-acetate (TPA) induces the PEL cell line into lytic phase (RENNE et al. 1996b). Electromicroscopy has revealed that many HHV-8 viral particles were found in the nucleus of TPA-treated cells (SAID et al. 1997) (Fig. 4). It has also been demonstrated that all of the PEL cells usually express latency-associated nuclear antigen (LANA, Fig. 2b), suggesting that they are in the latent phase. In addition, LANA and HHV-8 DNA have been shown to co-localize in dots of interphase nuclei and mitotic chromosomes. These data suggest that LANA tethers HHV-8 DNA to chromosomes during mitosis allowing the segregation of HHV-8 episomes onto progeny cells (BALLESTAS et al. 1999).

HHV-8 encodes some oncogenic proteins like G-protein-coupled receptor (GPCR) or cyclin D. Among them, ORF K9 (vIRF) is known to inhibit the interferon signaling pathway, suggesting that K9 works as an oncogene in PEL cells (GAO et al. 1997). However, expression of K9 protein is restricted in PEL cells

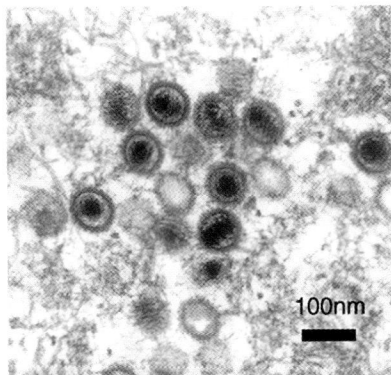

Fig. 4. Electromicroscopy of HHV-8-infected lymphoma cell. Many viral particles (virus capsids, diameter = 100nm) were observed in the nucleus of the TPA-induced HHV-8-positive cell line TY-1

regardless of TPA induction. Thus, further studies are needed to clarify the oncogenic role of HHV-8 in the pathogenesis of PEL.

5.3.3 HHV-8 and AIDS-ALCL

The existence of AIDS-ALCL has been stressed by some authors (CHADBURN et al. 1993; TIRELLI et al. 1995; NOSARI et al. 1996; BUSKE et al. 1997; DEPOND et al. 1997; NAKAMURA et al. 1999). However, the diagnostic criteria for ALCL are ambiguous and confusing in these reports. There is a general consensus that ALCL exhibits the following histopathological features: (1) anaplastic, large, blastic cell morphology with occasional horse-shoe-shaped/multiple nuclei and one or two prominent nucleoli, (2) much larger in size than the cells found in ordinary large cell lymphomas, with greater cytoplasmic volume, (3) growing in a cohesive sheet-like pattern (HARRIS et al. 1994). ALCL is classified into T/null cell-type lymphoma by the revised European-American lymphoma (REAL) classification (HARRIS et al. 1994). Meanwhile, reported AIDS-ALCL contained both B- and T-cell lymphoma.

We reported three cases of AIDS lymphomas that took an anaplastic large-cell morphology. Interestingly, all three were infected with HHV-8 (KATANO et al. 2000b). These three cases were complicated with Kaposi's sarcoma, PEL, and MCD. In addition, we demonstrated that inoculation of a PEL cell line to SCID mice produced HHV-8-positive and EBV-negative tumors in inoculated sites, while these tumor cells exhibited morphological characteristics of ALCL (Fig. 5). These findings suggest that HHV-8 can associate with solid lymphomas and that they can take an anaplastic large-cell morphology. These lymphomas should be distinguished from classical ALCLs, which are defined by the REAL classification even though their morphology and a part of their immunophenotype mimic classical ALCL.

5.3.4 HHV-8 and MCD

MCD is a rare disease characterized by plasmacytic lymphadenopathy with polyclonal hyperimmunoglobulinemia and high levels of IL-6 in the serum (FRIZZERA et al. 1983; CHEN 1984; YOSHIZAKI et al. 1989). Follicular hyperplasia with proliferation of plasma cells and hyaline vascular alterations in the lymph node are the histopathological hallmarks of MCD. In 1996, HHV-8 was first detected in some cases of MCD (SOULIER et al. 1995). Unlike KS and PEL, the HHV-8 genome is detected in only a minor portion of MCDs. However, HHV-8 DNA is frequently detected in tissues obtained from patients with MCD associated with HIV infection (SOULIER et al. 1995; OKSENHENDLER et al. 1996; PARRAVINCI et al. 1997).

In order to determine the association of MCD with HHV-8, we must clarify at least three questions. The first is whether HHV-8-encoded IL-6 (viral IL-6, vIL-6) plays a role in the proliferation of plasma cells: HHV-8 encodes a gene homologue of human IL-6 (hIL-6). This knowledge is attractive in considering the pathogenesis of HHV-8-associated MCDs, since the level of IL-6 is high in the sera of patients with MCD. In addition, vIL-6 was shown to be functional in the B9 cell prolifer-

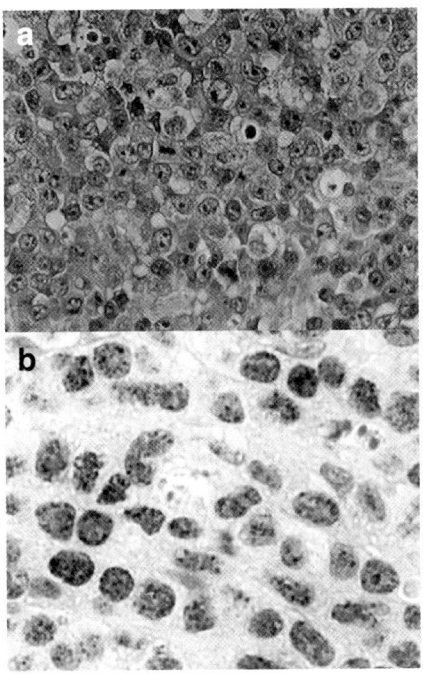

Fig. 5a,b. Engraftment of an HHV-8-positive lymphoma cell line in SCID mice. **a** Microscopic view of tumors in SCID mice transplanted with the HHV-8-positive cell line TY-1. Morphological features resembled those of an anaplastic large-cell lymphoma. **b** Immunohistochemistry revealed that these lymphoma cells were positive for HHV-8-encoded ORF73 protein (LANA)

ation assay (MOORE et al. 1996). However, the amount of vIL-6 in sera of patients with MCD has not been determined yet. Meanwhile, there are several points of interest in MCDs. The first point is whether vIL-6 can cause plasma cells to secrete hIL-6. Autocrine activation of vIL-6 and hIL-6 is expected, but such activation has not been demonstrated either in vivo, or in vitro.

The second point of interest is the localization and characteristics of HHV-8-infected cells in MCD. Immunohistochemical analysis revealed that HHV-8-infected cells are localized in the mantle zone of deformed lymph follicles (PARRAVINCI et al. 1997; DUPIN et al. 1999), and that the lytic proteins as well as latent proteins encoded by HHV-8 were expressed in cells (KATANO et al. 2000a). These data suggest that the role of HHV-8 in MCD differs from that in KS or PEL.

The third point is in serology. Anti-HHV-8 antibodies are detected in sera of HHV-8-infected individuals. However, there is little serological information indicating an association between MCD and HHV-8-infection so far.

Thus HHV-8 may at least play a role in pathogenesis of MCD. The role of cytokines such as IL-6 should be studied further.

5.3.5 HHV-8 Serology

Like other human herpes viruses, anti-HHV-8-antibodies are detected in sera of HHV-8-infected individuals. The seroprevalence against HHV-8 has been found to vary between studies, depending on the type of assay used and the countries where the investigations were carried out (SIMPSON et al. 1996; ANDRE et al. 1997; DAVIS et al. 1997; KEDES et al. 1997; CHATLYNNE et al. 1998; MAYAMA et al. 1998; MELBYE et al. 1998; RABKIN et al. 1998; REGAMEY et al. 1998).

To date, two methods [immunofluorescence assay (IFA) and enzyme-linked immunosorbent assay (ELISA)] have been used to detect antibodies to HHV-8. However, the target antigens differ among these methods. In IFA, latency-associated nuclear antigen (LANA), corresponding to ORF73 protein, is the major antigen, while in ELISA, ORF59 (PF-8), ORF65 (minor capsid protein), ORF26 (another possible minor capsid protein), or a lysate of whole viral particles have been used as antigens (SIMPSON et al. 1996; CHATLYNNE et al. 1998; DAVIS et al. 1997; PAU et al. 1998; RABKIN et al. 1998; REGAMEY et al. 1998). The use of these different antigens seems to have caused discrepancies in data reported to date (RABKIN et al. 1998). Recently, it was shown that lytic proteins such as ORFs 65, K8.1A and K8.1B, and the latent protein ORF73 exhibited high reactivity with sera of HHV-8-seropositive individuals on Western blot analysis (ZHU et al. 1999).

Thus, current HHV-8 antibody testing includes some problems in the accuracy, mainly in patients with asymptomatic HHV-8 infection (RABKIN et al. 1998). However, several groups have reported the seroprevalence to HHV-8 in the general population. Positivity in these studies has varied from 0% to 53%, depending on the assay methods used and the countries examined (CHATLYNNE et al. 1998; FUJII et al. 1998; MAYAMA et al. 1998; KATANO et al. 1999b; TEDESCHI et al. 1999). Infection with HHV-8 seems to be uncommon in the United States (0%–25%), Japan (0.2% or 2.2%) and Northern Europe (3%–5.1%), but more common in certain Mediterranean countries (4%–12%), and very prevalent in African countries such as Uganda (35%–53%).

The transmission modes of HHV-8 have not been clarified yet. In countries with endemic infection, horizontal transmission among children is suggested to be prevalent (MAYAMA et al. 1998), while sexual transmission appears to play an important role among homosexual men in non-endemic countries.

6 Conclusion

In this review, we described the association of AIDS lymphoma with two herpesviruses, EBV and HHV-8. While EBV and HHV-8 associate with the pathogenesis of AIDS-lymphomas, their mechanisms seem to be different. The discovery of HHV-8 contributed much in establishing the disease entity of PEL. HHV-6 and HHV-7 were identified from AIDS lymphoma, but consequently the associations of HHV-6 and HHV-7 with lymphoma were denied. However, the possibility of discovering new viruses that may change the concept of AIDS lymphoma dramatically cannot be ruled out. Probably, the mechanism of AIDS lymphomas will be more complex than we think at this moment.

Recent anti-retrovirus therapy has changed the pathology of AIDS. Whereas other AIDS-associated illnesses have decreased, AIDS lymphoma has not. It is our impression that AIDS lymphoma has changed to become more aggressive and fatal than before the development of HAART. Probably a longer life in an immunodeficient state may change the pathogenesis of AIDS lymphomas. In this sense, studies on the pathogenesis of AIDS lymphoma will become much more important.

References

Andre S, Schatz O, Bogner JR, Zeichhardt H, Stoffler MM, Jahn HU, Ullrich R, Sonntag AK, Kehm R, Haas J (1997) Detection of antibodies against viral capsid proteins of human herpesvirus 8 in AIDS-associated Kaposi's sarcoma. J Mol Med 75:145–152

Ballerini P, Gaidano G, Gong JZ, Tassi V, Saglio G, Knowles DM, Dalla-Favera R (1993) Multiple genetic lesions in acquired immunodeficiency syndrome-related non-Hodgkin's lymphoma. Blood 81:166–176

Ballestas ME, Chatis PA, Kaye KM (1999) Efficient persistence of extrachromosomal KSHV DNA mediated by latency-associated nuclear antigen. Science 284:641–644

Bashir R, Okano M, Kleveland K, Pirrucello S, Masih A, Sanger W, Fordyce-Boyer R, Purtilo D (1991) SCID/human mouse model of central nervous system lymphoproliferative disease. Lab Invest 65: 702–709

Beral V, Peterman T, Berkelman R, Jaffe H (1991) AIDS-associated non-Hodgkin lymphoma. Lancet 337:805–809

Billaud M, Rousset F, Calender A, Cordier M, Aubry JP, Laisse V, Lenoir GM (1990) Low expression of lymphocyte function-associated antigen (LFA)-1 and LFA-3 adhesion molecules is a common trait in Burkitt's lymphoma associated with and not associated with Epstein-Barr virus. Blood 75:1827–1833

Boshoff C, Gao SJ, Healy LE, Matthews S, Thomas AJ, Coignet L, Warnke RA, Strauchen JA, Matutes E, Kamel OW, Moore PS, Weiss RA, Chang Y (1998) Establishing a KSHV+ cell line (BCP-1) from peripheral blood and characterizing its growth in Nod/SCID mice. Blood 91:1671–1679

Buchbinder AQHY, Cockerell CJ, Friedman-Kien AE (1996) "Clinical manifestations and histopathologic features of classic, endemic African, and epidemic AIDS-associated Kaposi's sarcoma". In: Friedmann-Kien AE, Cockerell CJ (eds) Color atlas of AIDS, 2nd edn. Saunders WB Company, Philadelphia

Buske C, Hannig H, Hiddemann W, Bodemer W (1997) Human herpesvirus-8 (HHV-8) DNA associated with anaplastic large cell lymphoma of the B-cell type in an HIV-1-positive patient. Int J Cancer 73:303–304

Camilleri-Broet S, Davi F, Feuillard J, Bourgeois C, Seilhean D, Hauw JJ, Raphael M (1995) High expression of latent membrane protein 1 of Epstein-Barr virus and BCL-2 oncoprotein in acquired immunodeficiency syndrome-related primary brain lymphomas. Blood 86:432–435

Camilleri-Broet S, Davi F, Feuillard J, Seilhean D, Michiels JF, Brousset P, Epardeau B, Navratil E, Mokhtari K, Bourgeois C, Marelle L, Raphael M, Hauw JJ (1997) AIDS-related primary brain lymphomas: histopathologic and immunohistochemical study of 51 cases. The French Study Group for HIV-Associated Tumors. Hum Pathol 28:367–374

Carbone A, Cilia AM, Gloghini A, Canzonieri V, Pastore C, Todesco M, Cozzi M, Perin T, Volpe R, Pinto A, Gaidano G (1997a) Establishment of HHV-8-positive and HHV-8-negative lymphoma cell lines from primary lymphomatous effusions. Int J Cancer 73:562–569

Carbone A, Cilia AM, Gloghini A, Capello D, Todesco M, Quattrone S, Volpe R, Gaidano G (1998) Establishment and characterization of EBV-positive and EBV-negative primary effusion lymphoma cell lines harbouring human herpesvirus type-8. Br J Haematol 102:1081–1089

Carbone A, Gaidano G, Gloghini A, Pastore C, Saglio G, Tirelli U, Dalla-Favera R, Falini B (1997b) BCL-6 protein expression in AIDS-related non-Hodgkin's lymphomas: inverse relationship with Epstein-Barr virus-encoded latent membrane protein-1 expression. Am J Pathol 150:155–165

Centers for Disease Control and Prevention (1987) Revision of the CDC surveillance case definition for acquired immunodeficiency syndrome. Council of State and Territorial Epidemiologists; AIDS Program, Center for Infectious Diseases. MMWR Morb Mortal Wkly Rep 36 Suppl 1:1S–15S

Cesarman E, Chang Y, Moore PS, Said JW, Knowles DM (1995) Kaposi's sarcoma-associated herpesvirus-like DNA sequences in AIDS-related body-cavity-based lymphomas. N Engl J Med 332: 1186–1191

Chadburn A, Cesarman E, Jagirdar J, Subar M, Mir RN, Knowles DM (1993) CD30 (Ki-1) positive anaplastic large cell lymphomas in individuals infected with the human immunodeficiency virus. Cancer 72:3078–3090

Chaganti RS, Jhanwar SC, Koziner B, Arlin Z, Mertelsmann R, Clarkson BD (1983) Specific translocations characterize Burkitt's-like lymphoma of homosexual men with the acquired immunodeficiency syndrome. Blood 61:1265–1268

Chang Y, Cesarman E, Pessin MS, Lee F, Culpepper J, Knowles DM, Moore PS (1994) Identification of herpesvirus-like DNA sequences in AIDS-associated Kaposi's sarcoma. Science 266:1865–1869

Chatlynne LG, Lapps W, Handy M, Huang YQ, Masood R, Hamilton AS, Said JW, Koeffler HP, Kaplan MH, Friedman KA, Gill PS, Whitman JE, Ablashi DV (1998) Detection and titration of human herpesvirus-8-specific antibodies in sera from blood donors, acquired immunodeficiency syndrome patients, and Kaposi's sarcoma patients using a whole virus enzyme-linked immunosorbent assay. Blood 92:53–58

Chen KT (1984) Multicentric Castleman's disease and Kaposi's sarcoma. Am J Surg Pathol 8:287–293

Cossman J, Annunziata CM, Barash S, Staudt L, Dillon P, He WW, Ricciardi-Castagnoli P, Rosen CA, Carter KC (1999) Reed-Sternberg cell genome expression supports a B-cell lineage. Blood 94:411–416

Davis DA, Humphrey RW, Newcomb FM, O'Brien TR, Goedert JJ, Straus SE, Yarchoan R (1997) Detection of serum antibodies to a Kaposi's sarcoma-associated herpesvirus-specific peptide. J Infect Dis 175:1071–1079

Dent AL, Shaffer AL, Yu X, Allman D, Staudt LM (1997) Control of inflammation, cytokine expression, and germinal center formation by BCL-6. Science 276:589–592

DePond W, Said JW, Tasaka T, de Vos S, Kahn D, Cesarman E, Knowles DM, Koeffler HP (1997) Kaposi's sarcoma-associated herpesvirus and human herpesvirus 8 (KSHV/HHV8)-associated lymphoma of the bowel. Report of two cases in HIV-positive men with secondary effusion lymphomas. Am J Surg Pathol 21:719–724

Dupin N, Fisher C, Kellam P, Ariad S, Tulliez M, Franck N, van ME, Salmon D, Gorin I, Escande JP, Weiss RA, Alitalo K, Boshoff C (1999) Distribution of human herpesvirus-8 latently infected cells in Kaposi's sarcoma, multicentric Castleman's disease, and primary effusion lymphoma. Proc Natl Acad Sci USA 96:4546–4551

Dupin N, Grandadam M, Calvez V, Gorin I, Aubin JT, Havard S, Lamy F, Leibowitch M, Huraux JM, Escande JP, Agut H (1995) Herpesvirus-like DNA sequences in patients with Mediterranean Kaposi's sarcoma. Lancet 345:761–762

Falk S, Schmidts HL, Muller H, Berger K, Schneider M, Schlote W, Helm EB, Stille W, Hubner K, Stutte HJ (1987) Autopsy findings in AIDS – a histopathological analysis of fifty cases. Klin Wochenschr 65:654–663

Franceschi S, Dal Maso L, La Vecchia C (1999) Advances in the epidemiology of HIV-associated non-Hodgkin's lymphoma and other lymphoid neoplasms. Int J Cancer 83:481–485

Frizzera G, Banks PM, Massarelli G, Rosai J (1983) A systemic lymphoproliferative disorder with morphologic features of Castleman's disease. Pathological findings in 15 patients. Am J Surg Pathol 7:211–231

Fujii T, Taguchi H, Katano H, Mori S, Nakamura T, Nojiri N, Nakajima K, Tadokoro K, Juji T, Iwamoto A (1998) Seroprevalance of HHV-8 in HIV-1 positive and negative populations in Japan. J Med Virol 57:159–162

Gaidano G, Carbone A, Dalla-Favera R (1998) Pathogenesis of AIDS-related lymphomas: molecular and histogenetic heterogeneity. Am J Pathol 152:623–630

Gaidano G, Carbone A, Pastore C, Capello D, Migliazza A, Gloghini A, Roncella S, Ferrarini M, Saglio G, Dalla-Favera R (1997a) Frequent mutation of the 5' noncoding region of the BCL-6 gene in acquired immunodeficiency syndrome-related non-Hodgkin's lymphomas. Blood 89:3755–3762

Gaidano G, Cechova K, Chang Y, Moore PS, Knowles DM, Dalla FR (1996) Establishment of AIDS-related lymphoma cell lines from lymphomatous effusions. Leukemia 10:1237–1240

Gaidano G, Gloghini A, Gattei V, Rossi MF, Cilia AM, Godeas C, Degan M, Perin T, Canzonieri V, Aldinucci D, Saglio G, Carbone A, Pinto A (1997b) Association of Kaposi's sarcoma-associated herpesvirus-positive primary effusion lymphoma with expression of the CD138/syndecan-1 antigen. Blood 90:4894–4900

Gaidano G, Lo Coco F, Ye BH, Shibata D, Levine AM, Knowles DM, Dalla-Favera R (1994) Rearrangements of the BCL-6 gene in acquired immunodeficiency syndrome- associated non-Hodgkin's lymphoma: association with diffuse large-cell subtype. Blood 84:397–402

Gao SJ, Boshoff C, Jayachandra S, Weiss RA, Chang Y, Moore PS (1997) KSHV ORF K9 (vIRF) is an oncogene which inhibits the interferon signaling pathway. Oncogene 15:1979–1985

Groopman JE, Sullivan JL, Mulder C, Ginsburg D, Orkin SH, O'Hara CJ, Falchuk K, Wong-Staal F, Gallo RC (1986) Pathogenesis of B-cell lymphoma in a patient with AIDS. Blood 67:612–615

Grulich AE (1999) AIDS-associated non-Hodgkin's lymphoma in the era of highly active antiretroviral therapy. J Acquir Immune Defic Syndr 21 Suppl 1:S27–S30

Guarner J, del Rio C, Carr D, Hendrix LE, Eley JW, Unger ER (1991) Non-Hodgkin's lymphomas in patients with human immunodeficiency virus infection. Presence of Epstein-Barr virus by in situ hybridization, clinical presentation, and follow-up. Cancer 68:2460–2465

Hamilton-Dutoit SJ, Pallesen G, Franzmann MB, Karkov J, Black F, Skinhoj P, Pedersen C (1991) AIDS-related lymphoma. Histopathology, immunophenotype, and association with Epstein-Barr virus as demonstrated by in situ nucleic acid hybridization. Am J Pathol 138:149–163

Hamilton-Dutoit SJ, Raphael M, Audouin J, Diebold J, Lisse I, Pedersen C, Oksenhendler E, Marelle L, Pallesen G (1993a) In situ demonstration of Epstein-Barr virus small RNAs (EBER 1) in acquired immunodeficiency syndrome-related lymphomas: correlation with tumor morphology and primary site. Blood 82:619–624

Hamilton-Dutoit SJ, Rea D, Raphael M, Sandvej K, Delecluse HJ, Gisselbrecht C, Marelle L, van Krieken HJ, Pallesen G (1993b) Epstein-Barr virus-latent gene expression and tumor cell phenotype in acquired immunodeficiency syndrome-related non-Hodgkin's lymphoma. Correlation of lymphoma phenotype with three distinct patterns of viral latency. Am J Pathol 143:1072–1085

Harris NL, Jaffe ES, Stein H, Banks PM, Chan JK, Cleary ML, Delsol G, De W, Peeters C, Falini B, Gatter KC, Grogan TM, Isaacson PG, Knowles DM, Mason DY, Muller-Hermelink HK, Pileri SA, Piris MA, Ralfkiaer E, Warnke RA (1994) A revised European-American classification of lymphoid neoplasms: a proposal from the International Lymphoma Study Group. Blood 84:1361–1392

Henderson S, Rowe M, Gregory C, Croom-Carter D, Wang F, Longnecker R, Kieff E, Rickinson A (1991) Induction of bcl-2 expression by Epstein-Barr virus latent membrane protein 1 protects infected B cells from programmed cell death. Cell 65:1107–1115

Herndier BG, Kaplan LD, McGrath MS (1994) Pathogenesis of AIDS lymphomas. AIDS 8:1025–1049

Herndier BG, Sanchez HC, Chang KL, Chen YY, Weiss LM (1993) High prevalence of Epstein-Barr virus in the Reed-Sternberg cells of HIV-associated Hodgkin's disease. Am J Pathol 142:1073–1079

Itoh T, Shiota M, Takanashi M, Hojo I, Satoh H, Matsuzawa A, Moriyama T, Watanabe T, Hirai K, Mori S (1993) Engraftment of human non-Hodgkin lymphomas in mice with severe combined immunodeficiency. Cancer 72:2686–2694

Kaposi M (1872) Idiopatiches multiples pigment sarcom der Haut. Arch Dermatol Syphil 4:265–272

Katano H, Hoshino Y, Morishita Y, Nakamura T, Satoh H, Iwamoto A, Herndier B, Mori S (1999a) Establishing and characterizing a CD30-positive cell line harboring HHV-8 from a primary effusion lymphoma. J Med Virol 58:394–401

Katano H, Morishita Y, Cui LX, Watanabe T, Hirai K, Mori S (1996) Expression of latent membrane protein 1 in clinically isolated cases and animal models of AIDS-associated non-Hodgkin's lymphomas. Pathol Int 46:568–574

Katano H, Sata T, Suda T, Nakamura T, Tachikawa N, Nishizumi H, Sakurada S, Hayashi Y, Koike M, Iwamoto A, Kurata T, Mori S (1999b) Expression and antigenicity of human herpesvirus 8 encoded ORF59 protein in AIDS-associated Kaposi's sarcoma. J Med Virol 59:346–355

Katano H, Sato Y, Kurata T, Mori S, Sata T (1999c) High expression of HHV-8-encoded ORF73 protein in spindle-shaped cells of Kaposi's sarcoma. Am J Pathol 155:47–52

Katano H, Sato Y, Kurata T, Mori S, Sata T (2000a) Expression and Localization of Human Herpesvirus 8-Encoded Proteins in Primary Effusion Lymphoma, Kaposi's Sarcoma and Multicentric Castleman's Disease. Virology 266:335–344

Katano H, Suda T, Morishita Y, Yamamoto K, Hoshino Y, Nakamura K, Tachikawa N, Sata T, Hamaguchi H, Iwamoto A, Mori S (2000b) HHV-8-associated solid lymphomas occurring in AIDS patients take anaplastic large cell morphology. Mod Pathol 13:77–85

Kedes DH, Ganem D, Ameli N, Bacchetti P, Greenblatt R (1997) The prevalence of serum antibody to human herpesvirus 8 (Kaposi sarcoma-associated herpesvirus) among HIV-seropositive and high-risk HIV-seronegative women. JAMA 277:478–481

Khanna R, Burrows SR, Nicholls J, Poulsen LM (1998) Identification of cytotoxic T cell epitopes within Epstein-Barr virus (EBV) oncogene latent membrane protein 1 (LMP1): evidence for HLA A2 supertype-restricted immune recognition of EBV-infected cells by LMP1-specific cytotoxic T lymphocytes. Eur J Immunol 28:451–458

Klatt EC, Nichols L, Noguchi TT (1994) Evolving trends revealed by autopsies of patients with the acquired immunodeficiency syndrome. 565 autopsies in adults with the acquired immunodeficiency syndrome, Los Angeles, Calif, 1982–1993. Arch Pathol Lab Med 118:884–890

Knowles DM (1999) Immunodeficiency-associated lymphoproliferative disorders. Mod Pathol 12: 200–217

Komano J, Maruo S, Kurozumi K, Oda T, Takada K (1999) Oncogenic role of epstein-barr virus-encoded RNAs in Burkitt's lymphoma cell line akata. J Virol 73:9827–9831

MacMahon EM, Glass JD, Hayward SD, Mann RB, Becker PS, Charache P, McArthur JC, Ambinder RF (1991) Epstein-Barr virus in AIDS-related primary central nervous system lymphoma. Lancet 338:969–973

Mayama S, Cuevas LE, Sheldon J, Omar OH, Smith DH, Okong P, Silvel B, Hart CA, Schulz TF (1998) Prevalence and transmission of Kaposi's sarcoma-associated herpesvirus (human herpesvirus 8) in Ugandan children and adolescents. Int J Cancer 77:817–820

Meeker TC, Shiramizu B, Kaplan L, Herndier B, Sanchez H, Grimaldi JC, Baumgartner J, Rachlin J, Feigal E, Rosenblum M, McGrath MS (1991) Evidence for molecular subtypes of HIV-associated lymphoma: division into peripheral monoclonal, polyclonal and central nervous system lymphoma. AIDS 5: 669–674

Melbye M, Cook PM, Hjalgrim H, Begtrup K, Simpson GR, Biggar RJ, Ebbesen P, Schulz TF (1998) Risk factors for Kaposi's-sarcoma-associated herpesvirus (KSHV/HHV-8) seropositivity in a cohort of homosexual men, 1981–1996. Int J Cancer 77:543–548

Migliazza A, Martinotti S, Chen W, Fusco C, Ye BH, Knowles DM, Offit K, Chaganti RS, Dalla-Favera R (1995) Frequent somatic hypermutation of the 5′ noncoding region of the BCL6 gene in B-cell lymphoma. Proc Natl Acad Sci USA 92:12520–12524

Miles SA, Rezai AR, Salazar GJ, Vander MM, Stevens RH, Logan DM, Mitsuyasu RT, Taga T, Hirano T, Kishimoto T, Martinez-Maza O (1990) AIDS Kaposi sarcoma-derived cells produce and respond to interleukin 6. Proc Natl Acad Sci USA 87(11):4068–4072

Mohar A, Romo J, Salido F, Jessurun J, Ponce de Leon S, Reyes E, Volkow P, Larraza O, Peredo MA, Cano C, Gomez G, Sepulveda J, Mueller N (1992) The spectrum of clinical and pathological manifestations of AIDS in a consecutive series of autopsied patients in Mexico. AIDS 6:467–473

Moore PS, Boshoff C, Weiss RA, Chang Y (1996) Molecular mimicry of human cytokine and cytokine response pathway genes by KSHV. Science 274:1739–1744

Moore PS, Chang Y (1995) Detection of herpesvirus-like DNA sequences in Kaposi's sarcoma in patients with and without HIV infection. N Engl J Med 332:1181–1185

Mori S, Koike M, Hondo R, Wakabayashi T, Maeda Y, Kazuyama Y, Kataoka I, Watanabe T (1991) Malignant lymphomas in Japanese AIDS patients. Acta Pathol Jpn 41:744–750

Mosialos G, Birkenbach M, Yalamanchili R, VanArsdale T, Ware C, Kieff E (1995) The Epstein-Barr virus transforming protein LMP1 engages signaling proteins for the tumor necrosis factor receptor family. Cell 80:389–399

Murray RJ, Kurilla MG, Brooks JM, Thomas WA, Rowe M, Kieff E, Rickinson AB (1992) Identification of target antigens for the human cytotoxic T cell response to Epstein-Barr virus (EBV): implications for the immune control of EBV-positive malignancies. J Exp Med 176:157–168

Nador RG, Cesarman E, Chadburn A, Dawson DB, Ansari MQ, Sald J, Knowles DM (1996) Primary effusion lymphoma: a distinct clinicopathologic entity associated with the Kaposi's sarcoma-associated herpes virus. Blood 88:645–656

Nakamura K, Katano H, Hoshino Y, Nakamura T, Hosono O, Masunaga A, Mori S, Iwamoto A, Tamaki K (1999) Human herpesvirus type 8 (HHV-8) and Epstein-Barr virus (EBV)-associated cutaneous lymphoma taking anaplastic large cell morphology in an HIV-positive patient. Br J Dermatol 141:141–145

Nosari A, Cantoni S, Oreste P, Schiantarelli C, Landonio G, Alexiadis S, Gargantini L, Caggese L, Gambacorta M, Morra E (1996) Anaplastic large cell (CD30/Ki-1+) lymphoma in HIV+ patients: clinical and pathological findings in a group of ten patients. Br J Haematol 95:508–512

Oksenhendler E, Duarte M, Soulier J, Cacoub P, Welker Y, Cadranel J, Cazals HD, Autran B, Clauvel JP, Raphael M (1996) Multicentric Castleman's disease in HIV infection: a clinical and pathological study of 20 patients. AIDS 10:61–67

Onizuka T, Moriyama M, Yamochi T, Kuroda T, Kazama A, Kanazawa N, Sato K, Kato T, Ota H, Mori S (1995) BCL-6 gene product, a 92- to 98-kD nuclear phosphoprotein, is highly expressed in germinal center B cells and their neoplastic counterparts. Blood 86:28–37

Parravinci C, Corbellino M, Paulli M, Magrini U, Lazzarino M, Moore PS, Chang Y (1997) Expression of a virus-derived cytokine, KSHV vIL-6, in HIV-seronegative Castleman's disease. Am J Pathol 151:1517–1522

Pau CP, Lam LL, Spira TJ, Black JB, Stewart JA, Pellett PE, Respess RA (1998) Mapping and serodiagnostic application of a dominant epitope within the human herpesvirus 8 ORF 65-encoded protein. J Clin Microbiol 36:1574–1577

Pelicci PG, Knowles DMd, Arlin ZA, Wieczorek R, Luciw P, Dina D, Basilico C, Dalla-Favera R (1986) Multiple monoclonal B cell expansions and c-myc oncogene rearrangements in acquired immune deficiency syndrome-related lymphoproliferative disorders. Implications for lymphomagenesis. J Exp Med 164:2049–2060

Rabkin CS, Schulz TF, Whitby D, Lennette ET, Magpantay LI, Chatlynne L, Biggar RJ (1998) Interassay correlation of human herpesvirus 8 serologic tests. HHV-8 Interlaboratory Collaborative Group. J Infect Dis 178:304–309

Regamey N, Cathomas G, Schwager M, Wernli M, Harr T, Erb P (1998) High human herpesvirus 8 seroprevalence in the homosexual population in Switzerland. J Clin Microbiol 36:1784–1786

Renne R, Lagunoff M, Zhong W, Ganem D (1996a) The size and conformation of Kaposi's sarcoma-associated herpesvirus (human herpesvirus 8) DNA in infected cells and virions. J Virol 70:8151–8154

Renne R, Zhong W, Herndier B, McGrath M, Abbey N, Kedes D, Ganem D (1996b) Lytic growth of Kaposi's sarcoma-associated herpesvirus (human herpesvirus 8) in culture. Nat Med 2:342–346

Rowe M, Lear AL, Croom-Carter D, Davies AH, Rickinson AB (1992) Three pathways of Epstein-Barr virus gene activation from EBNA1-positive latency in B lymphocytes. J Virol 66:122–131

Russo JJ, Bohenzky RA, Chien MC, Chen J, Yan M, Maddalena D, Parry JP, Peruzzi D, Edelman IS, Chang Y, Moore PS (1996) Nucleotide sequence of the Kaposi sarcoma-associated herpesvirus (HHV8). Proc Natl Acad Sci USA 93:14862–14867

Said JW, Chien K, Tasaka T, Koeffler HP (1997) Ultrastructural characterization of human herpesvirus 8 (Kaposi's sarcoma-associated herpesvirus) in Kaposi's sarcoma lesions: electron microscopy permits distinction from cytomegalovirus (CMV). J Pathol 182:273–281

Shiramizu B, Herndier B, Meeker T, Kaplan L, McGrath M (1992) Molecular and immunophenotypic characterization of AIDS-associated, Epstein-Barr virus-negative, polyclonal lymphoma. J Clin Oncol 10:383–389

Shiramizu B, Herndier BG, McGrath MS (1994) Identification of a common clonal human immunodeficiency virus integration site in human immunodeficiency virus-associated lymphomas. Cancer Res 54:2069–2072

Simpson GR, Schulz TF, Whitby D, Cook PM, Boshoff C, Rainbow L, Howard MR, Gao SJ, Bohenzky RA, Simmonds P, Lee C, de RA, Hatzakis A, Tedder RS, Weller IV, Weiss RA, Moore PS (1996) Prevalence of Kaposi's sarcoma associated herpesvirus infection measured by antibodies to recombinant capsid protein and latent immunofluorescence antigen. Lancet 348:1133–1138

Soulier J, Grollet L, Oksenhendler E, Cacoub P, Cazals HD, Babinet P, d'Agay MF, Clauvel JP, Raphael M, Degos L, et al. (1995) Kaposi's sarcoma-associated herpesvirus-like DNA sequences in multicentric Castleman's disease. Blood 86:1276–1280

Subar M, Neri A, Inghirami G, Knowles DM, Dalla-Favera R (1988) Frequent c-myc oncogene activation and infrequent presence of Epstein-Barr virus genome in AIDS-associated lymphoma. Blood 72:667–671

Tedeschi R, De PP, Schulz TF, Dillner J (1999) Human serum antibodies to a major defined epitope of human Herpesvirus 8 small viral capsid antigen. J Infect Dis 179:1016–1020

Tirelli U, Vaccher E, Zagonel V, Talamini R, Bernardi D, Tavio M, Gloghini A, Merola MC, Monfardini S, Carbone A (1995) CD30 (Ki-1)-positive anaplastic large-cell lymphomas in 13 patients with and 27 patients without human immunodeficiency virus infection: the first comparative clinicopathologic study from a single institution that also includes 80 patients with other human immunodeficiency virus-related systemic lymphomas. J Clin Oncol 13:373–380

Wang D, Liebowitz D, Kieff E (1985) An EBV membrane protein expressed in immortalized lymphocytes transforms established rodent cells. Cell 43:831–840
Welch K, Finkbeiner W, Alpers CE, Blumenfeld W, Davis RL, Smuckler EA, Beckstead JH (1984) Autopsy findings in the acquired immune deficiency syndrome. JAMA 252:1152–1159
Yoshizaki K, Matsuda T, Nishimoto N, Kuritani T, Taeho L, Aozasa K, Nakahata T, Kawai H, Tagoh H, Komori T, Kishimoto S, Hirano T, Kishimoto T (1989) Pathogenic significance of interleukin-6 (IL-6/BSF-2) in Castleman's disease. Blood 74:1360–1367
Zheng H, Mori S (1997) Epstein-Barr virus-infected lymphocytes in non-neoplastic lymph nodes of patients infected with human immunodeficiency virus. Pathol Int 47:217–221
Zhu L, Wang R, Sweat A, Goldstein E, Horvat R, Chandran B (1999) Comparison of human sera reactivities in immunoblots with recombinant human herpesvirus (HHV)-8 proteins associated with the latent (ORF73) and lytic (ORFs 65, K8.1A, and K8.1B) replicative cycles and in immunofluorescence assays with HHV-8-infected BCBL-1 cells. Virology 256:381–392

III
Molecular Mechanisms of Oncogenesis

Role of Epstein-Barr Virus in Burkitt's Lymphoma

K. Takada

1 Introduction... 141
2 BL-Derived Akata Cell Line that Retains BL-Type EBV Expression after Prolonged
   In Vitro Cultivation... 142
3 Isolation of EBV-Negative Subclones from the BL-Derived Akata Line............ 142
4 EBV Reinfection of EBV-Negative Akata Cells: A Model for BL-Type EBV Infection...... 143
5 Dependence of Malignant Phenotypes of EBV-Positive Akata Cells on the Presence of EBV... 144
6 Dependence of Apoptosis Resistance of EBV-Positive Akata Cells
   on the Presence of EBV... 144
7 EBERs' Role on Malignant Phenotype and Apoptosis Resistance in Akata Cells......... 145
8 Summary... 149
References... 150

1 Introduction

Epstein-Barr virus (EBV) is associated with more than 90% of Burkitt's lymphoma (BL) in the African regions of endemicity and less frequently (15%–20%) with sporadic BL occurring worldwide (reviewed by Rickinson and Kieff 1996). The most consistent finding in BL, whether EBV-infected or not, is the juxtaposition of the c-*myc* locus to the immunoglobulin (Ig) H enhancer through a reciprocal translocation t(8;14) and, more rarely, to the Igκ locus t(2;8) or the Igλ locus t(8;22). This feature led to the hypothesis that this translocation results in constitutive activation of the c-*myc* gene, which seems to be a key step in the development of BL (Klein 1981). On the other hand, the role of EBV is still obscure.

EBV infects primary B lymphocytes in vitro and transforms them into blasts that can proliferate indefinitely. Such EBV-transformed lymphoblastoid cells maintain the entire viral genome in a plasmid form and express a limited number of EBV gene products, including six nuclear antigens (EBNA1, EBNA2, EBNA3A, EBNA3B, EBNA3C, and EBNALP), three membrane proteins (LMP1, LMP2A, and LMP2B), EBV-encoded small RNAs (EBERs, specifically EBER1 and EBER2), and trans-

---

Department of Tumor Virology, Institute for Genetic Medicine, Hokkaido University, N15 W7, Kita-ku, Sapporo 060-8638, Japan

cripts from the *Bam*HI A region (BARF0). Among them, EBNA2 and LMP1 are particularly important for the immortalization of primary lymphocytes (KIEFF 1996). However, these proteins are not expressed in BL. These observations raise a question about the role of EBV in oncogenesis. It has been difficult to answer this question because we did not have an in vitro system of Burkitt's-type EBV expression that was characterized by expression of EBNA1, EBER, BARF0 and a slight amount of LMP2A. We overcame this problem and proved that EBV is required for the maintenance of malignancies (SHIMIZU et al. 1994; KOMANO et al. 1998). Similar results were also reported by CHODOSH et al. (1998) and RUF et al. (1999).

Here, we summarize a series of our works. EBV contributes to the malignant phenotype and resistance to apoptosis in BL line Akata (SHIMIZU et al. 1994; KOMANO et al. 1998). The EBERs are responsible for these phenotypes (KOMANO et al. 1999). Transfection of the EBER genes into EBV-negative Akata clones restores the capacity for the growth in soft agar, tumorigenicity in SCID mice, resistance to apoptotic inducers, and upregulated expression of bcl-2 oncoproteins that are originally retained in parental EBV-positive Akata cells and are lost in EBV-negative subclones. BL cells, including Akata, possess a chromosomal translocation involving the c-*myc* locus, which results in constitutive activation of the c-*myc* gene (KLEIN 1981). Our results suggest that EBERs upregulate expression of bcl-2 protein to protect cells from c-myc-induced apoptosis, and to allow c-myc to exert its oncogenic function in BL.

2 BL-Derived Akata Cell Line that Retains BL-Type EBV Expression after Prolonged In Vitro Cultivation

Most BL lines maintain BL-type EBV expression in vivo, but change the EBV expression to that of immortalized lymphocytes during in vitro cultivation (ROWE et al. 1987). The Akata cell line is unique in this respect because it retains BL-type EBV expression after prolonged in vitro cultivation (SHIMIZU et al. 1994). It is derived from a Japanese BL, possesses a Burkitt's type c-*myc* translocation [t(8;14)], and expresses surface Ig of the G(κ) class (TAKADA et al. 1991). The Akata line is now commonly used as a virus source, a targeting host to generate recombinant EBV, and a tool to study viral replication since crosslinking of surface IgG with anti-IgG antibody induces viral replication synchronously and efficiently (TAKADA 1984; TAKADA and ONO 1989).

3 Isolation of EBV-Negative Subclones from the BL-Derived Akata Line

Initially, Akata cells were 100% positive for EBNA. However, after serial passage for about 2 years, we found that approximately half of the cells were negative for

EBNA. From that culture, we could isolate EBV-positive and EBV-negative cell clones by limiting dilution (SHIMIZU et al. 1994). Southern blot analysis with the BamHI K fragment of EBV DNA used as a probe, and polymerase chain reaction (PCR) analysis with primer pairs specific for coding regions of EBNA1, EBNA2, and LMP1 confirmed that EBNA-negative clones do not contain EBV DNA.

Since parental Akata cells were almost 100% positive for EBNA, it was most probable that the EBV plasmid was lost in some Akata cells during cultivation. To further confirm this possibility, one of the EBV-positive Akata cell clones was maintained in culture. After 6 months, the EBNA positivity decreased from the initial 100% to 83%. Again, from that culture, EBV-negative cell clones were isolated by limiting dilution. Both EBV-positive and -negative clones were virtually 100% positive for surface Ig of the G, kappa class and possessed chromosome markers characteristic of the parental Akata cells, and so were clearly derived from Akata cells and not from contaminated unrelated cells. Thus, it was confirmed that the EBV plasmid was lost in some Akata cells during cultivation.

Among the many EBV-positive BL lines, the Akata line is the first case from which EBV-negative cell clones were isolated. Why were EBV-negative cell clones isolated from the Akata line but not from other BL lines? It has been reported that the mechanism of plasmid maintenance is not perfect and 4% of cells lose EBV plasmids per generation (KIRCHMAIER and SUGDEN 1995). If so, theoretically, the EBV-negative population should increase during cultivation, although we cannot see such a phenomenon. The most probable explanation is that all EBV-infected cell lines except Akata depend on the presence of EBV for their survival. Accumulation of mutations of cellular genes during cultivation of Akata cells made some fraction of the cells independent of EBV under the ordinary culture condition, and thus, we could isolate EBV-negative subclones.

4 EBV Reinfection of EBV-Negative Akata Cells: A Model for BL-Type EBV Infection

Next, we examined whether EBV-reinfection of EBV-negative Akata cells resulted in BL-type EBV expression (KOMANO et al. 1998). For this purpose, we generated and used a recombinant Akata EBV-knockout viral thymidine kinase gene, and replaced it with a neomycin resistance gene (neoR) (YOSHIYAMA et al. 1995; SHIMIZU et al. 1996), because we could not isolate EBV-reinfected Akata clones with either wild-type Akata EBV or B95-8 EBV since the population of infected cells that could stably retain EBV was extremely low. We successfully isolated EBV-reinfected cell clones in the selection medium. All the G418-resistant clones were positive for EBNA.

From more than a hundred clones, we randomly selected 12 clones that were subjected to immunoblotting and reverse transcription (RT)-PCR to identify viral latency. EBNA1 was detected by Western blotting, and BARF0, EBER1, and LMP2A were detected by RT-PCR. The clones utilized the Q promoter for EBNA1 transcription and were negative for other latent proteins. These data were consistent with the conclusion that reinfected Akata cells possessed BL-type EBV expression.

5 Dependence of Malignant Phenotypes of EBV-Positive Akata Cells on the Presence of EBV

Isolation of EBV-positive and EBV-negative cell clones with the same origin makes it possible to examine the effects of EBV in BL cells. Both proliferated at nearly the same rate in RPMI 1640 medium supplemented with 10% fetal calf serum. However, phenotypic differences were found in two assays, the soft agar colony assay and that for tumorigenicity in SCID mice, which are commonly used as markers of malignancies (SHIMIZU et al. 1994; KOMANO et al. 1998). EBV-positive cell clones formed colonies in soft agar, but EBV-negative clones scarcely did. EBV-positive clones produced tumors in SCID mice, but EBV-negative clones did not. Similar results were obtained in comparison between EBV-reinfected Akata cell clones and cell clones that were transfected with the *neoR* gene into the same EBV-negative Akata cell clones and selected for G418 resistance. These results indicate that malignant phenotypes are dependent on the presence of EBV.

6 Dependence of Apoptosis Resistance of EBV-Positive Akata Cells on the Presence of EBV

Akata cells underwent apoptosis after treatment of cells with various inducers such as cycloheximide, UV light, and glucocorticoid. We examined whether there was any difference in the susceptibility to apoptotic cell death between EBV-infected and -uninfected Akata clones (KOMANO et al. 1998). After treatment with apoptotic stimuli, EBV-negative Akata cells showed more prominent DNA laddering than EBV-positive Akata cells. The survival rate was examined by a colorimetric assay that measured the activity of mitochondrial dehydrogenases. As a result, EBV-positive clones from parental Akata cells were found to be more resistant to cell death than EBV-negative clones. In addition, restoration of the resistance to apoptosis was observed in the EBV-reinfected Akata clones. These findings made it clear that resistance to apoptosis of Akata cells was dependent on the presence of EBV. It should be noted that EBV-infected Akata cells were negative for LMP1 and BHRF1, which are known to be anti-apoptotic proteins. They were expressed only when virus replication was induced by incubation of cells with anti-IgG.

We then examined expression of bcl-2, an oncoprotein having an anti-apoptotic function, in EBV-infected and uninfected cells. In Western blotting, EBV-positive clones from the parental Akata cells were shown to express a higher level of bcl-2 than EBV-negative clones. Furthermore, expression of bcl-2 was higher in reinfected clones than in *neoR*-transfected clones.

7 EBERs' Role on Malignant Phenotype and Apoptosis Resistance in Akata Cells

The next question is what EBV product is responsible for malignant phenotypes and resistance to apoptosis in Akata cells. Four EBV genes are expressed in Akata cells, EBNA1, EBER, BARF0, and a slight amount of LMP2A. Our conclusion is that EBER is responsible for these phenotypes (KOMANO et al. 1998, 1999).

We constructed a plasmid, pEK, that carried a single copy of EBER and *neoR* and isolated G418-resistant clones by transfecting pEK into an EBV-negative Akata cell clone (Fig. 1). Compared with the level of EBV-reinfected cell clones, the

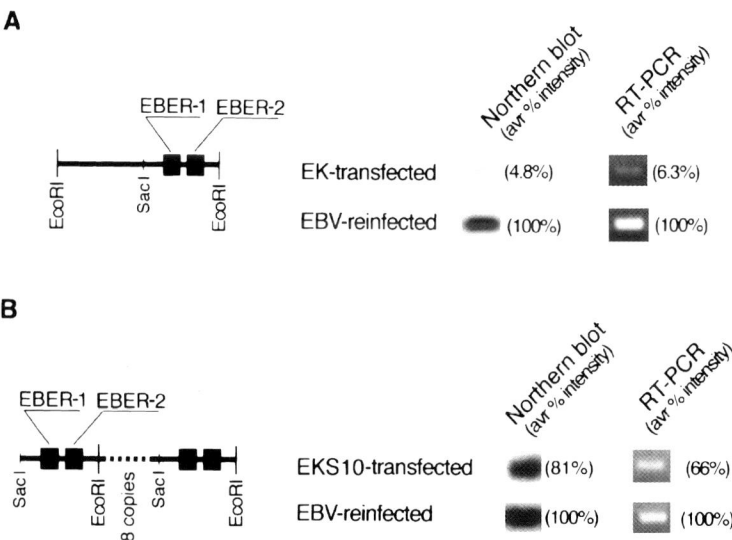

Fig. 1A,B. Relative EBER expression in EK- and EKS10-transfected Akata cells compared with EBV-reinfected Akata cell. For each EBER transfectant derived from different EBV-negative clones, more than 50 cell clones were examined for EBER expression and four to six clones with highest EBER expression were chosen for further studies. This figure shows a representative result. Four EBER-transfected cell clones from an EBV-negative Akata cell clone were mixed and subjected to quantitative assays for EBER. **A** Schematic drawing of the *Eco*RI K fragment of Akata EBV DNA (*left*) and relative EBER expression in EK-transfected cells (*right*). **B** Schematic drawing of the EKS10 unit (*left*) and relative EBER expression in EKS10-transfected cells (*right*)

average signal intensity for EBERs in four EK-transfected cell clones was 4.8% by Northern blot analysis and 6.3% by RT-PCR, even though clones expressing the highest level were analyzed. To obtain a higher level of EBER expression in transfected cells, we constructed a plasmid, pEKS10, that contained 10 tandem repeats of the SacI/EcoRI subfragment from the EcoRI K fragment of Akata EBV DNA (Fig. 1). We isolated G418-resistant clones by transfecting pEKS10 into an EBV-negative Akata cell clone. The average EBER expression in four EKS10-transfected cell clones was 81% by Northern blot analysis and 66% by RT-PCR compared with the level of EBV-reinfected cell clones. It was difficult to obtain levels of EBER expression in transfected cells equivalent to those in EBV-reinfected cells. The amount of the EBER-1 transcript was approximately twofold greater than that of EBER-2 in EBV-positive, EBV-reinfected, and EBER-transfected Akata cell clones by semiquantitative RT-PCR assay.

The growth characteristics of EK- and EKS10-transfected Akata cell clones were compared with EBV-reinfected cell clones and neoR-transfected cell clones. First, we performed a soft agar colony assay using cell clones derived from three independent EBV-negative clones (Fig. 2). EBV-reinfected cell clones formed colonies but neoR-transfected cell clones scarcely did. The absolute number of colonies differed among experiments, presumably due to clonal variation. EK-transfected cell clones formed a few colonies, whereas EKS10-transfected cell clones formed significantly more colonies than neoR-transfected cell clones, but formed fewer colonies than EBV-reinfected cell clones did. Second, we examined whether EKS10-transfected cells were tumorigenic in SCID mice. In three separate experiments, tumors developed in mice (1/5, 1/5, 7/15) injected with EKS10-transfected cells and in those (2/4, 2/4, 8/9) receiving EBV-reinfected cells, but not in those (0/4, 0/4, 0/15) with neoR-transfected cells. In an in situ hybridization study, almost all the tumor cells were positive for EBER-1 expression.

To test whether expression of EBERs restored the resistance to apoptosis, we performed an apoptosis assay using a set of cell clones (Fig. 3). Apoptosis was induced by exposing cells to inducers such as cycloheximide (CHX), glucocorticoid, and hypoxic stress, and surviving cells were measured with the MTT assay. The survival rate (%SR) was the highest in EBV-reinfected cells and the lowest in neoR-transfected cells. The difference of %SR between neoR-transfected and EK-transfected cells was not significant. In contrast, the %SR of EKS10-transfected cells was significantly higher than that of neoR-transfected cells for all stimuli.

We then examined whether upregulation of bcl-2 oncoprotein had been restored in EKS10-transfected cells (Fig. 4). The level of bcl-2 protein expression in EBV-reinfected cells was found to be higher than that in neoR-transfected cells. The bcl-2 expression in EKS10-transfected cells was increased compared with neoR-transfected cells. We could not find significant upregulation of bcl-2 protein in EK-transfected cells.

How could EBERs, RNAs, contribute to malignant phenotypes and resistance to apoptosis? EBERs were reported to bind to some cellular proteins, La (LERNER et al. 1981), EAP/L22 (TOCZYSKI et al. 1991; TOCZYSKI et al. 1994), and PKR (CLARKE et al. 1991). Above all, association of EBERs with PKR is noteworthy.

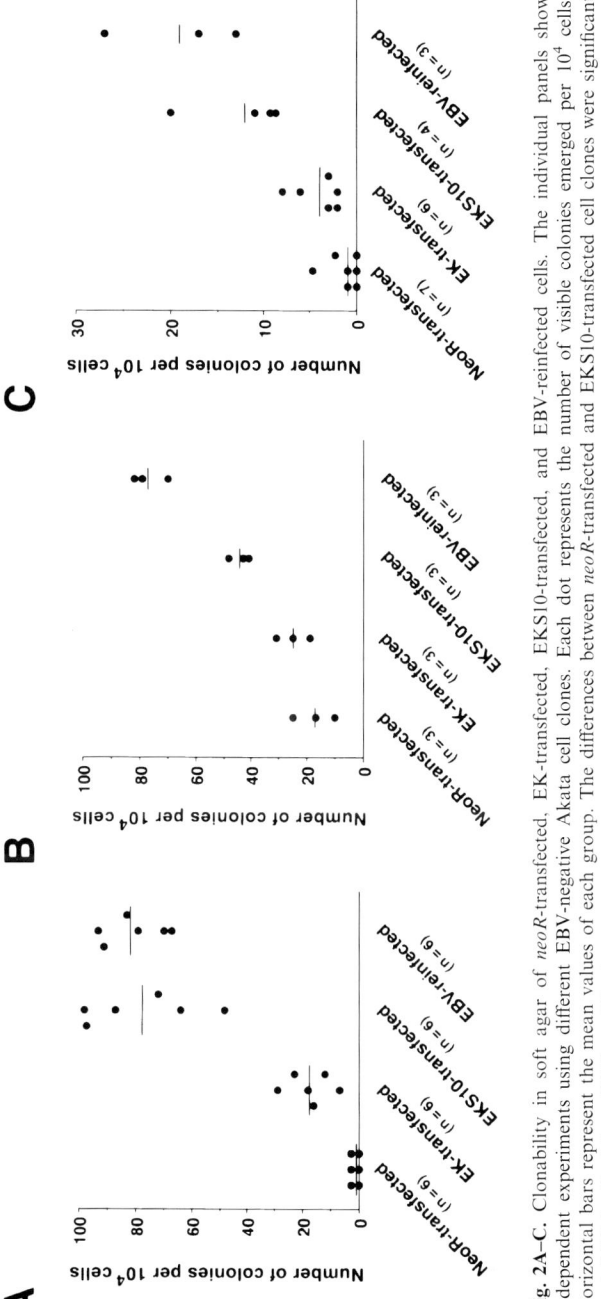

Fig. 2A–C. Clonability in soft agar of *neoR*-transfected, EK-transfected, EKS10-transfected, and EBV-reinfected cells. The individual panels show independent experiments using different EBV-negative Akata cell clones. Each dot represents the number of visible colonies emerged per 10^4 cells. Horizontal bars represent the mean values of each group. The differences between *neoR*-transfected and EKS10-transfected cell clones were significant ($p < 0.01$) by the Mann–Whitney U test

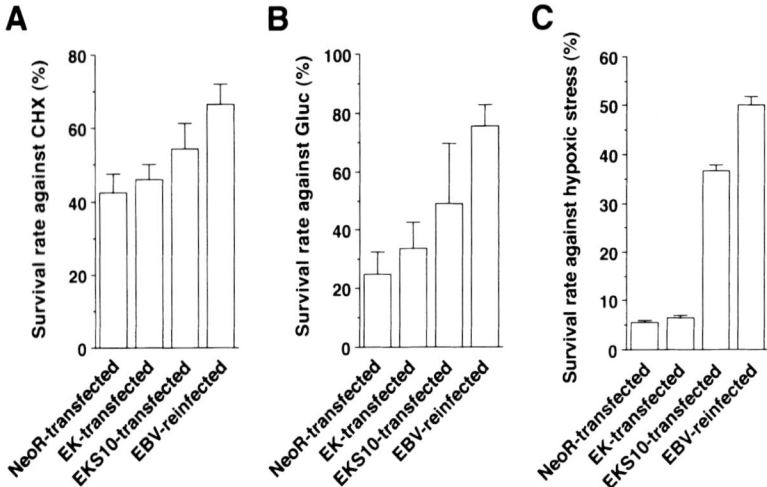

Fig. 3. Resistance to apoptosis of *neoR*-transfected, EK-transfected, EKS10-transfected, and EBV-reinfected cells against cycloheximide (CHX; **A**), glucocorticoid (Gluc; **B**), and hypoxic stress (**C**). The bar shows the mean values ± standard deviation of six clones. The data presented here are typical results from three independent experiments. By *t*-test analysis, the differences between mean values from *neoR*- and EKS10-transfected cells were significant at $p < 0.01$ (CHX), $p < 0.05$ (Gluc), and $p < 0.001$ (hypoxic stress) levels

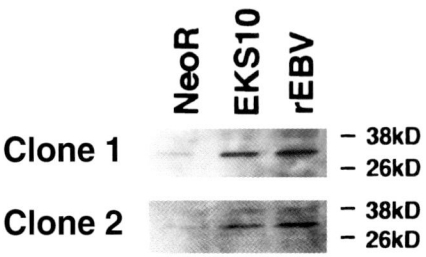

Fig. 4. Expression of bcl-2 protein in *neoR*-transfected, EKS10-transfected, and EBV-reinfected cells derived from two independent EBV-negative Akata cell clones. The data presented here are typical results of several experiments

PKR, previously described as an interferon-induced protein kinase, is induced by interferons (IFs) and activated by double-stranded RNAs to phosphorylate eIF-2α, a translation initiation factor (MEURS et al. 1990). This phosphorylation results in blockage of protein synthesis, through which interferons are thought to exert their anti-viral effects. EBERs were shown to bind to PKR and block activation of PKR as well as adenovirus-associated RNAs (VA)I and VAII, which might result in enhancement of protein synthesis of general messenger (m)RNAs (SHARP et al. 1993). The dominant negative form of PKR has been shown to predispose cells to malignant transformation or even initiate it (KOROMILAS et al. 1992; MEURS et al. 1993). In addition, GCN4, a yeast homologue of PKR, is known to regulate specific gene expression via translational control (ABASTADO et al. 1991). Since the expression levels of β-actin, c-myc, bax, and PKR in EBV-positive Akata cells were equal to those in EBV-negative Akata cells (data not shown), whereas the expression of bcl-2 protein was specifically upregulated in EBV-positive Akata cells, it is possible that inactivation of PKR by EBERs may enhance translation of bcl-2 mRNA more efficiently than other mRNAs to confer the malignant phenotype and resistance to apoptosis.

8 Summary

We established an in vitro system representing BL-type EBV infection, which is characterized by expression of EBNA1, EBER, BARF0, and LMP2A, and absence of EBNA2 and LMP1 expression (SHIMIZU et al. 1994; KOMANO et al. 1998). Comparison of EBV-positive and -negative Akata cell clones revealed that EBV contributes to the malignant phenotype and resistance to apoptosis. This is clear evidence that EBV is not a passenger and plays a role in BL. Moreover, we found that EBERs are responsible for these phenotypes (KOMANO et al. 1999). In the transfection study, EBER-expressing Akata cell clones restored the malignant phenotype, resistance to apoptosis and upregulated expression of bcl-2 protein to a level comparable to the restoration rate of EBER expression compared with EBV-reinfected cell clones. Many RNAs are known to have catalytic functions; however, there has been no report describing an oncogenic RNA. This is the first paper that provides evidence that RNA polymerase III-transcribed virus-encoded small RNAs affect the malignant phenotype and resistance to apoptosis.

Like Akata cells (TAKADA et al. 1991), all the BL cells possess a chromosomal translocation involving the c-*myc* locus, which results in constitutive activation of the c-*myc* gene (KLEIN 1981). In mammalian cells, deregulated expression of c-myc has been shown to contribute not only to tumorigenesis (LAND et al. 1983) but also to induce apoptosis (ASKEW et al. 1991; EVAN et al. 1992; MILNER et al. 1993). Therefore, BL cells are predisposed to c-myc-induced apoptosis. Our data imply that EBV infection would upregulate expression of bcl-2 protein to protect cells from c-myc-induced apoptosis, and to allow c-myc to exert its oncogenic functions (VAUX et al. 1988; BRITO-BABAPULLE et al. 1991; BISSONNETTE et al. 1992; FANIDI et al. 1992; KARSAN et al. 1993; MOHAMMAD et al. 1993; OLTVAI et al. 1993; MARIN et al. 1995). In this way bcl-2 might cooperate with c-myc in the development of BL (Fig. 5).

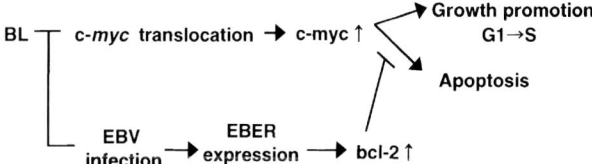

Fig. 5. Myc-EBER synergy in the development of BL (hypothesis)

References

Abastado JP, Miller PF, Jackson BM, Hinnebusch AG (1991) Suppression of ribosomal reinitiation at upstream open reading frames in amino acid-starved cells forms the basis for GCN4 translational control. Mol Cell Biol 11:486–496

Askew DS, Ashmun RA, Simmons BC, Cleveland JL (1991) Cell cycle arrest and accelerates apoptosis. Oncogene 6:1915–1922

Bissonnette RP, Echeverri F, Mahboubi A, Green DR (1992) Apoptotic cell death induced by c-myc is inhibited by bcl-2. Nature 359:552–554

Brito-Babapulle V, Crawford A, Khokhar T, Laffan M, Matutes E, Fairhead S, Catovsky D (1991) Translocations t(14;18) and t(8;14) with rearranged bcl-2 and c-myc in a case presenting as B-ALL (L3). Leukemia 5:83–87

Chodosh J, Holder VP, Gan Y, Belgaumi A, Sample J, Sixbey JW (1998) Eradication of latent Epstein-Barr virus by hydroxyurea alters the growth-transformed cell phenotype. J Infect Dis 177:1194–1201

Clarke PA, Schwemmle M, Schickinger J, Hilse K, Clemens MJ, Abastado JP, Miller PF, Jackson BM, Hinnebusch AG (1991) Binding of Epstein-Barr virus small RNA EBER-1 to the double-stranded RNA-activated protein kinase DAI. Nucleic Acids Res 19:243–248

Evan GI, Wyllie AH, Gilbert CS, Littlewood TD, Land H, Brooks M, Waters CM, Penn LZ, Hancock DC (1992) Induction of apoptosis in fibroblasts by c-myc protein. Cell 69:119–128

Fanidi A, Harrington EA, Evan GI (1992) Cooperative interaction between c-myc and bcl-2 proto-oncogenes. Nature 359:554–556

Karsan A, Gascoyne RD, Coupland RW, Shepherd JD, Phillips GL, Horsman DE (1993) Combination of t(14;18) and a Burkitt's type translocation in B-cell malignancies. Leukemia Lymphoma 10: 433–441

Kieff E (1996) Epstein-Barr virus and its replication. In: Fields BN, Knipe DM, Howley PM (eds) Fields Virology, 3rd edn. Lippincott-Raven, Philadelphia

Kirchmaier AL, Sugden B (1995) Plasmid maintenance of derivatives of oriP of Epstein-Barr virus. J Virol 69:1280–1283

Klein G (1981) The role of gene dosage and genetic transpositions in carcinogenesis. Nature 294:313–318

Komano J, Maruo S, Kurozumi K, Oda T, Takada K (1999) Oncogenic role of Epstein-Barr virus-encoded RNAs in Burkitt's lymphoma cell line Akata. J Virol 73:9827–9831

Komano J, Sugiura M, Takada K (1998) Epstein-Barr virus contributes to the malignant phenotype and to apoptosis resistance in Burkitt's lymphoma cell line Akata. J Virol 72:9150–9156

Koromilas AE, Roy S, Barber GN, Katze MG, Sonenberg N (1992) Malignant transformation by a mutant of the IFN-inducible dsRNA-dependent protein kinase. Science 257:1685–1689

Land H, Parada LF, Weinberg RA (1983) Cellular oncogenes and multistep carcinogenesis. Science 222:771–778

Lerner MR, Andrews NC, Miller G, Steitz JA (1981) Two small RNAs encoded by Epstein-Barr virus and complexed with protein are precipitated by antibodies from patients with systemic lupus erythematosus. Proc Natl Acad Sci USA 78:805–809

Marin MC, Hsu B, Stephens LC, Brisbay S, McDonnell TJ (1995) The functional basis of c-myc and bcl-2 complementation during multistep lymphomagenesis in vivo. Exp Cell Res 217:240–247

Meurs E, Chong K, Galabru J, Thomas NS, Kerr IM, Williams BR, Hovanessian AG (1990) Molecular cloning and characterization of the human double-stranded RNA-activated protein kinase induced by interferon. Cell 62:379–390

Meurs EF, Galabru J, Barber GN, Katze MG, Hovanessian AG (1993) Tumor suppressor function of the interferon-induced double-stranded RNA-activated protein kinase. Proc Natl Acad Sci USA 90: 232–236

Milner AE, Grand RJ, Waters CM, Gregory CD (1993) Apoptosis in Burkitt lymphoma cells is driven by c-myc. Oncogene 8:3385–3391

Mohammad RM, Mohamed AN, Smith MR, Jawadi NS, al-Katib A (1993) A unique EBV-negative low-grade lymphoma line (WSU-FSCCL) exhibiting both t(14;18) and t(8;11). Cancer Genet Cytogen 70:62–67

Oltvai ZN, Milliman CL, Korsmeyer SJ (1993) Bcl-2 heterodimerizes in vivo with a conserved homolog, Bax, that accelerates programmed cell death. Cell 74:609–619

Rickinson AB, Kieff E (1996) Epstein-Barr virus. In: Fields BN, Knipe DM, Howley PM (eds) Fields Virology, 3rd edn. Lippincott-Raven, Philadelphia

Rowe M, Rowe DT, Gregory CD, Young LS, Farrell PJ, Rupani H, Rickinson AB (1987) Differences in B cell growth phenotype reflect novel patterns of Epstein-Barr virus latent gene expression in Burkitt's lymphoma cells. EMBO J 6:2743–2751

Ruf IK, Rhyne PW, Yang H, Hutt-Fletcher CM, Cleveland JL, Sample JT (1999) Epstein-Barr virus regulates c-MYC, apoptosis, and tumorigenicity in Burkitt lymphoma. Mol Cell Biol 19:1651–1660

Sharp TV, Schwemmle M, Jeffrey I, Laing K, Mellor H, Proud CG, Hilse K, Clemens MJ (1993) Comparative analysis of the regulation of the interferon-inducible protein kinase PKR by Epstein-Barr virus RNAs EBER-1 and EBER-2 and adenovirus VAI RNA. Nucleic Acids Res 21:4483–4490

Shimizu N, Tanabe-Tochikura A, Kuroiwa Y, Takada K (1994) Isolation of Epstein-Barr virus (EBV)-negative cell clones from the EBV-positive Burkitt's lymphoma (BL) line Akata: malignant phenotypes of BL cells are dependent on EBV. J Virol 68:6069–6073

Shimizu N, Yoshiyama H, Takada K (1996) Clonal propagation of Epstein-Barr virus (EBV) recombinants in EBV-negative Akata cells. J Virol 70:7260–7263

Takada K (1984) Cross-linking of cell surface immunoglobulins induces Epstein-Barr virus in Burkitt lymphoma lines. Int J Cancer 33:27–32

Takada K, Horinouchi K, Ono Y, Aya T, Osato T, Takahashi M, Hayasaka S (1991) An Epstein-Barr virus-producer line Akata: establishment of the cell line and analysis of viral DNA. Virus Genes 5:147–156

Takada K, Ono Y (1989) Synchronous and sequential activation of latently infected Epstein-Barr virus genomes. J Virol 63:445–449

Toczyski DP, Matera AG, Ward DC, Steitz JA (1994) The Epstein-Barr virus (EBV) small RNA EBER1 binds and relocalizes ribosomal protein L22 in EBV-infected human B lymphocytes. Proc Natl Acad Sci USA 91:3463–3467

Toczyski DP, Steitz JA (1991) EAP, a highly conserved cellular protein associated with Epstein-Barr virus small RNAs (EBERs). EMBO J 10:459–466

Vaux DL, Cory S, Adams JM (1988) Bcl-2 gene promotes haemopoietic cell survival and cooperates with c-myc to immortalize pre-B cells. Nature 335:440–442

Yoshiyama H, Shimizu N, Takada K (1995) Persistent Epstein-Barr virus infection in a human T-cell line: unique program of latent virus expression. EMBO J 14:3706–3711

EBV Regulates c-MYC, Apoptosis, and Tumorigenicity in Burkitt's Lymphoma

I.K. Ruf[1], P.W. Rhyne[1], H. Yang[2], C.M. Borza[3], L.M. Hutt-Fletcher[3], J.L. Cleveland[2,4], and J.T. Sample[1,5]

1 Introduction	153
2 Results	154
2.1 The Type I EBV Latency Program Promotes Cell Survival in BL Following Withdrawal of Survival Factors	154
2.2 EBV- and Growth-Phase-Associated Expression of c-MYC and Bcl-2	155
2.3 EBV Infection, but not EBNA-1 alone, Regulates c-MYC Levels, Sensitivity to Apoptosis, and Restores Tumorigenic Potential to EBV-Negative BL Cells	156
3 Conclusions	159
References	160

1 Introduction

Epstein-Barr Virus (EBV) latency-associated gene expression in Burkitt's lymphoma (BL) cells is limited to a small set of gene products, none of which are known to overtly influence cell growth potential. Designated type I latency, this program of viral gene expression is characteristic of BL tumor cells in vivo and is restricted to expression of the EBNA-1 protein, the *Bam*HI-A rightward transcripts (BARTs), and two small non-coding RNAs, EBER-1 and EBER-2 (Arrand and Rymo 1982; Rowe et al. 1986; Gregory et al. 1990; Brooks et al. 1993; Fries et al. 1997). Notably, with the exception of EBNA-1, which is required for viral genome maintenance, none of these latency-associated gene products is essential for EBV-mediated immortalization of B lymphocytes in vitro (Swaminathan et al. 1991; Robertson et al. 1994). Consequently, a direct contribution of EBV to tumorigenic potential in BL has not been established. Thus, the dominant factor presumed to be responsible for maintenance of tumorigenicity in BL is the deregulated expression of the c-*MYC* proto-oncogene, which occurs as a conse-

---

[1] Program in Viral Oncogenesis and Tumor Immunology, Department of Virology and Molecular Biology, St. Jude Children's Research Hospital, Memphis, TN 38105
[2] Department of Biochemistry, St. Jude Children's Research Hospital, Memphis, TN 38105
[3] School of Biological Sciences, University of Missouri-Kansas City, Kansas City, MO 64110
[4] Department of Biochemistry, University of Tennessee Health Science Center, Memphis, TN 38163
[5] Department of Pathology, University of Tennessee Health Science Center, Memphis, TN 38163

quence of chromosomal translocation (reviewed in MAGRATH 1990). This assumption, however, was challenged by the work of Takada and colleagues (SHIMIZU et al. 1994) who found that spontaneous loss of the EBV genome in the BL cell line Akata is associated with a concomitant loss of tumorigenic potential. The main objectives of this study were to define the contribution of EBV and individual latent gene products to tumorigenicity in BL using the Akata BL cell system.

2 Results

2.1 The Type I EBV Latency Program Promotes Cell Survival in BL Following Withdrawal of Survival Factors

EBV-negative Akata BL cells have previously been shown to be non-tumorigenic relative to their EBV-positive counterparts when assayed for growth in soft agar and tumor induction in athymic nude mice (SHIMIZU et al. 1994). Despite this observed difference in tumorigenicity between EBV-positive and EBV-negative Akata cells, we did not observe any difference in the growth properties of these cells under standard culture conditions (Fig. 1A). This suggested that EBV may confer a growth advantage which is primarily manifest under growth-limiting conditions, such as would be encountered in soft agar assays or in vivo. To address this possibility, BL cells from the logarithmic and stationary phases of the growth cycle

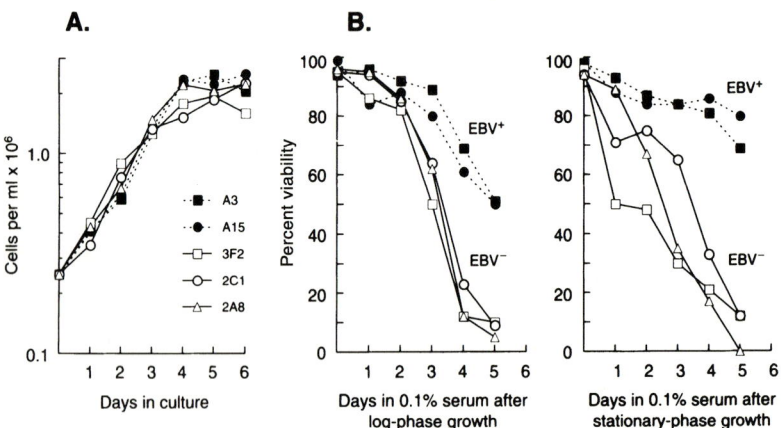

Fig. 1A,B. Type I latency promotes BL cell survival following serum deprivation. **A** Clonal EBV-positive (*shaded symbols*) and EBV-negative (*open symbols*) Akata cells were seeded at 2.5×10^5 cells/ml in standard growth medium containing 10% serum, and cell concentration was monitored daily for 6 days to assess growth rate. **B** Cells from the log (day 2) and stationary (day 5) phases of growth were washed and transferred to growth medium containing 0.1% serum, in triplicate, at a density of 5×10^5 cells/ml. Cell viability was determined daily by trypan blue dye exclusion

were analyzed for survival in low serum (0.1%) over a 4-day period. As demonstrated in Fig. 1B (left panel), cells from log-phase growth, regardless of EBV status, were sensitive to the reduced-serum conditions, resulting in the death of the majority of cells over the 4-day course of the assay. By contrast, EBV-positive BL cells from stationary-phase growth exhibited a marked delay in cell death when compared to stationary-phase EBV-negative cells, which died at rates comparable to log-phase cells. We confirmed that the death of these cells in our assay was due to apoptosis through the examination of cytospin preparations of stationary-phase EBV-positive and EBV-negative cells cultured in low serum for three days (RUF et al. 1999). EBV-negative cells displayed the morphologic changes characteristic of apoptotic death (condensed chromatin and vacuolation), while the majority of EBV-positive cells appeared relatively normal. Thus, EBV-positive Akata cells were found to be more resistant to induction of apoptosis than their EBV-negative counterparts and, as predicted, this difference in susceptibility to apoptosis was most apparent under growth-limiting conditions.

2.2 EBV- and Growth-Phase-Associated Expression of c-MYC and Bcl-2

Given that c-MYC is the primary mediator of apoptosis in BL (MILNER et al. 1992), we next investigated the possibility that altered c-MYC expression might account for the observed EBV-dependent resistance to apoptosis. To do so, we examined the level of c-MYC protein in EBV-positive and EBV-negative Akata cells throughout the cell growth cycle. As shown in Fig. 2, no detectable difference in the level of c-MYC protein was observed between EBV-positive and EBV-negative cells in the log phase of growth (days 1–3). However, a dramatic decrease in c-MYC levels was observed in EBV-positive cells as they entered the stationary phase of growth (days 4 and 5), in agreement with the increased resistance to apoptosis observed in EBV-positive stationary-phase cells. By contrast, EBV-negative cells maintained a constant level of c-MYC expression throughout the cell growth cycle (Fig. 2). We also examined three additional (non-Akata) EBV-positive BL cell lines and found a similar decrease in c-MYC expression as cells entered stationary phase (RUF et al. 1999). Thus, EBV-mediated downregulation of c-MYC under growth-limiting conditions appeared to be a general property of EBV-positive BL cells. In addition to c-MYC, we examined the expression of several pro- and anti-apoptotic proteins of the Bcl-2 family. We consistently found that the level of the anti-apoptotic protein Bcl-2 increased in both EBV-positive and EBV-negative cells as they progressed through the cell growth cycle. Furthermore, EBV-positive Akata cells generally expressed a higher steady-state level of Bcl-2 protein than EBV-negative Akata cells (Fig. 2). No EBV- or growth-phase-associated differences in expression levels were observed for Mcl-1, Bcl-X_L, or Bax (data not shown).

The potential mechanism of c-MYC regulation was initially investigated by examining the effects of growth phase on c-*myc* messenger (m)RNA levels. In

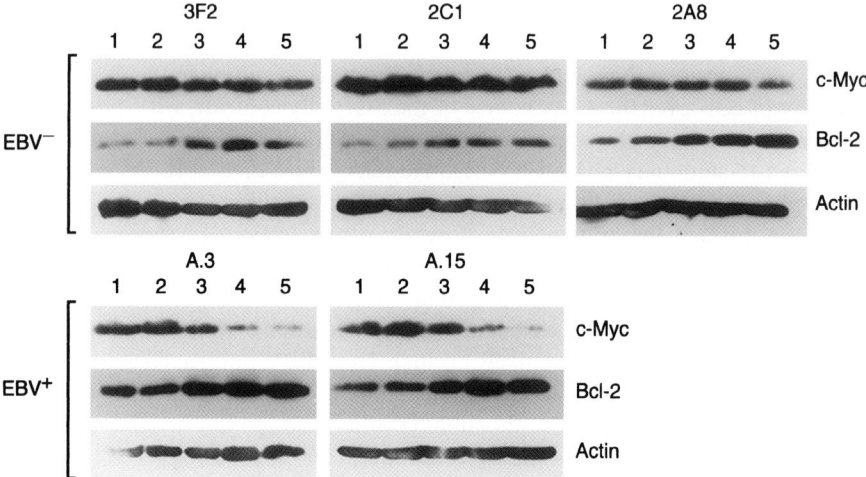

Fig. 2. EBV- and growth-phase-associated expression of c-MYC and Bcl-2. Clonal EBV-positive (A.3 and A.15) and EBV-negative (3F2, 2C1 and 2A8) Akata BL cells were seeded (2.5×10^5 cells/ml) in complete growth medium, and 2×10^6 cells were harvested daily from the experiment presented in Fig. 1 for the first 5 days in culture to monitor expression of c-MYC and Bcl-2 by immunoblot analysis. Detection of actin served as a loading control

contrast to c-MYC protein levels in EBV-positive cells, no changes in c-*myc* transcripts were detected as cells progressed into stationary-phase (Fig. 3). This suggests that EBV-induced regulation of c-*MYC* expression occurs at the post-transcriptional level. We subsequently examined the half-life of c-MYC in EBV-negative and EBV-positive Akata cells from both log- and stationary-phase growth by cyclohexamide- and pulse-chase experiments and found no difference in c-MYC $t_{1/2}$ between EBV-positive and EBV-negative cells, regardless of cell growth phase (Ruf et al. 1999). Consequently, we propose that EBV-mediated downregulation of c-MYC likely occurs via a translational mechanism.

2.3 EBV Infection, but not EBNA-1 alone, Regulates c-MYC Levels, Sensitivity to Apoptosis, and Restores Tumorigenic Potential to EBV-Negative BL Cells

The correlation between EBV infection and resistance to apoptosis, c-MYC regulation, and tumorigenic potential suggests that EBV is essential for these phenotypes. However, it was plausible that loss of the EBV genome was a consequence of loss of tumorigenic potential, rather than the underlying cause of the loss of tumorigenicity. To test this, as well as begin to investigate the contribution of individual latent gene products, we established several reinfected EBV-negative Akata cell lines (2A8.1, 2A8.2, and 2A8.3) and EBV-negative Akata lines constitutively expressing EBNA-1 (EBV$^-$/EBNA-1$^+$). We then evaluated the ability of

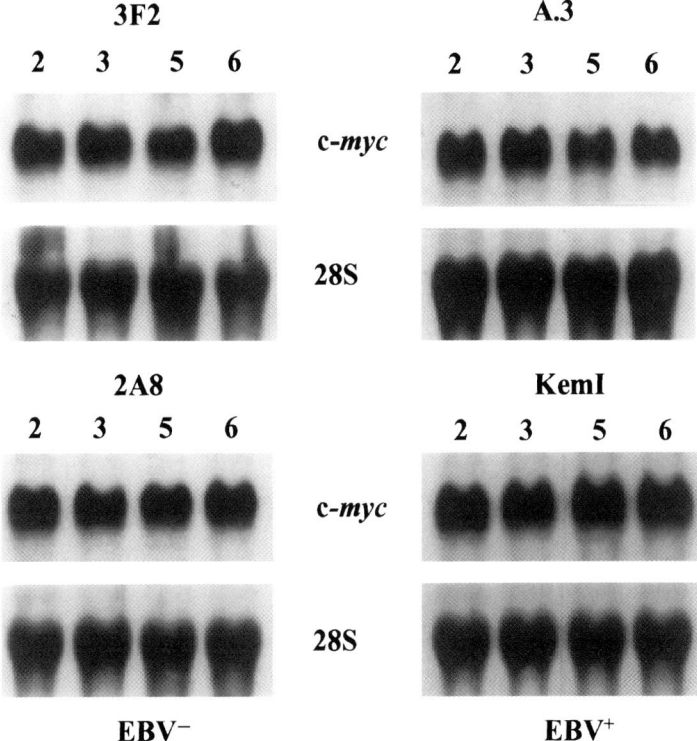

Fig. 3. EBV down-regulates c-MYC expression through a post-transcriptional mechanism. Northern blot analysis of c-*myc* mRNA and 28S rRNA (loading control) levels is EBV-negative (3F2 and 2A8) and EBV-positive (A.3 and KemI) BL cells at 2, 3, 5, and 6 days after seeding in complete growth medium at 2.5×10^5 cells/ml. Each lane contained 10μg of total cellular RNA

these reinfected and EBV⁻/EBNA-1⁺ cell lines to mediate c-MYC downregulation and sensitivity to apoptosis. Reinfection of EBV-negative Akata cells was sufficient to restore downregulation of c-MYC in stationary-phase cells (Fig. 4A, left panel). However, as also illustrated in Fig. 4A (right panel), enforced expression of EBNA-1 in the context of EBV-negative Akata cells was insufficient to mediate downregulation of c-MYC. Consistent with these results, the reinfected cells displayed less susceptibility to apoptosis than their EBV-negative counterparts when deprived of serum, while EBNA-1 expression provided no significant survival advantage (Fig. 4B).

We next assessed the tumorigenic potential of both the reinfected cells and EBV⁻/EBNA-1⁺ cells by tumor induction assays in SCID mice. As shown in Table 1, reinfected 2A8 cells produced tumors in all mice injected with these cells, whereas the parental 2A8 line and an additional EBV-negative Akata line (3F2) failed to yield any tumors in the 20 weeks post-injection during which mice were observed for tumor formation. All tumors obtained in this study were analyzed for latent gene expression patterns, including expression of the EBV oncoprotein

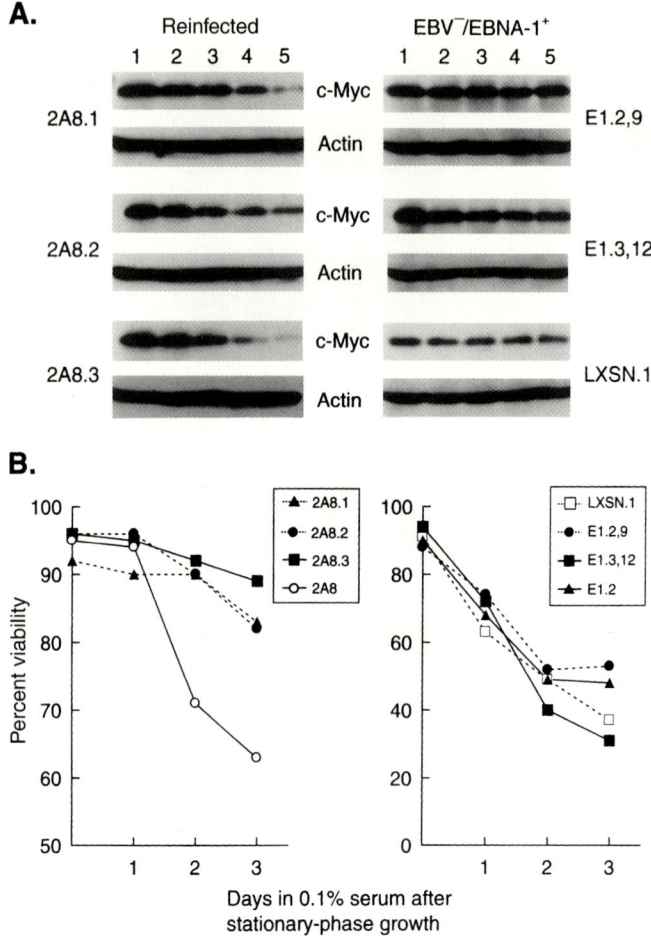

Fig. 4A,B. EBV infection, but not EBNA-1 alone, regulates c-MYC levels and sensitivity to apoptosis. **A** Immunoblot analysis of c-MYC throughout the cell growth cycle in reinfected and EBV-negative/EBNA-1-positive Akata cells. Cells were seeded as described in Fig. 1 and harvested at the indicated day post-seeding. **B** EBV infection, but not EBNA-1 alone, protects Akata cells from apoptosis during stationary-phase growth. After reaching stationary phase (day 5 post-seeding), indicated cells were deprived of serum, and percent viability was determined at the indicated intervals by trypan blue dye exclusion

LMP-1, and were found to accurately reflect the phenotype of the injected cell line (Table 1 and data not shown). Consistent with the inability of EBNA-1 to provide a survival advantage under growth-limiting conditions, EBV$^-$/EBNA-1$^+$ cell lines failed to produce tumors in SCID mice. Thus, EBV infection of Akata cells, but not EBNA-1 expression alone, is sufficient to mediate the downregulation of c-MYC and resistance to apoptosis under growth-limiting conditions, as well as enhance tumorigenic potential.

Table 1. EBV, but not EBNA-1, restores tumorigenicity to EBV-negative Akata cells

Akata cell line(s)[a]	EBV/EBNA-1 status	No. of mice that developed tumors	No. of tumors in which LMP-1 was detected[b]
A.15	+/+	3/3	0/3
3F2(3) and 2A8(9)	−/−	0/12	NA
Reinfected	+/+	10/10	0/10
pLXSN	−/−	0/12	NA
EBNA-1	−/+	0/8	NA

NA, not applicable.

[a] Numbers in parentheses are number of mice injected with each cell line; three clones of re-infected 2A8 cells were injected in triplicate and one mouse received a pool of reinfected 3F2 cells; each of three clones of pLXSN-containing 3F2 cells (vector control) were injected in quadruplicate; two pools of 3F2 lines expressing EBNA-1 were each injected in triplicate, and two cloned 3F2 lines expressing EBNA-1 were each injected into individual mice.

[b] Expression of the EBV oncoprotein LMP-1 in tumor cells, relative to an LCL-positive control, was determined by immunoblot analysis; all tumors were EBNA-1$^+$.

3 Conclusions

We have investigated the contributory role of EBV and an individual latent gene product (EBNA-1) to tumorigenicity in Akata BL cells and have shown that reestablishment of type I latency in EBV-negative Akata cells is sufficient to restore tumorigenic potential. Similar results were recently reported by others (KOMANO et al. 1998). Additionally, we have shown that EBV-positive BL cells display an increased resistance to apoptosis under growth-limiting conditions, which correlates with a modest upregulation of the anti-apoptotic protein Bcl-2 and a concomitant post-transcriptional decrease in c-MYC expression. Although the EBNA-1 protein is expressed in all EBV-associated tumors and is reported to have oncogenic potential, enforced expression of the EBNA-1 protein in EBV-negative Akata cells failed to restore the observed EBV-dependent regulation of c-MYC and tumorigenic potential (WILSON et al. 1996; KUBE et al. 1999). These results provide direct evidence that EBV can significantly contribute to the tumorigenic potential of BL, and suggest a novel mechanism whereby the restricted latency program of EBV promotes long-term survival of latently infected cells, and thus virus persistence. Additionally, recently we have shown that stable expression of the EBERs in EBV-negative Akata cells can significantly enhance the tumorigenic potential of these cells. Interestingly, this enhancement of tumorigenicity occurs independently of an effect on apoptosis or regulation of c-MYC and Bcl-2. Thus, EBV likely promotes the tumorigenic potential of BL cells through at least two avenues: an EBER-dependent mechanism that enhances proliferative potential, and a second mechanism, mediated by an as yet unidentified EBV gene, that counters the pro-apoptotic consequences of deregulated expression of c-MYC in BL cells.

Acknowledgements. The data presented herein have been published previously (RUF et al. 1999). This work was supported by United States Public Health Services (PHS) grants DE-11116 (to L.M.H.-F.), CA-76379, and DK44158 (to J.L.C.), and CA-73544 and CA-56639 (to J.T.S.), Cancer Center (CORE) grant CA-21756, and the American Lebanese Syrian Associated Charities. I.K.R. was supported by PHS grant T32-AI-07372.

References

Arrand JR, Rymo L (1982) Characterization of the major Epstein-Barr virus-specific RNA in Burkitt lymphoma-derived cells. J Virol 41:376–389
Brooks LA, Lear AL, Young LS, Rickinson AB (1993) Transcripts from the Epstein-Barr virus BamHI A fragment are detectable in all three forms of virus latency. J Virol 67:3182–3190
Fries KL, Sculley TB, Webster-Cyriaque J, Rajadurai P, Sadler RH, Raab-Traub N (1997) Identification of a novel protein encoded by the BamHI A region of the Epstein-Barr virus. J Virol 71:2765–2771
Gregory CD, Rowe M, Rickinson AB (1990) Different Epstein-Barr virus-B cell interactions in phenotypically distinct clones of a Burkitt's lymphoma cell line. J Gen Virol 71:1481–1495
Komano J, Sugiura M, Takada K (1998) Epstein-Barr virus contributes to the malignant phenotype and to apoptosis resistance in Burkitt's lymphoma cell line Akata. J Virol 72:9150–9156
Kube D, Vockerodt M, Weber O, Hell K, Wolf J, Haier B, Grasser FA, Muller-Lantzsch N, Kieff E, Diehl V, Tesch H (1999) Expression of Epstein-Barr virus nuclear antigen 1 is associated with enhanced expression of CD25 in the Hodgkin cell line L428. J Virol 73:1630–1636
Magrath I (1990) The pathogenesis of Burkitt's lymphoma. Adv Cancer Res 55:133–270
Milner AE, Johnson GD, Gregory CD (1992) Prevention of programmed cell death in Burkitt's lymphoma cell lines by bcl-2-dependent and -independent mechanisms. Int J Cancer 52:636–644
Robertson ES, Tomkinson B, Kieff E (1994) An Epstein-Barr virus with a 58-kilobase-pair deletion that includes BARF0 transforms B lymphocytes in vitro. J Virol 68:1449–1458
Rowe DT, Rowe M, Evan GI, Wallace LE, Farrell PJ, Rickinson AB (1986) Restricted expression of EBV latent genes and T-lymphocyte-detected membrane antigen in Burkitt's lymphoma cells. EMBO J 5:2599–2607
Ruf IK, Rhyne PW, Yang H, Borza CM, Hutt-Fletcher LM, Cleveland JL, Sample JT (1999) Epstein-Barr virus regulates c-MYC, apoptosis, and tumorigenicity in Burkitt lymphoma. Mol Cell Biol 19:1651–1660
Shimizu N, Tanabe-Tochikura A, Kuroiwa Y, Takada K (1994) Isolation of Epstein-Barr virus (EBV)-negative cell clones from the EBV-positive Burkitt's lymphoma (BL) line Akata: malignant phenotypes of BL cells are dependent on EBV. J Virol 68:6069–6073
Swaminathan S, Tomkinson B, Kieff E (1991) Recombinant Epstein-Barr virus with small RNA (EBER) genes deleted transforms lymphocytes and replicates in vitro. Proc Natl Acad Sci USA 88:1546–1550
Wilson JB, Bell JL, Levine AJ (1996) Expression of Epstein-Barr virus nuclear-antigen-1 induces B cell neoplasia in transgenic mice. EMBO J 15:3117–3126

Epstein-Barr Virus Infection of Human Epithelial Cells

S. IMAI[1], J. NISHIKAWA[2], M. KURODA[1], and K. TAKADA[3]

1 Introduction	161
2 In Vitro Infection Model	163
2.1 Attempts to Infect Epithelial Cells by EBV in Culture	163
2.2 Establishment of an Efficient Infection System	164
2.3 Mechanism of Infection	164
3 Expression and Function of EBV Genes in Epithelial Cells	168
3.1 Viral Latent Gene Expression	168
3.2 Latent Gene Functions for Malignant Cell Growth	170
3.3 Virus Productive Infection	174
4 Immunological Aspects of EBV-Associated Epithelial Tumors	175
References	177

1 Introduction

Much attention has been given to the close association of Epstein-Barr virus (EBV) with human epithelial cancers following the first detection of the virus in nasopharyngeal carcinoma (NPC) tissues (ZUR HAUSEN et al. 1970). Related examinations confirmed the regular presence of the viral genome in poorly differentiated and undifferentiated subtypes of NPC, irrespective of its local and ethnic incidence rate, thereby suggesting a strong etiological link of EBV to the tumor (reviewed in KLEIN 1979). Additionally, oral hairy leukoplakia (OHL) in AIDS patients and lymphoepithelioma-like carcinoma histologically resembling EBV-positive NPC at sites other than nasopharynx, such as the lung or salivary gland, have also been categorized as EBV-associated human epithelial proliferative lesions (for review, see RICKINSON and KIEFF 1996). The first report by BURKE

---

[1] Department of Microbiology, Kochi Medical School, Kohasu, Okoh-cho, Nankoku, Kochi 783-8505, Japan
[2] First Department of Internal Medicine, Yamaguchi University School of Medicine, 1144 Kogushi, Ube, Yamaguchi 755-8505, Japan
[3] Department of Tumor Virology, Institute for Genetic Medicine, Hokkaido University, N15 W7, Kita-ku, Sapporo 060-8638, Japan

et al. (1990) suggesting that EBV infection was a pathogenetic factor in primary gastric carcinoma of the lymphoepithelioma type evoked further studies on this subject (MIN et al. 1991; SHIBATA et al. 1991; NIEDOBITEK et al. 1992). The subsequent accumulation of evidence has emphasized a novel etiological link of EBV to gastric carcinoma, including a common type of the tumors (SHIBATA and WEISS 1992; TOKUNAGA et al. 1993; FUKAYAMA et al. 1994; IMAI et al. 1994; OTT et al. 1994; UEMURA et al. 1994). Based on a number of worldwide epidemiological surveys, it was estimated that gastric carcinoma is the most frequent of the EBV-associated human malignancies (for reviews, see OSATO and IMAI 1996; LEVIN and LEVINE 1998).

Recent studies also suggested that EBV is related to breast and hepatocellular carcinoma (LABRECQUE et al. 1995; BONNET et al. 1999; SUGAWARA et al. 1999). In the latter, EBV may bear the oncogenic outcome indirectly through transactivation of hepatitis C virus replication (SUGAWARA et al. 2000). Apparently, in light of the above observations, the extent of EBV-positive epithelial malignancy is still expanding. Apart from these newly found epithelial tumors, where a link with EBV remains to be investigated, all tumor cells in individual NPCs and gastric carcinomas are infected exclusively and clonally with EBV and express, though limited in number, latent EBV genes (DESGRANGES et al. 1975; RAAB-TRAUB and FLYNN 1986; YOUNG et al. 1988; WEISS et al. 1989; IMAI et al. 1994; OTT et al. 1994). This suggests a specific and active role for EBV in carcinogenesis. Of greater significance, clonal EBV infection was demonstrated in almost all precursor cells (dysplasia) and early tumor cells (carcinoma in situ) of each individual pre-invasive lesion related to NPC as well, implying a decisive role for the virus in the initiation or early phase of NPC development (PATHMANATHAN et al. 1995). The detection of EBV genome in early gastric carcinoma and in normal, or metaplastic, gastric epithelium likewise suggests virus involvement in gastric carcinogenesis (UEMURA et al. 1994; FUKAYAMA et al. 1994; HAYASHI et al. 1996; ARIKAWA et al. 1997; YANAI et al. 1997).

In contrast to B cells, however, either the mechanism of EBV infection or of the EBV-induced pathological events in epithelial cells, is less-well understood. This is due mainly to the unavailability of an in vitro infection model for investigating EBV activity in epithelial cells. This lack of progress may lead some to question a putative oncogenic role of EBV in epithelial cells, and thereby support the notion that EBV is merely "a passenger" or "a silent bystander". Such a negative concept, particularly for gastric carcinoma, has been supported by some epidemiological and pathological data obtained with a small number of specimens (ROWLANDS et al. 1993; HORIUCHI et al. 1994) or by limited testing of differences in clinicopathologic findings and viral genetic variations between EBV-positive and -negative gastric carcinomas (TAKANO et al. 1999). Nevertheless, the present chapter will investigate and review the recent impact of EBV on human epithelial cancers, as reported in some major studies. We will stress those studies that may provide a breakthrough in elucidating the role of the virus in epithelial tumor development.

2 In Vitro Infection Model

2.1 Attempts to Infect Epithelial Cells by EBV in Culture

A few classical, but important, studies reported the successful experimental infection of epithelial cells by EBV. Artificial implantation of membrane receptors with subsequent exposure to EBV led to the entry and expression of EBV in epithelial cells, indicating that epithelial cells can inherently permit EBV infection, providing that they are supplemented with sufficient numbers of receptor molecules for EBV (SHAPIRO and VOLSKY 1982). Evidence for direct infection of some primary cultures of NPC tumor cells and a specific cultured epithelial cell line was also found, signifying that certain epithelial cells possess a naturally occurring EBV receptor (GLASER et al. 1976; BEN-BASSAT et al. 1982). A further interesting finding was that primary explant cultures of human ectocervical epithelia could be infected only by wild-type EBV strains, but not by laboratory strains such as B95-8 or P3HR-1 EBV, which might have adapted to lymphoid cells during long-term in vitro propagation (SIXBEY et al. 1983). This finding raised the possibility that the biochemical difference of the viral envelope among the virus strains may be a key component as the ligand to the naturally occurring epithelial receptor(s) for EBV (see also Sect. 2.3).

Another success in EBV infection of epithelial cells in vitro was deduced from a well-known serological finding commonly seen in patients with NPC. The majority of NPC patients mount immunoglobulin A (IgA) antibodies to the EBV viral capsid antigen (VCA) and to the major envelope glycoprotein gp350/220, or more specifically, to the early antigen (EA) complex in their sera, all of which are usually measured as a diagnostic and/or prognostic parameter for NPC (HENLE and HENLE 1976). Detection of these virus-specific IgAs sometimes presages the clinical onset of NPC, and this may be the case also with EBV-associated gastric carcinoma (LEVINE et al. 1995). Based on this finding, an in vitro transient assay has been conducted, which demonstrated that coexistence of EBV virions and polymeric IgA (pIgA) antibody against virus gp350/220, but not virions alone, were necessary for the entry of EBV into colon carcinoma-derived epithelial cells HT-29 (SIXBEY and YAO 1992). Immune complexes between EBV and anti-gp350/220 pIgA bind to the pIg receptor (pIgR), i.e., the secretory component (SC) protein, expressed on the epithelial surfaces. These complexes are subsequently endocytosed into the cells. This phenomenon was reproduced later by others, using NPC-derived cell lines from which the EBV genome had been lost during in vitro passages; however, in this study, examination of NPC biopsy samples documented the restricted expression of SC protein in a fraction of NPC tumor cells, but not in untransformed metaplastic oropharyngeal epithelia, indicating that EBV infection through this pathway may require previous certain malignant changes of epithelial cells (LIN et al. 1997). Neither this pIgR-mediated nor the known EBV receptor (complement receptor type 2, CR2; CD21)-mediated experimental infection succeeded in producing epithelial cells persistently infected with the virus in vitro (SIXBEY and YAO

1992; Li et al. 1992, 1997). Overall, these early attempts to infect epithelial cells would have been carried out with considerable difficulty, and thus any successful achievements would have been few and far between. Recently however, a new, powerful analytical tool for investigation has been introduced into this research field (see below).

2.2 Establishment of an Efficient Infection System

The genetic engineering required to produce recombinant EBV has contributed greatly to the identification of the latent EBV genes essential for B-cell immortalization by the virus (HAMMERSCHMIDT and SUGDEN 1988). Subsequently, a more organized system for the large-scale production of clonal EBV recombinants was established, based on a Burkitt's lymphoma (BL) cell line, designated Akata (TAKADA and ONO 1989; SHIMIZU et al. 1996). This has opened a way to investigate the complex interactions in the EBV-epithelium field.

The first series of experiments was undertaken using a recombinant virus clone carrying a neomycin-resistant gene (neo^r) as a selective marker (neo^r-rEBV), which had been inserted in the *Bam*HI X leftward reading frame 1 (BXLF1) locus of the virus genome. This region codes for viral thymidine kinase, which is dispensable for B-cell immortalization by EBV and production of progeny virus. This unique and powerful system has yielded new insights into EBV-epithelial cell interactions. Exposure to the neo^r-rEBV preparation, followed by G418 selection in culture, has resulted in the successful isolation of many EBV-converted cell clones from three virus-negative gastric carcinoma cell lines (YOSHIYAMA et al. 1997). A related study further extended the range of epithelial cells used and has confirmed that many epithelial cell lines from various tissues are susceptible to EBV infection in vitro (IMAI et al. 1998). This efficient epithelial infection system was also applied to a study using NPC cell lines that had suffered EBV genomic loss caused by long-term in vitro culture, and has established stably reinfected clones of the tumor cells as well (CHANG et al. 1999). Interestingly, almost all the isolated carcinoma cell clones infected with neo^r-rEBV have been shown to reproduce EBV gene expression, similar to the behavior of epithelial tumors in vivo (YOUNG et al. 1988; BROOKS et al. 1992; IMAI et al. 1994; SUGIURA et al. 1996; YOSHIYAMA et al. 1997; IMAI et al. 1998; CHANG et al. 1999), thereby providing a useful in vitro model.

2.3 Mechanism of Infection

The CR2 molecule expressed on B cells serves as a functional EBV receptor (FINGEROTH et al. 1984). The viral ligand for CR2 is the major EBV glycoprotein gp350/220 coded by the BLLF1 locus, which has a similar sequence to the natural ligand C3d (NEMEROW et al. 1989). It is controversial, however, whether normal epithelial cells express CR2 and utilize it normally as a functional receptor for EBV in vivo. There have been histochemical findings suggestive of actual CR2 expression, both in normal and malignant epithelial tissues, in various anatomical sites (BILLAUD et al. 1989; TIMENS et al. 1991). In gastric tissues, both detectable CR2 expression (LEONCINI et al. 1993) and the converse (ODA et al. 1993; HARN et al. 1995; SHIN et al. 1996; CHAPEL et al. 2000; our own observations) have been found. A molecule immunostainable and/or immunoprecipitable by the well-defined anti-CR2 monoclonal antibody anti-B2 or HB-5 was reported to be expressed in the normal oropharyngeal mucosal epithelia and in uterine cervical epithelia (YOUNG et al. 1986; SIXBEY et al. 1987). However, this molecule (~195kDa) was shown to clearly differ in molecular weight (MW) from reported MW of the genuine CR2 (approximately 145kDa) and was unable to interact with viral gp350/220. It is still uncertain whether this CR2-like molecule is actually functional in EBV infection, partly because subsequent observations revealed its rapid disappearance on the cell surface following explantation of cells to in vitro culture (YOUNG et al. 1989). A transcriptional variant of CR2 messenger (m)RNA with additional 5' untranslated sequences was identified in some human epithelial cell lines, RHEK-1 and HeLa, which have weak capacity for EBV binding; however, this transcript was proven to have a coding sequence identical to CR2 mRNA transcribed in B cells (BIRKENBACH et al. 1992). Conclusively, these data indicate that some epithelial cells can express CR2 molecule, but in general, the abundance is substantially lower, the level insufficient for mediating infection.

Experimental infection of epithelial cells using neo^r-rEBV, described in the previous section, has shown that carcinoma cell lines, which are clearly negative for CR2, could still be infected with EBV. This therefore implicates a CR2-independent entry of the virus into human epithelia (YOSHIYAMA et al. 1997; IMAI et al. 1998; CHANG et al. 1999; NISHIKAWA et al. 1999). A similar experimental approach, on the other hand, has also proved that CR2-mediated infection of EBV can occur in another human epithelial cell line, 293, without exogenous transduction of the CR2 gene (FINGEROTH et al. 1999). Accordingly, it is likely that EBV exploits several different mechanisms for entry into epithelial cells; thus, in the next step of the investigation, it is necessary to determine the nature of this CR2-independent infection. We have found that cell-to-cell contact increases markedly the transmission efficiency of EBV from virus-producing B cells to virus-recipient epithelial cells in vitro, possibly reflecting a common mode for virus transmission in epithelia in vivo (IMAI et al. 1998). A mechanism involving cell fusion between virus-donor and recipient cells, as had been proposed previously (BAYLISS and WOLF 1980), can be readily excluded in our co-cultivation experiment (IMAI et al. 1998). A more precise re-examination of the original finding was carried out by others, who have confirmed that direct cell-to-cell contact indeed enhances EBV infection efficiency in NPC cell lines (CHANG et al. 1999). Moreover, their data led these authors to conclude that cell-to-cell contact not only provides a CR2-independent epithelial route of EBV infection, but also synergistically reinforces CR2-mediated infection. Cell-to-cell contact, therefore, seems to be particularly attractive as a comprehensive mechanism for EBV infection in epithelial cells, because it may be able to explain the in vivo spread of EBV over diverse epithelial tissues, even though CR2 is not expressed there. This concept may also be upheld by the detection of lytically infected intramucosal B cells as a probable virus source (TAO et al. 1995). Although such an augmented viral transmission by close cell-to-cell contact was observed also in human T-cell leukemia virus type I (HTLV-I), the exact mechanism is still unknown (CANN and CHEN 1996). These recent in vitro studies regarding EBV, however, indicate at least that epithelial cells are presumed to utilize both CR2-dependent and CR2-independent "devices" for EBV infection, the latter of which may shed light on the existence of an unidentified receptor-ligand system.

As stated in the previous section of this chapter, the pIgA-mediated entry of EBV into epithelial cells is another conceivable infection mechanism (SIXBEY and YAO 1992). A subsequent study performed by the same group has suggested that the epithelial polarization status is a key determinant for the infectious outcome of EBV, mediated through the pIgR process (GAN et al. 1997). Based on the in vivo results obtained from a mouse model, they prepared the pIgR-transfected Madin-Darby canine kidney (MDCK/pIgR) epithelial cells, whose polarity can be manipulated in vitro, and demonstrated that, when MDCK/pIgR cells, but not untransfected MDCK cells, retain their proper polarity, basolaterally internalized viruses (as a pIgA-EBV complex) are eliminated apically from the cells by transcytosis without evidence of infection (e.g., viral gene expression). In contrast, nonpolarized MDCK/pIgR cells, as most carcinoma cells have disorganized polarity or apolarity, permitted infection after pIgA-mediated entry of EBV. This also involves postulating that a loss of polarity, associated with certain premalignant or dysplastic changes in cells, may be essential for the successful infection of epithelium by EBV (GAN et al. 1997; also see Sect. 3.3). Another approach utilizing CR2-transfected MDCK cells, however, has demonstrated that EBV enters into the polarized MDCK/CR2 cells preferentially from the apical surface rather than from the basolateral surface and establishes infection, as ascertained by the expression of viral BZLF1, despite the relatively dense distribution of CR2 on the basolateral site (CHODOSH et al. 2000). The lytic cycle initiated in BZLF1-positive MDCK/CR2 cells largely ends in abortive infection, as shown by the bare detection of viral late lytic antigens such as gp350/220 and the undetectability of infectious virion in culture, perhaps due to the inherent low permissiveness of MDCK cells per se. Nonetheless, a small fraction of infected cells displayed the basolateral accumulation of gp350/220 indicative of newly synthesized virions. This apical EBV entry followed by the possible basolateral release of progeny may illustrate well the initial step of systemic EBV dissemination in primary infection in vivo; EBV can overcome an epithelial barrier of the normal mucosa (generally oral mucosa) by polarity-dependent virus translocation, consequently reaching submucosal tissues containing B cells, a natural reservoir for EBV. To explain the disparity of consequences between pIgR-mediated and CR2-mediated viral trafficking seen in the same cell system, the authors speculated the existence of a cellular cofactor in polarized MDCK cells, which is segregated mainly on the apical side and enhances apical infectivity synergistically with CR2 (CHODOSH et al. 2000).

Genetic construction of mutated EBV recombinants, in which several genes encoding envelope glycoproteins are deleted from the entire virus genome, has provided significant progress toward understanding the infection mechanism. A clonal EBV recombinant, lacking gp42 coded by the BZLF2 locus, was shown to be capable of binding to B cells, perhaps through the classical gp350/220-CR2 interaction; however, this recombinant was unable to penetrate the B-cell membrane (WANG and HUTT-FLETCHER 1998). Nevertheless, this mutant virus was still capable of infecting epithelial cells. This same group previously discovered the binding property of gp42 to the HLA-DR β chain (SPRIGGS et al. 1996), and further demonstrated the absolute requirement of HLA-DR as a cofactor for the infection of B cells by EBV (LI et al. 1997). The authors have speculated that the gp42-HLA-DR interaction facilitates virions to obtain closer access to the cell membrane than does the initial gp350/220-CR2 interaction and to catalyze subsequent membrane fusion through an interaction with an additional major glycoprotein complex, gp85-gp25 (the EBV homolog of herpes simplex virus gH-gL) (LI et al. 1995). Similarly, another recent study has demonstrated that exogenous introduction of the HLA-DP or -DQ gene, as well as HLA-DR, conferred susceptibility to EBV infection on resistant cells that express CR2 but not HLA class II (HAAN et al. 2000). The authors, therefore, concluded that CR2 and HLA class II are sufficient for EBV infection in cells of various origins. However, as these authors also stated, CR2 gene-transfected epithelial cells (SVK-CR2), which lack HLA class II expression as most epithelial cells do, can be efficiently infected by EBV (LI et al. 1992), further implying that a coreceptor(s) different from the HLA class II molecules naturally exists in epithelial cells.

A most recent experiment using recombinant EBV lacking gp85 coded by the BXLF2 locus has shown that there is a drastically reduced binding capacity, and, of course, loss of infectivity as well, of EBV to both epithelial cells and B cells (ODA et al. 2000). The data suggested that gp85 is not only a virion component essential for the formation of the tripartite gH-gL-gp42 complex and subsequent fusion of the viral envelope with the cell membrane as demonstrated before (MILLER and HUTT-FLETCHER 1988), but that gp85 is also another ligand for an unidentified, possibly epithelial-specific binding receptor. WANG et al. (1998) described a model in which wild-type EBV virions are proposed to contain two types of envelope glycoprotein complexes, the tripartite gH-gL-gp42 and the bipartite gH-gL complexes, which can mediate the infection of B cells and epithelial cells, respectively, in a mutually exclusive manner. The gH-gL complex is deemed to interact with the putative epithelial receptor. With respect to this concept, some other recent results by others are of special interest. JANZ et al. (2000) have shown that a gp350/220-disrupted mutant virus can still infect both B cells and epithelial cells, thereby supporting the existence of not only an additional viral ligand, but also a functional receptor different from the CR2-gp350/220 and HLA-DR-gp42 combinations. The reason for this is, again, that epithelial cells sometimes express little or no CR2 and are usually negative for HLA class II expression. A relevant recent finding has also suggested that EBV entry into primary human monocytes could occur through a CR2-independent process (SAVARD et al. 2000). Taken together, these data support the existence of a novel, generally used (co)receptor for EBV on epithelial cells. A future topic for investigation is the identification of a receptor, other than CR2, expressed preferentially on epithelial cells and the determination of its ligand on the virion.

3 Expression and Function of EBV Genes in Epithelial Cells

3.1 Viral Latent Gene Expression

Thus far, EBV-associated epithelial diseases have been examined extensively for the expression of latent EBV genes in vivo and in vitro. There are three representative forms of viral latency, i.e., latency I, II and III. Cells of latency III, represented by EBV-immortalized lymphoblastoid B-cell lines (LCLs), express all virus latent genes: six EBV-determined nuclear antigens (EBNAs), two EBV-encoded small RNAs (EBERs), three latent membrane proteins (LMPs), and BamHI A rightward transcripts (BARTs), which are also referred to as complementary strand transcripts (CSTs), or BARF0 RNAs. Latency I and latency II cells share in common the expression of EBNA1, EBERs, and BARTs/CSTs; cells of latency II additionally express LMP1, LMP2A, and LMP2B (reviewed in RICKINSON and KIEFF 1996).

In general, latency II is displayed by NPC cells. Furthermore, latency II was reproduced experimentally by hybridomas between latency III LCL and virus-negative nonlymphoid cells (CONTRERAS-SALAZAR et al. 1989; CONTRERAS-BRODIN et al. 1991; KERR et al. 1992). Latency I, typically seen in BL, is observed in gastric carcinoma cells. However, some variants of latency I/II are also recognized in these same epithelial tumors. About 30% of NPC cases are negative for LMP1 and LMP2B at the transcriptional (single-round PCR amplification) and protein levels (FAHRAEUS et al. 1988; YOUNG et al. 1988; BROOKS et al. 1992), and about 40% of gastric carcinoma cases are positive for LMP2A mRNA (SUGIURA et al. 1996). Such latency I-like gene expression was further reported in transplantable EBV-associated gastric carcinoma cells, which can be propagated through severe combined immunodeficient (SCID) mice (IWASAKI et al. 1998). Several nonmalignant gastric epithelial cell lines harboring EBV (designated GT38 and GT39) have been established in vitro by another laboratory (TAJIMA et al. 1998). Surprisingly, these cell lines exhibited latent gene expression consistent with latency III, with a high proportion being lytic cycle gene-expressing cells (TAKASAKA et al. 1998). This difference in latency type between in vivo gastric carcinoma cells and the above cell lines may simply reflect the different genetic cellular background – fully malignant vs non-malignant. The possible reason for this is that latency III with partial spontaneous, productive infection similar to GT38 and GT39 cell lines is seen in OHL, another EBV-related non-malignant epithelial lesion (WEBSTER-CYRIAQUE and RAAB-TRAUB 1998). The isolation of

GT38 and GT39 lines from noncancerous tissues provides direct evidence for the EBV infection of a gastric epithelium in vivo, although it may be a rare event.

It has been demonstrated that another EBV gene, BARF1, is possibly involved in the oncogenic process of NPC. BARF1 mRNA is organized with open reading frames (ORFs), located in a large fragment (~40kb) spanning from BamHI D to A of the genome. Although the BARF1 gene was considered initially as an early gene in B cells, subsequent experiments revealed its transforming and immortalizing capacities on both primate and rodent cells; thus, BARF1 is now defined as a latency-associated gene (GRIFFIN and KARRAN 1984; WEI and OOKA 1989; KARRAN et al. 1992; WEI et al. 1997). Using biopsy specimens, some transcriptional studies have detected BARF1 expression preferentially in an EBV-associated malignant epithelial lesion (i.e., NPC), compared to a benign epithelial lesion (i.e., OHL), and only rarely in other lymphoid neoplasias, such as Hodgkin's disease and T-cell lymphoma (BRINK et al. 1998; HAYES et al. 1999). In addition, more than 80% of EBV-associated gastric carcinomas have recently turned out to express BARF1 (ZUR HAUSEN et al. 2000). Together, these data indicate that the BARF1 gene is most likely expressed in an epithelial-specific manner and has a potential oncogenic role (also see next section), though the expression is generally low at the mRNA level (SBIH-LAMMALI et al. 1996).

Another recent study identified two novel rightward transcripts (termed A73 and RPMS1 according to their cloned complementary (c)DNAs), which are classified as members of the BARTs/CSTs family (HITT et al. 1989; KARRAN et al. 1992), from a cDNA library of the nude mouse-passaged NPC cell line C15 (SMITH et al. 2000; also see Sect. 3.2). In the same article, the RNase protection assay indicated that, of all the transcripts of the BARTs/CSTs family capable of encoding proteins, these two transcripts occur more abundantly and, furthermore, that they have a remarkable diversity in their amino acid sequences, compared with the hitherto well-analyzed BARF0 and RK-BARF0 proteins (FRIES et al. 1997; KIENZLE et al. 1999).

Finally, recent discoveries of a novel latency pattern of virus gene expression are worthy of mention. Detection of the expression of EBERs by in situ hybridization has been commonly used to examine EBV infection, due to the high copy numbers in an infected cell. However, the transcription of EBERs has been found to be suppressed in certain epithelial tumors latently infected with EBV, namely in a subset of NPCs (TAKEUCHI et al. 1997), in hepatocellular carcinoma (SUGAWARA et al. 1999), and in breast cancer (BONNET et al. 1999). Although the pathogenetic significance of EBV in the last two malignancies is largely unknown at present, such variable, or occasional lack of, EBERs' expression in epithelial neoplasms may account for, to some degree, the conflicting results in the literature concerning the association of EBV with breast carcinoma, especially when the association itself is attributed to the expression of EBERs in certain breast carcinomas (IEZZONI et al. 1995; GLASER et al. 1998; BONNET et al. 1999). These case-oriented studies indicate that in situ detection of EBER expression cannot always be used as a reliable and sensitive diagnostic tool of EBV infection.

3.2 Latent Gene Functions for Malignant Cell Growth

The fact that epithelial cells can remain in the latency II viral program, such as NPC, has lead to studies focusing mainly on a unique virus oncogene, LMP1. Many studies have documented the multi-potency of LMP1 implicated in epithelial oncogenesis. When singly transduced, LMP1 transforms rodent fibroblasts (Wang et al. 1985; Baichwal and Sugden 1988), alters keratin gene expression (Dawson et al. 1990), induces consistent morphological changes, and inhibits cell differentiation in several immortalized human keratinocytes (Fahraeus et al. 1990). LMP1 also induces expression of the epidermal growth factor receptor (Miller et al. 1995), blocks p53-mediated apoptosis via activation of the A20 gene (Fries et al. 1996), and activates matrix metalloproteinase 9 (MMP-9) which may be involved in the highly invasive and metastatic capacity of NPC (Yoshizaki et al. 1998). Moreover, the development of epidermal hyperplasia was reported in LMP1-transgenic mice (Kulwichit et al. 1998). These pleiotropic biological activities of LMP1 are based on its molecular associations with the tumor necrosis factor receptor-associated proteins (TRAFs) (Mosialos et al. 1995; Miller et al. 1997; Devergne et al. 1998), as well as nuclear factor (NF)-κB signaling (Mitchell and Sugden 1995; Huen et al. 1995; Paine et al. 1995; Sylla et al. 1998), and c-Jun N-terminal kinase (JNK) signaling (Eliopoulos and Young 1998). These associations are mediated through the longer cytoplasmic domains of LMP1, called C-terminal activating regions (CTARs) 1 and/or 2. In addition, a recent publication has shown that LMP1 is able to transform MDCK epithelial cells with concomitant induction of the cellular Ets-1 transcription factor, suggesting a regulatory role for Ets-1 in LMP1-inducible urokinase-type plasminogen activator (UPA) activity (Kim et al. 2000).

There has been no clear evidence that EBV provides any continuous contributions to the malignant phenotype of EBV-positive (but LMP1-negative) gastric carcinoma cells, or to the promotion of growth in the gastric epithelium. For the first time, we have recently shown that after EBV infection, cultured gastric epithelial cells derived from a nonmalignant gastric tissue (designated PGE-5) gain the capacity of enhanced population doubling and a higher saturation density, together with the acquisition of clonability in agar, whereas the virus-negative parental PGE-5 cells have a limited lifespan (Nishikawa et al. 1999; also see Figs. 1 and 2). Furthermore, a few of the EBV-converted clones of PGE-5 formed tumors subcutaneously in SCID mice (Fig. 2). The EBV-convertants of PGE-5 displayed a gene expression pattern similar to that of latency I, the same as that seen in gastric carcinoma cells in vivo (Fig. 1) A similar, growth-promoting effect by EBV and viral latency-type has also been observed in the gastric carcinoma cell line NU-GC-3 (Imai et al. 1998; Nishikawa et al. 1999). EBV-induced immortalization of gastric epithelium may be inferred also from the establishment of nonmalignant gastric epithelial cell lines infected with EBV, namely, the GT38 and GT39 lines mentioned in the previous section (Tajima et al. 1998; Takasaka et al. 1998).

A lower frequency of tumor cells undergoing apoptosis in EBV-positive gastric carcinomas, compared with in EBV-negative ones (Ohfuji et al. 1996), may predict

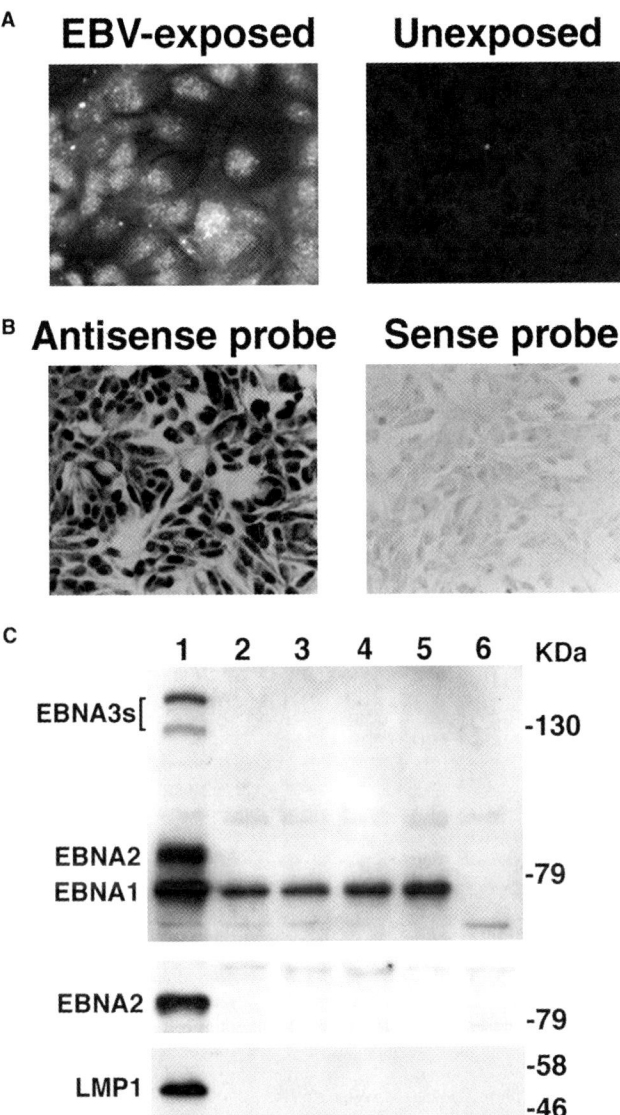

Fig. 1A–C. EBV latent gene expression in nonmalignant gastric epithelial cells. **A** Immunofluorescence staining of EBNA with EBV-seropositive human serum. Results of a neo^r-rEBV-infected, G418-resistant PGE-5 cell clone (*left*) and the parental PGE-5 cells (*right*) are shown, ×400. **B** In situ hybridization for EBER1. An EBV-infected PGE-5 clone was hybridized with the antisense oligoprobe (*left*) and with the sense oligoprobe (*right*). Intense nuclear signals are evident only in the case using the antisense probe, ×200. **C** Immunoblotting for detection of EBNAs and LMP1. The blots were probed with pooled human sera for EBNA1, -2, -3A, -3B and 3C (*top blot*), PE2 monoclonal antibody for EBNA2 (*middle blot*), and with CS1-4 monoclonal antibodies for LMP1 (*bottom blot*). Lane 1: LCL immortalized with neo^r-rEBV; lanes 2 to 5: neo^r-rEBV-infected PGE-5 cell clones; lane 6: parent PGE-5 cells. Cell lysates extracted from 10^5 cells were loaded per slot

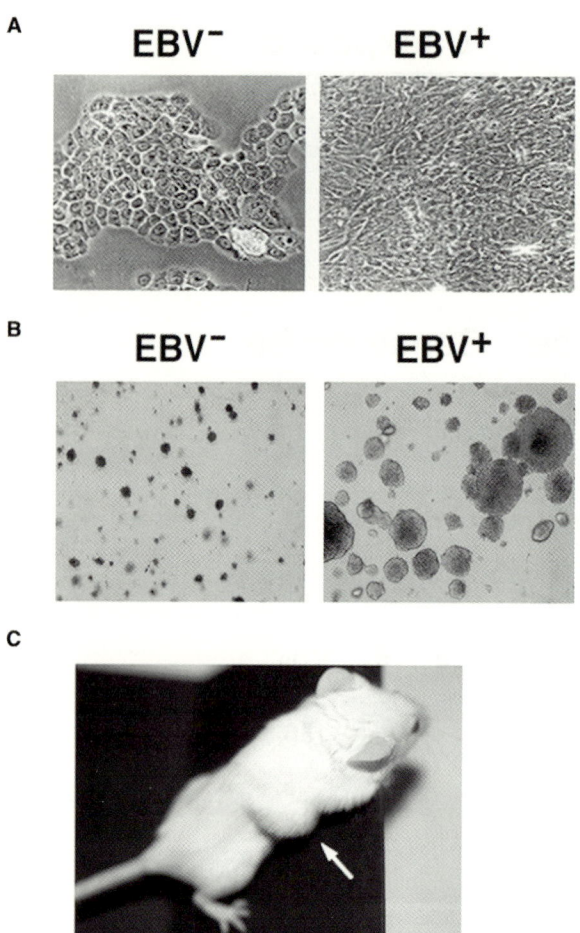

Fig. 2A–C. Malignant growth of EBV-infected PGE-5 clones. **A** Growth in fluid culture. A neo^r-transfected PGE-5 clone (*left*) and a neo^r-rEBV-infected PGE-5 clone (*right*) are representatively shown, ×100. Both clones were trypsinized, reseeded into culture wells under the same culture conditions, and photographed 5 days after passage. **B** Growth in semisolid agarose. A neo^r-transfected PGE-5 clone as a control (*left*) and a neo^r-rEBV-infected PGE-5 clone (*right*) are representatively shown, ×40. 10^4 cells were seeded per well in 6-well plates and photographed 4 weeks after seeding. Almost all colonies in the control culture consisted of dead cells, whereas the virus-infected clone formed many large colonies of viable cells. **C** Tumorigenesis in SCID mice. 10^7 cells of neo^r-rEBV-infected PGE-5 clone were injected subcutaneously per mouse and photographed 72 days after inoculation

an undefined anti-apoptotic virus gene expressed in epithelial tumors representing latency I, such as gastric carcinoma. In the study, as well as another study (GULLEY et al. 1996), there were no significant differences in both p53 and Bcl-2 positivity between the two groups of gastric carcinomas. Another recent examination also produced the similar results indicative of a significantly lower rate of apoptosis in EBV-positive gastric carcinoma than EBV-negative one; by contrast, the low susceptibility to apoptosis was found to closely correlate with overexpression of Bcl-2, but not with p53 expression, suggesting the virus-induced, *bcl*-2-mediated anti-apoptotic effect (KUME et al. 1999). Although BHRF-1 (HENDERSON et al. 1993) and, more recently, BALF1 (MARSHALL et al. 1999) have been identified as a functional EBV homolog of the *bcl*-2 proto-oncogene so far, and in fact, the BHRF-1 product can suppress epithelial apoptosis induced by tumor necrosis factor alpha (TNF-α) or anti-Fas antibody (KAWANISHI 1997), both genes are expressed during the virus lytic cycle. A recent study using a latency I BL line, Akata, has demonstrated that optimal EBERs expression may be responsible for the malignant phenotype and the resistance to apoptosis related to upregulation of Bcl-2 expression (KOMANO et al. 1999). By contrast, except for LMP1, little knowledge concerning the functional roles for the viral latent genes known to be expressed commonly in latency I and latency II (i.e., EBNA1, EBERs, BARTs/CSTs, and LMP2A/2B) has been obtained in the epithelial cell environment. At present, since our data have indicated that an upregulation of Bcl-2 associated with EBV or EBERs is not observed in epithelial cell lines (S. Imai et al., unpublished observation), it appears that mechanisms other than those for BL must be involved in EBV-induced malignant changes in epithelial cells. In this context, analogous to Akata BL cells, our PGE-5 cells may be a useful tool for investigating the role for the "latency I genes" in epithelial cells, including an anti-apoptotic effect. Interestingly, a paradoxical biological function of LMP1 possibly associated with gastric carcinogenesis has been reported recently: LMP1 expression reversibly deprives gastric carcinoma cells of their malignant phenotypes and, furthermore, renders the cells more susceptible to apoptotic cell death (SHEU et al. 1998). These results can explain a possible in vivo event that the selective elimination of LMP1-positive clones of gastric carcinoma cells may occur, thereby facilitating the emergence of more malignant and immune-resistant clones. The upregulation of *bcl*-2 was observed also in most NPC cases, but there has been no correlation between Bcl-2 expression and the presence of EBV or LMP1 positivity (LU et al. 1993; AGATHANGGELOU et al. 1995).

The EBV BARF1 latent gene, preferentially expressed in epithelial lesions with latent EBV infection, encodes a 33-kDa protein homologous to the human colony-stimulating factor 1 receptor (hCSF-1R); the predicted degree of similarity is approximately 38% at the amino acid level (STROCKBINE et al. 1998). Furthermore, this BARF1-encoded 33-kDa protein has been shown to be functional, as ascertained by the neutralizing activity of hCSF-1 in vitro (STROCKBINE et al. 1998). This strongly suggests a biological activity of the BARF1 protein to promote EBV-infected cell growth directly (WEI and OOKA 1989; KARRAN et al. 1990; WEI et al. 1997) and/or through inhibition of the virus-specific immune response, because hCSF-1 is known to act as a potent immunomodulator that induces cytokine release from mononuclear cells, such as interferon alpha (COHEN et al. 1999). Interestingly as well, among the products of the newly identified BARTs/CSTs family transcripts, A73 has been shown to interact with RACK1, a repressor protein involved in the signaling pathway via the *src* kinase and protein kinase C families (SMITH et al. 2000). The other product, RPMS1, has been proven to bind to a cellular factor, CBF1 (RBP-Jκ), resulting in inhibition of the CBF1-mediated transcription associated with activated Notch (NotchIC) (SMITH et al. 2000).

Based on these recent achievements in this research area, it is now more important than ever to analyze and clarify the roles of latent EBV genes including such newly identified ones in epithelial oncogenesis.

3.3 Virus Productive Infection

Latent EBV infection concomitant with virus replicative gene expression has been reported to occur in EBV-associated human epithelial lesions. However, except for OHL, infected cells entering the lytic cycle, if any, generally form only a very minor population in the whole tumor mass or occasionally express the immediate early protein BZLF1 only (COCHET et al. 1993; PATHMANATHAN et al. 1995; SELVES et al. 1996). Although productive infection may play an important role in virus spread in vivo, it is probably disadvantageous for tumor cell survival and growth. This has suggested a therapeutic approach specific for EBV-infected cancer cells, whereby induction of the lytic cycle is enforced by delivery of EBV genes BZLF1 or BRLF1, with the parallel use of antiherpesvirus drugs such as ganciclovir (WESTPHAL et al. 2000). Nevertheless, in terms of epithelial tumor development, several experimental findings regarding productive infection in epithelial cells seem to be intriguing. An experimental EBV infection system using SVK-CR2 epithelial cell lines revealed previously that most acutely infected cells underwent a shift to the lytic cycle in a differentiation-dependent manner (LI et al. 1992). The subsequent study conducted by the same group successfully isolated stable, latently infected clones of SVK-CR2 through long-term cloning culture (KNOX et al. 1996). These stable infected clones were found to become extremely refractory to stimulation for both differentiation and virus lytic cycle induction, and they displayed latent gene expression positive for EBNA1, EBERs, and BARTs/CSTs only (LMP1 and LMP2 were detected at an extremely low level), which resembles in vivo tumor cells of LMP1-negative NPC. Thus, these observations strongly support again the former assumption (BECKER et al. 1991) that sustained, intrinsic inability of keratinocytes to differentiate, which correlates strictly with the latent infection state, is the absolute cellular prerequisite for the occurrence of this malignancy. In addition, documented capability of LMP1 to inhibit epithelial cell differentiation may explain a direct role for EBV in another possible in vivo scenario for LMP1-positive NPC development; LMP1 blocks further differentiation of partially differentiated keratinocytes or those predisposed to differentiation, resulting in establishment of stable latent EBV infection. It is still unknown whether the phenomenon mimics exactly in vivo events: the clonal selection of cells with intrinsic nonpermissiveness from the heterogeneous, infected cell population, or a rare, special precursor clone of keratinocytes, which will not permit productive infection (i.e., differentiation-resistant), is preferentially targeted by EBV, thereby endowing the precursor with growth potential. In any event, however, EBV-infected clones susceptible to productive infection will presumably fail to proceed into a more malignant stage. Rather, they are reminiscent of a highly productive but nonmalignant disease, OHL (BECKER et al. 1991). Given that the biological substrate for the future appearance of the NPC tumor is the stratified squamous epithelium, which gained the inability to differentiate prior to EBV infection, it is possible that naturally existing cofactors (dietary substances) may also be capable of playing a role in the initial neoplastic changes in NPC (KNOX et al. 1996; POIRIER et al. 1987), in addition to a well-established genetic factor (LU et al. 1990). In this context, chemical inducers of EBV replication are also assumed to represent possible environmental risk factors for the known, local endemic nature of NPC (ITO et al. 1981; MIZUNO et al. 1983; ZENG et al. 1984; TOMEI et al. 1987). Certain types of naturally occurring phorbol esters like 12-O-tetradecanoyl phorbol-13-acetate (TPA), known as a tumor-promoting compound, may contribute to tumor development through the induction of the lytic cycle (LI et al. 1992; KNOX et al. 1996). This would then lead to an increase in the size of the virus-infected cell pool and to the selective survival and growth promotion of nonpermissive clones as tumor progenitors. The TPA enhancement of lytic cycle was observed also in in vitro EBV-converted gastric carcinoma cell lines and EBV-infected gastric epithelial cell lines derived from stomach biopsy (YOSHIYAMA et al. 1997; TAKASAKA et al. 1998), though any epidemiological evidence for the possible linkage of such nonviral environmental risk factors or genetic predisposition to EBV-associated gastric carcinoma has not yet appeared (QIU et al. 1997; TASHIRO et al. 1998).

4 Immunological Aspects of EBV-Associated Epithelial Tumors

The representative EBV-associated epithelial tumor NPC exhibits a distinctive pathological feature of intense lymphocyte infiltration in the stroma (i.e., so-called lymphoepithelioma-like histology). This is also the case with EBV-associated gastric carcinoma; the undifferentiated subgroup of gastric carcinoma with lymphoepithelioma-like histology has extremely high EBV positivity (over 90%) (MIN et al. 1991; SHIBATA et al. 1991; TOKUNAGA et al. 1993; IMAI et al. 1994). Furthermore, most of the usual histological subgroups, including early gastric carcinoma, are frequently accompanied by prominent lymphocytic infiltration (SHIBATA and WEISS 1992; TOKUNAGA et al. 1993; UEMURA et al. 1994). This clearly differentiates such tumors from those gastric carcinomas unrelated to EBV. Such tumor-infiltrating lymphocytes (TILs) are known to predominantly comprise T cells (HERAIT et al. 1987; MIN et al. 1991; TOKUNAGA et al. 1993). This characteristic histology presumably represents host immune response directed to the tumor cells, which may also be reflected by a relatively preferable prognosis in TIL-rich gastric carcinoma cases (so-called gastric carcinoma with lymphoid stroma), compared with TIL-less cases (WATANABE et al. 1976). A correlation between the degree of lymphocyte infiltration and prolonged survival rate of patients is observed also in NPC cases (GIANNINI et al. 1991; ZONG et al. 1993). In NPC biopsies, TILs with CD4-positive and CD8-positive ($CD8^+$) phenotypes are present in varying proportions (HERAIT et al. 1987). Immunophenotypic analysis of EBV-associated gastric carcinomas, on the other hand, revealed that most TILs in the tumor bear the $CD8^+$ cytotoxic T-lymphocyte (CTL) phenotype together with several markers indicative of active proliferation (SAIKI et al. 1996). However, it has been obscure whether the TILs are making any contribution to operative host defense machinery against EBV-carrying malignant cells or, conversely, to tumor cell growth. Functional characterization of the infiltrating $CD8^+$ T cells was performed for EBV-associated gastric carcinomas and demonstrated that these T cells are able to recognize and lyse EBV-infected autologous and HLA-matched allogeneic LCLs, but not mitogen-activated autologous B lymphoblasts (KUZUSHIMA et al. 1999). However, both relevant antigenic peptide-pulse and vaccinia recombinant virus vector-mediated gene transfer failed to identify the target EBV gene for CTL. The authors speculate that some cellular proteins, which provide epitopes effective for massive TIL stimulation but insufficient for eliciting CTL-mediated killing, may be involved in the strong T-cell response seen in EBV-associated gastric carcinoma. Since this study utilized HLA class I (A24)-matched B-cell and fibroblast system, it is unclear which EBV gene is attributable to the expression of such cellular proteins in gastric epithelia. In this regard, emerging experimental evidence (also see Sect. 3.1) raises another possibility that novel EBV-coded latent proteins, which are preferentially expressed and suggested to be functional in the epithelial environment in vivo, may provide other antigenic viral epitopes recognizable by the tumor-infiltrating $CD8^+$ T cells (ZUR HAUSEN et al. 2000; SMITH et al. 2000). Additionally, another group has recently identified CTLs specific to a latent EBV protein, BARF0; however, its expression level in infected cells is trickily regulated by differential splicing, thereby silencing or minimizing the CTL-mediated lysis of target cells (KIENZLE et al. 1998, 1999). Further research is to be expected also in this area.

It is known that NPC cells are highly proliferative (very short doubling time) and invasive in vivo, whereas the tumor cells are prone to apoptotic cell death in vitro as represented by their poor adaptation to in vitro culture and their low transplantability into immunodeficient mice (BUSSON et al. 1988). This biological gap of NPC cell growth between in vivo and in vitro conditions strongly suggests the importance of tumor tissue microenvironment in situ, which is advantageous for the tumor cell survival and growth. SBIH-LAMMALI et al. (1999) have provided clear evidence for this recently. The authors have demonstrated, consistent with a previous study (AGATHANGGELOU et al. 1995), that the majority of NPC cells express both CD95 (Fas/APO-1) and CD40 at the high level in vivo, raising possible pathogenetic significance of the immunoregulatory molecules. Indeed in the same examination, NPC-derived cell lines have been shown to enter the apoptotic process when the CD95 pathway alone is activated, while pretreatment with CD40 ligand (CD40L; CD154), which is often presented by TILs in vivo, can prevent the cells from apoptosis induced by CD95 activation (SBIH-LAMMALI et al. 1999). This finding would validate a role for TILs in tumor cell survival in vivo, thereby enhancing the tumor growth associated with EBV infection.

Immunological characterization of gastric carcinoma cells revealed that CD54 (intercellular adhesion molecule-1, ICAM-1) and CD95 are expressed more abundantly in EBV-positive tumor cells than in EBV-negative tumor cells, although the expression of CD95 ligand (CD95L) on TILs was not examined (SAIKI et al. 1996). In some cancers such as malignant melanoma, tumor cells are known to express CD95L on their surfaces or release its soluble form, thereby inducing apoptotic cell death of TILs (HAHNE et al. 1996). TILs are substantially expressing upregulated CD95 relative to peripheral blood T cells (CARDI et al. 1998). This "Fas counterattack" mechanism is suggested to enable tumor cells to evade tumor-specific immune surveillance through CD95-mediated killing of tumor-infiltrating T cells (O'CONNELL et al. 1996). To understand the better prognosis in EBV-associated gastric carcinoma by this phenomenon, the CD95L status should be verified in comparison with the virus-negative tumors.

Acknowledgements. Our own work described in this chapter was supported by scientific research grants from the Japanese Ministry of Education, Science, Sports and Culture, and from the President Research Fund of Kochi Medical School Hospital.

References

Agathanggelou A, Niedobitek G, Chen R, Nicholls J, Yin W, Young LS (1995) Expression of immune regulatory molecules in Epstein-Barr virus-associated nasopharyngeal carcinomas with prominent lymphoid stroma. Evidence for a functional interaction between epithelial tumor cells and infiltrating lymphoid cells. Am J Pathol 147:1152–1160

Arikawa J, Tokunaga M, Tashiro Y, Tanaka S, Sato E, Haraguchi K, Yamamoto A, Toyohira O, Tsuchimochi A (1997) Epstein-Barr virus-positive multiple early gastric cancers and dysplastic lesions: a case report. Pathol Int 47:730–734

Baichwal VR, Sugden B (1988) Transformation of Balb 3T3 cells by the BNLF-1 gene Epstein-Barr virus. Oncogene 2:461–467

Bayliss GJ, Wolf H (1980) Epstein-Barr virus-induced cell fusion. Nature 287:164–165

Becker J, Leser U, Marschall M, Langford A, Jilg W, Gelderblom H, Reichart P, Wolf H (1991) Expression of proteins encoded by Epstein-Barr virus trans-activator genes depends on the differentiation of epithelial cells in oral hairy leukoplakia. Proc Natl Acad Sci USA 88:8332–8336

Ben-Bassat H, Mitrani-Rosenbaum S, Goldblum N (1982) Induction of Epstein-Barr virus nuclear antigen and DNA synthesis in a human epithelial cell line after Epstein-Barr virus infection. J Virol 41:703–708

Billaud M, Busson P, Huang D, Mueller-Lantzch N, Rousselet G, Pavlish O, Wakasugi H, Seigneurin JM, Tursz T, Lenoir GM (1989) Epstein-Barr virus (EBV)-containing nasopharyngeal carcinoma cells express the B-cell activation antigen blast2/CD23 and low levels of the EBV receptor CR2. J Virol 63:4121–4128

Birkenbach M, Tong X, Bradbury LE, Tedder TF, Kieff E (1992) Characterization of an Epstein-Barr virus receptor on human epithelial cells. J Exp Med 176:1405–1414

Bonnet M, Guinebretiere J-M, Kremmer E, Grunewald V, Benhamou E, Contesso G, Joab I (1999) Detection of Epstein-Barr virus in invasive breast cancers. J Natl Cancer Inst 91:1376–1381

Brink AA, Vervoort MB, Middeldorp JM, Meijer CJ, van den Brule AJ (1998) Nucleic acid sequence-based amplification, a new method for analysis of spliced and unspliced Epstein-Barr virus latent transcripts, and its comparison with reverse transcriptase PCR. J Clin Microbiol 36:3164–3169

Brooks L, Yao QY, Rickinson AB, Young LS (1992) Epstein-Barr virus latent gene transcription in nasopharyngeal carcinoma cells: Coexpression of EBNA1, LMP1, and LMP2 transcripts. J Virol 66:2689–2697

Burke AP, Yen TSB, Shekitka KM, Sobin LH (1990) Lymphoepithelial carcinoma of the stomach with Epstein-Barr virus demonstrated by polymerase chain reaction. Mod Pathol 3:377–380

Busson P, Ganem G, Flores P, Mugneret F, Clausse B, Caillou B, Braham K, Wakasugi H, Lipinski M, Tursz T (1988) Establishment and characterization of three transplantable EBV-containing nasopharyngeal carcinomas. Int J Cancer 42:599–606

Cann AJ, Chen ISY (1996) Human T-cell leukemia virus type I and II. In: Fields BN, Knipe DM, Howley PM (eds) Virology, 3rd edn. Lippincott-Raven, Philadelphia, New York

Cardi G, Heaney JA, Schned AR, Ernstoff MS (1998) Expression of Fas(APO-1/CD95) in tumor-infiltrating and peripheral blood lymphocytes in patients with renal cell carcinoma. Cancer Res 58:2078–2080

Chang Y, Tung C-H, Huang Y-T, Lu J, Chen J-Y, Tsai C-H (1999) Requirement for cell-to-cell contact in Epstein-Barr virus infection of nasopharyngeal carcinoma cells and keratinocytes. J Virol 73:8857–8866

Chapel F, Fabiani B, Davi F, Raphael M, Tepper M, Champault G, Guettier C (2000) Epstein-Barr virus and gastric carcinoma in western patients: comparison of pathological parameters and p53 expression in EBV-positive and negative tumours. Histopathology 36:252–261

Chodosh J, Gan Y-J, Holder VP, Sixbey JW (2000) Patterned entry and egress by Epstein-Barr virus in polarized CR2-positive epithelial cells. Virology 266:387–396

Cochet C, Martel-Renoir D, Grunewald V, Bosq J, Cochet G, Schwaab G, Bernaudin JF, Joab I (1993) Expression of the Epstein-Barr virus immediate early gene, BZLF1, in nasopharyngeal carcinoma tumor cells. Virology 197:358–365

Cohen JI, Lekstrom (1999) Epstein-Barr virus BARF1 protein is dispensable for B-cell transformation and inhibits alpha interferon secretion from mononuclear cells. J Virol 73:7627–7632

Contreras-Brodin BA, Anvret M, Imreh S, Altiok E, Klein G, Masucci M (1991) B cell phenotype-dependent expression of the Epstein-Barr virus nuclear antigens EBNA-2 to EBNA-6: studies with somatic cell hybrids. J Gen Virol 72:3025–3033

Contreras-Salazar B, Klein G, Masucci MG (1989) Host cell-dependent regulation of growth transformation associated Epstein-Barr virus antigens in somatic cell hybrids. J Virol 63:2768–2772

Dawson CW, Rickinson AB, Young LS (1990) Epstein-Barr virus latent membrane protein inhibits human epithelial cell differentiation. Nature 344:777–780

Desgranges C, Wolf H, De-The G, Shanmugaratnam K, Cammoun N, Ellouz R, Klein G, Lennert K, Munoz N, zur Hausen H (1975) Nasopharyngeal carcinoma. X. Presence of Epstein-Barr genomes in separated epithelial cells of tumors in patients from Singapore, Tunisia and Kenya. Int J Cancer 16: 7–15

Devergne O, McFarland EC, Mosialos G, Izumi KM, Ware CF, Kieff E (1998) Role of the TRAF binding site and NF-κB activation in Epstein-Barr virus latent membrane protein 1-induced cell gene expression. J Virol 72:7900–7908

Eliopoulos AG, Young LS (1998) Activation of the cJun N-terminal kinase (JNK) pathway by the Epstein-Barr virus-encoded latent membrane protein 1 (LMP1). Oncogene 16:1731–1742

Fahraeus R, Fu HL, Ernberg I, Finke J, Rowe M, Klein G, Falk K, Nilsson E, Yadav M, Busson P, Tursz T, Kallin B (1988) Expression of Epstein-Barr virus-encoded proteins in nasopharyngeal carcinoma. Int J Cancer 42:329–338

Fahraeus R, Rymo L, Rhim JS, Klein G (1990) Morphological transformation of human keratinocytes expressing the LMP gene of Epstein-Barr virus. Nature 345:447–449

Fingeroth JD, Weis JJ, Tedder TF, Strominger JL, Biro PA, Fearon DT (1984) Epstein-Barr virus receptor of human B lymphocytes is the C3d receptor CR2. Proc Natl Acad Sci USA 81:4510–4514

Fingeroth JD, Diamond ME, Sage DR, Hayman J, Yates J (1999) CD21-dependent infection of an epithelial cell line, 293, by Epstein-Barr virus. J Virol 73:2115–2125

Fries KL, Miller WE, Raab-Traub N (1996) Epstein-Barr virus latent membrane protein 1 blocks p53-mediated apoptosis through the induction of the A20 gene. J Virol 70:8653–8659

Fries KL, Sculley TB, Webster-Cyriaque J, Rajadurai P, Sadler RH, Raab-Traub N (1997) Identification of a novel protein encoded by the BamHI A region of the Epstein-Barr virus. J Virol 71: 2765–2771

Fukayama M, Hayashi Y, Iwasaki Y, Chong J, Ooba T, Takizawa T, Koike M, Mizutani S, Miyaki M, Hirai K (1994) Epstein-Barr virus-associated gastric carcinoma and Epstein-Barr virus infection of the stomach. Lab Invest 71:73–81

Gan Y-J, Chodosh J, Morgan A, Sixbey JW (1997) Epithelial cell polarization is a determinant in the infectious outcome of immunoglobulin A-mediated entry by Epstein-Barr virus. J Virol 71: 519–526

Giannini A, Bianchi S, Messerini L, Gallo O, Asprella-Libonati G, Olmi P, et al (1991) Prognostic significance of accessory cells and lymphocytes in nasopharyngeal carcinoma. Path Res Pract 187:496–502

Glaser R, de The G, Lenoir G, Ho JHC (1976) Superinfection of epithelial nasopharyngeal carcinoma cells with Epstein-Barr virus. Proc Natl Acad Sci USA 73:960–963

Glaser SL, Ambinder RF, DiGiuseppe JA, Horn-Ross PL, Hsu JL (1998) Absence of Epstein-Barr virus EBER-1 transcripts in an epidemiologically diverse group of breast cancers. Int J Cancer 75:555–558

Griffin BE, Karran L (1984) Immortalization of monkey epithelial cells by specific fragments of Epstein-Barr virus DNA. Nature 309:78–82

Gulley ML, Pulitzer DR, Eagan PA, Schneider BG (1996) Epstein-Barr virus infection is an early event in gastric carcinogenesis and is independent of bcl-2 expression and p53 accumulation. Hum Pathol 27:20–27

Haan KM, Kwok WW, Longnecker R, Speck P (2000) Epstein-Barr virus entry utilizing HLA-DP or HLA-DQ as a coreceptor. J Virol 74:2451–2454

Hahne M, Rimoldi D, Schroter M, Romero P, Schreier M, French LE, Schneider P, Bornand T, Fontana A, Lienard D, Cerottini J, Tschopp J (1996) Melanoma cell expression of Fas(Apo-1/CD95) ligand: implications for tumor immune escape. Science 274:1363–1366

Hammerschmidt W, Sugden B (1989) Genetic analysis of immortalizing functions of Epstein-Barr virus in human B lymphocytes. Nature 340:393–397

Harn H-J, Chang J-Y, Wang M-W, Ho L-I, Lee H-S, Chiang J-H, Lee W-H (1995) Epstein-Barr virus-associated gastric adenocarcinoma in Taiwan. Hum Pathol 26:267–271

zur Hausen A, Brink AATP, Craanen ME, Middeldorp JM, Meijer CJLM, van den Brule AJC (2000) Unique transcription pattern of Epstein-Barr virus (EBV) in EBV-carrying gastric adenocarcinomas: expression of the transforming BARF1 gene. Cancer Res 60:2745–2748

zur Hausen H, Schulte-Holthauzen H, Klein G, Henle W, Henle G, Clifford P, Santesson L (1970) EBV DNA in biopsies of Burkitt tumours and anaplastic carcinomas of the nasopharynx. Nature 228:1956–1958

Hayashi K, Teramoto N, Akagi T, Sasaki Y, Suzuki T (1996) In situ detection of Epstein-Barr virus in the gastric glands with intestinal metaplasia. Am J Gastroenterol 91:1481

Hayes DP, Brink AATP, Vervoort MBHJ, Middeldorp JM, Meijer CJLM, van den Brule AJC (1999) Expression of Epstein-Barr virus (EBV) transcripts encoding homologues to important human proteins in diverse EBV associated diseases. J Clin Pathol: Mol Pathol 52:97–103

Henderson S, Huen D, Rowe M, Dawson C, Johnson G, Rickinson A (1993) Epstein-Barr virus-coded BHRF1 protein, a viral homologue of Bcl-2, protects human B cells from programmed cell death. Proc Natl Acad Sci USA 90:8479–8483

Henle G, Henle W (1976) Epstein-Barr virus specific IgA serum antibodies as an outstanding feature of nasopharyngeal carcinoma. Int J Cancer 17:1–7

Herait P, Ganem G, Lipinski M, Carlu C, Micheau C, Schwaab G, de-The G, Tursz T (1987) Lymphocyte subsets in tumour of patients with undifferentiated nasopharyngeal carcinoma: presence of lymphocytes with the phenotype of activated T cells. Br J Cancer 55:135–139

Hitt MM, Allday MJ, Hara T, Karran L, Jones MD, Busson P, Tursz T, Ernberg I, Griffin BE (1989) EBV gene expression in an NPC-related tumour. EMBO J 8:2639–2651

Horiuchi K, Mishima K, Ohsawa M, Aozasa K (1994) Carcinoma of stomach and breast with lymphoid stroma: localisation of Epstein-Barr virus. J Clin Pathol 47:538–540

Huen DS, Henderson SA, Croom-Carter D, Rowe M (1995) The Epstein-Barr virus latent membrane protein-1 (LMP1) mediates activation of NF-kappa B and cell surface phenotype via two effector regions in its carboxy-terminal cytoplasmic domain. Oncogene 10:549–560

Iezzoni JC, Gaffey MJ, Weiss LM (1995) The role of Epstein-Barr virus in lymphoepithelioma-like carcinomas. Am J Clin Pathol 103:308–315

Imai S, Koizumi S, Sugiura M, Tokunaga M, Uemura Y, Yamamoto N, Tanaka S, Sato E, Osato T (1994) Gastric carcinoma: Monoclonal epithelial malignant cells expressing Epstein-Barr virus latent infection protein. Proc Natl Acad Sci USA 91:9131–9135

Imai S, Nishikawa J, Takada K (1998) Cell-to-cell contact as an efficient mode of Epstein-Barr virus infection of diverse human epithelial cells. J Virol 72:4371–4378

Ito Y, Kawanishi M, Hirayama T, Takabayashi S (1981) Combined effect of the extracts from Croton tiglium, Euphorbia lathyris or *Euphorbia tirucalli* and n-butyrate on Epstein-Barr virus expression in human lymphoblastoid P3HR-1 and Raji cells. Cancer Lett 12:175–180

Iwasaki Y, Chong J-M, Hayashi Y, Ikeno R, Arai K, Kitamura M, Koike M, Hirai K, Fukayama M (1998) Establishment and characterization of a human Epstein-Barr virus-associated gastric carcinoma in SCID mice. J Virol 72:8321–8326

Janz A, Oezel M, Kurzeder C, Mautner J, Pich D, Kost M, Hammerschmidt W, Delecluse H-J (2000) Infectious Epstein-Barr virus lacking the major glycoprotein BLLF1 (gp350/220) demonstrates the existence of additional viral ligands. J Virol 74:10142–10152

Karran L, Gao Y, Smith PR, Griffin BE (1992) Expression of a family of complementary-strand transcripts in Epstein-Barr virus-infected cells. Proc Natl Acad Sci USA 89:8058–8062

Karran L, Teo CG, King D, Hitt MM, Gao Y, Griffin BE (1990) Establishment of immortalized primate epithelial cells with EBV DNA. Int J Cancer 45:763–772

Kawanishi M (1997) Epstein-Barr virus BHRF1 protein protects intestine 407 epithelial cells from apoptosis induced by tumor necrosis factor alpha and anti-Fas antibody. J Virol 71:3319–3322

Kerr BM, Lear AL, Rowe M, Croom-Carter D, Young LS, Rookes SM, Gallimore PH, Rickinson AB (1992) Three transcriptionally distinct forms of Epstein-Barr virus latency in somatic cell hybrids: Cell phenotype dependence of virus promoter usage. Virology 187:189–201

Kienzle N, Sculley TB, Poulsen L, Buck M, Cross S, Raab-Traub N, Khanna R (1998) Identification of a cytotoxic T-lymphocyte response to the novel BARF0 protein of Epstein-Barr virus: A critical role for antigen expression. J Virol 72:6614–6620

Kienzle N, Sculley TB, Greco S, Khanna R (1999) Silencing virus-specific cytotoxic T cell-mediated immune recognition by differential splicing: A novel implication of RNA processing for antigen presentation. J Immunol 162:6963–6966

Kienzle N, Buck M, Greco S, Krauer K, Sculley T (1999) Epstein-Barr virus-encoded RK-BARF0 protein expression. J Virol 73:8902–8906

Kim KR, Yoshizaki T, Miyamori H, Hasegawa K, Horikawa T, Furukawa M, Harada S, Seiki M, Sato H (2000) Transformation of Madin-Darby canine kidney (MDCK) epithelial cells by Epstein-Barr virus latent membrane protein 1 (LMP1) induces expression of Ets1 and invasive growth. Oncogene 19:1764–1771

Klein G (1979) The relationship of the virus to nasopharyngeal carcinoma. In: Epstein MA, Achong BG (eds) The Epstein-Barr virus. Springer-Verlag Berlin, pp 339–350

Knox PG, Li Q-X, Rickinson AB, Young LS (1996) In vitro production of stable Epstein-Barr virus-positive epithelial cell clones which resemble the virus:cell interaction observed in nasopharyngeal carcinoma. Virology 215:40–50

Komano J, Maruo S, Kurozumi K, Oda T, Takada K (1999) Oncogenic role of Epstein-Barr virus-encoded RNAs in Burkitt's lymphoma cell line Akata. J Virol 73:9827–9831

Kulwichit W, Edwards RH, Davenport EM, Baskar JF, Godfrey V, Raab-Traub N (1998) Expression of the Epstein-Barr virus latent membrane protein 1 induces B cell lymphoma in transgenic mice. Proc Natl Acad Sci USA 95:11963–11968

Kume T, Oshima K, Shinohara T, Takeo H, Yamashita Y, Shirakusa T, Kikuchi M (1999) Low rate of apoptosis and overexpression of bcl-2 in Epstein-Barr virus-associated gastric carcinoma. Histopathology 34:502–509

Kuzushima K, Nakamura S, Nakamura T, Yamamura Y, Yokoyama N, Fujita M, Kiyono T, Tsurumi T (1999) Increased frequency of antigen-specific CD8+ cytotoxic T lymphocytes infiltrating an Epstein-Barr virus-associated gastric carcinoma. J Clin Invest 104:163–171

Labrecque LG, Barnes DM, Fentiman IS, Griffin BE (1995) Epstein-Barr virus in epithelial cell tumors: a breast cancer study. Cancer Res 55:39–45

Leoncini L, Vindigni C, Megha T, Funto I, Pacenti L, Musaro M, Renieri A, Seri M, Anagnostopoulos J, Tosi P (1993) Epstein-Barr virus and gastric cancer: data and unanswered questions. Int J Cancer 53:898–901

Levin LI, Levine PH (1998) The epidemiology of Epstein-Barr virus-associated human cancer. In: Osato T, Takada K, Tokunaga M (eds) Epstein-Barr virus and human cancer. Gann Monograph in Cancer Res No. 45, Karger, Basel Tokyo, pp 51–74

Levine PH, Stemmermann GS, Lennette ET, Hildesheim A, Shibata D, Nomura A (1995) Elevated antibody titers to Epstein-Barr virus prior to the diagnosis of Epstein-Barr virus-associated gastric adenocarcinoma. Int J Cancer 60:642–644

Lin C-T, Lin C-R, Tan G-K, Chen W, Dee AN, Chan W-Y (1997) The mechanism of Epstein-Barr virus infection in nasopharyngeal carcinoma cells. Am J Pathol 150:1745–1756

Li QX, Young LS, Niedobitek G, Dawson CW, Birkenbach M, Wang F, Rickinson AB (1992) Epstein-Barr virus infection and replication in a human epithelial system. Nature 356:347–350

Li Q, Turk SM, Hutt-Fletcher LM (1995) The Epstein-Barr virus (EBV) BZLF2 gene product associates with the gH and gL homologs of EBV and carries an epitope critical to infection of B cells but not of epithelial cells. J Virol 69:3987–3994

Li Q, Spriggs MK, Kovats S, Turk SM, Comeau MR, Nepom B, Hutt-Fletcher LM (1997) Epstein-Barr virus uses HLA class II as a cofactor for infection of B lymphocytes. J Virol 71:4657–4662

Lu QL, Elia G, Lucas S, Thomas JA (1993) Bcl-2 proto-oncogene expression in Epstein-Barr-virus-associated nasopharyngeal carcinoma. Int J Cancer 53:29–35

Lu S-J, Day NE, Degos L, Lepage V, Wang P-C, Chan S-H, Simons M, McKnight B, Easton D, Zeng Y, de-The G (1990) Linkage of a nasopharyngeal carcinoma susceptibility locus to the HLA region. Nature 346:470–471

Marshall WL, Yim C, Gustafson E, Graf T, Sage DR, Hanify K, Williams L, Fingeroth J, Finberg RW (1999) Epstein-Barr virus encodes a novel homolog of the bcl-2 oncogene that inhibits apoptosis and associates with bax and bak. J Virol 73:5181–5185

Miller N, Hutt-Fletcher LM (1988) A monoclonal antibody to glycoprotein gp85 inhibits fusion but not attachment of Epstein-Barr virus. J Virol 62:2366–2372

Miller WE, Earp HS, Raab-Traub N (1995) The Epstein-Barr virus latent membrane protein 1 induces expression of the epidermal growth factor receptor. J Virol 69:4390–4398

Miller WE, Mosialos G, Kieff E, Raab-Traub N (1997) Epstein-Barr virus LMP1 induction of the epidermal growth factor receptor is mediated through a TRAF signaling pathway destinct from NF-κB activation. J Virol 71:586–594

Min K-W, Holmquist S, Peiper SC, Oileary TJ (1991) Poorly differentiated adenocarcinoma with lymphoid stroma (lymphoepithelioma-like carcinomas) of the stomach. Report of three cases with Epstein-Barr virus genome demonstrated by the polymerase chain reaction. Am J Clin Pathol 96:219–227

Mitchell T, Sugden B (1995) Stimulation of NF-kappa B-mediated transcription by mutant derivatives of the latent membrane protein of Epstein-Barr virus. J Virol 69:2968–2976

Mizuno F, Koizumi S, Osato T, Kokwaro JO, Ito Y (1983) Chinese and African Euphorbiaceae plant extracts: markedly enhancing effect on Epstein-Barr virus-induced transformation. Cancer Lett 19:199–205

Mosialos G, Birkenbach M, Yalamanchili R, VanArsdale T, Ware C, Kieff E (1995) The Epstein-Barr virus transforming protein LMP1 engages signaling proteins for the tumor necrosis factor receptor family. Cell 80:389–399

Nemerow GR, Houghten RA, Moore MD, Cooper NR (1989) Identification of an epitope in the major envelope protein of Epstein-Barr virus that mediates viral binding to the B lymphocyte EBV receptor (CR2). Cell 56:369–377

Niedobitek G, Herbst H, Young LS, Rowe M, Dienemann D, Germer C, Stein H (1992) Epstein-Barr virus and carcinomas. Expression of the viral genome in an undifferentiated gastric carcinoma. Diagnostic Mol Pathol 1:103–108

Nishikawa J, Imai S, Takada K (1999) Epstein-Barr virus promotes epithelial cell growth in the absence of EBNA2 and LMP1 expression. J Virol 73:1286–1292

O'Connell J, O'Sullivan GC, Collins JK, Shanahan F (1996) The Fas counterattack: Fas-mediated T cell killing by colon cancer cells expressing Fas ligand. J Exp Med 184:1075–1082

Oda K, Tamaru J, Takenouchi T, Mikata A, Nunomura M, Saitoh N, Sarashina H, Nakajima N (1993) Association of Epstein-Barr virus with gastric carcinoma with lymphoid stroma. Am J Pathol 143:1063–1071

Oda T, Imai S, Chiba S, Takada K (2000) Epstein-Barr virus lacking glycoprotein gp85 cannot infect B cells and epithelial cells. Virology 276:52–58

Ohfuji S, Osaki M, Tsujitani S, Ikeguchi M, Sairenji T, Ito H (1996) Low frequency of apoptosis in Epstein-Barr virus-associated gastric carcinoma with lymphoid stroma. Int J Cancer 68:710–715

Osato T, Imai S (1996) Epstein-Barr virus and gastric carcinoma. Seminars in Cancer Biol 7:175–182

Ott G, Kirchner TH, Muller-Hermelink HK (1994) Monoclonal Epstein-Barr virus genomes but lack of EBV-related protein expression in different types of gastric carcinoma. Histopathology 25:323–329

Paine E, Scheinman RI, Baldwin AS JR, Raab-Traub N (1995) Expression of LMP1 in epithelial cells leads to the activation of a select subset of NF-κB/Rel family proteins. J Virol 69:4572–4576

Pathmanathan R, Prasad U, Sadler R, Flynn K, Raab-Traub N (1995) Clonal proliferations of cells infected with Epstein-Barr virus in preinvasive lesions related to nasopharyngeal carcinoma. N Engl J Med 333:693–698

Poirier S, Ohshima H, de-The G, Hubert A, Bourgade MC, Bartsch H (1987) Volatile nitrosamine levels in common foods from Tunisia, South China and Greenland, high-risk areas for nasopharyngeal carcinoma. Int J Cancer 39:293–296

Qiu K, Tomita Y, Hashimoto M, Ohsawa M, Kawano K, Wu D-M, Aozasa K (1997) Epstein-Barr virus in gastric carcinoma in Suzhou, China and Osaka, Japan: association with clinico-pathologic factors and HLA-subtype. Int J Cancer 71:155–158

Raab-Traub N, Flynn K (1986) The structure of the termini of the Epstein-Barr virus as a marker of clonal cellular proliferation. Cell 47:883–889

Rickinson AB, Kieff E (1996) Epstein-Barr virus. In: Fields BN, Knipe DM, Howley PM (eds) Virology, 3rd edn. Lippincott-Raven, Philadelphia, New York

Rowlands DC, Ito M, Mangham DC, Reynolds G, Herbst H, Hallissey MT, Fielding JWL, Newbold KM, Jones EL, Young LS, Niedobitek G (1993) Epstein-Barr virus and carcinomas: rare association of the virus with gastric adenocarcinomas. Br J Cancer 68:1014–1019

Saiki Y, Ohtani H, Naito Y, Miyazawa M, Nagura H (1996) Immunophenotypic characterization of Epstein-Barr virus-associated gastric carcinoma: massive infiltration by proliferating CD8+ T-lymphocytes. Lab Invest 75:67–76

Savard M, Belanger C, Tardif M, Gourde P, Flamand L, Gosselin J (2000) Infection of primary human monocytes by Epstein-Barr virus. J Virol 74:2612–2619

Sbih-Lammali F, Djennaoui D, Belaoui H, Bouguermouh A, Decaussin G, Ooka T (1996) Transcriptional expression of Epstein-Barr virus genes and proto-oncogenes in north African nasopharyngeal carcinoma. J Med Virol 49:7–14

Sbih-Lammali F, Clausse B, Ardila-Osorio H, Guerry R, Talbot M, Havouis S, Ferradini L, Bosq J, Tursz T, Busson P (1999) Control of apoptosis in Epstein-Barr virus-positive nasopharyngeal carcinoma cells: Opposite effects of CD95 and CD40 stimulation. Cancer Res 59:924–930

Selves J, Bibeau F, Brousset P, Meggetto F, Mazerolles C, Voigt JJ, Pradere B, Chiotasso P, Delsol G (1996) Epstein-Barr virus latent and replicative gene expression in gastric carcinoma. Histopathology 28:121–127

Shapiro IM, Volsky DJ (1982) Infection of normal human epithelial cells by Epstein-Barr virus. Science 219:1225–1228

Sheu L-F, Chen A, Wei Y-H, Ho K-C, Cheng J-Y, Meng C-L, Lee W-H (1998) Epstein-Barr virus LMP1 modulates the malignant potential of gastric carcinoma cells involving apoptosis. Am J Pathol 152:63–74

Shibata DS, Tokunaga M, Uemura Y, Sato E, Tanaka S, Weiss LM (1991) Association of Epstein-Barr virus with undifferentiated gastric carcinomas with intense lymphoid infiltration. Am J Pathol 139:469–474

Shibata D, Weiss LM (1992) Epstein-Barr virus-associated gastric adenocarcinoma. Am J Pathol 140:769–774

Shimizu N, Yoshiyama H, Takada K (1996) Clonal propagation of Epstein-Barr virus (EBV) recombinants in EBV-negative Akata cells. J Virol 70:7260–7263

Shin WS, Kang MW, Kang JH, Choi MK, Ahn BM, Kim JK, Sun HS, Min K-W (1996) Epstein-Barr virus-associated gastric adenocarcinomas among Koreans. Am J Clin Pathol 105:174–181

Sixbey JW, Vesterinen EH, Nedrud JG, Raab-Traub N, Walton LA, Pagano JS (1983) Replication of Epstein-Barr virus in human epithelial cells infected in vitro. Nature 306:480–483

Sixbey JW, Davis DS, Young LS, Hutt-Fletcher L, Tedder TF, Rickinson AB (1987) Human epithelial cell expression of an Epstein-Barr virus receptor. J Gen Virol 68:805–811

Sixbey JW, Yao Q-Y (1992) Immunoglobulin A-induced shift of Epstein-Barr virus tissue tropism. Science 255:1578–1580

Smith PR, De Jesus O, Turner D, Hollyoake M, Karstegl CE, Griffin BE, Karran L, Wang Y, Hayward SD, Farrell PJ (2000) Structure and content of CST (BART) family RNAs of Epstein-Barr virus. J Virol 74:3082–3092

Spriggs MK, Armitage RJ, Comeau MR, Strockbine L, Farrah T, Macduff B, Ulrich D, Alderson MR, Mullberg J, Cohen JI (1996) The extracellular domain of the Epstein-Barr virus BZLF2 protein binds the HLA-DR beta chain and inhibits antigen presentation. J Virol 70:5557–5563

Strockbine LD, Cohen JI, Farrah T, Lyman SD, Wagener F, DuBose RF, Armitage RJ, Spriggs MK (1998) The Epstein-Barr virus BARF1 gene encodes a novel, soluble colony-stimulating factor-1 receptor. J Virol 72:4015–4021

Sugawara Y, Mizugaki Y, Uchida T, Torii T, Imai S, Makuuchi M, Takada K (1999) Detection of Epstein-Barr virus (EBV) in hepatocellular carcinoma tissue: a novel EBV latency characterized by the absence of EBV-encoded small RNA expression. Virology 256:196–202

Sugawara Y, Makuuchi M, Kato N, Shimotohno K, Takada K (2000) Enhancement of hepatitis C virus replication by Epstein-Barr virus-encoded nuclear antigen 1. EMBO J 18:5755–5760

Sugiura M, Imai S, Tokunaga M, Koizumi S, Uchizawa M, Okamoto K, Osato T (1996) Transcriptional analysis of Epstein-Barr virus gene expression in EBV-positive gastric carcinoma. Unique viral latency in the tumour cells. Br J Cancer 74:625–631

Sylla BS, Hung SC, Davidson DM, Hatzivassiliou E, Malinin NL, Wallach D, Gilmore TD, Kieff E, Mosialos G (1998) Epstein-Barr virus-transforming protein latent infection membrane protein 1 activates transcription factor NF-κB through a pathway that includes the NF-κB-inducing kinase and the IκB kinases IKKα and IKKβ. Proc Natl Acad Sci USA 95:10106–10111

Tajima M, Komuro M, Okinaga K (1998) Establishment of Epstein-Barr virus-positive human gastric epithelial cell lines. Jpn J Cancer Res 89:262–268

Takada K, Ono Y (1989) Synchronous and sequential activation of latently infected Epstein-Barr virus genomes. J Virol 63:445–449

Takano Y, Kato Y, Saegusa M, Mori S, Shiota M, Masuda M, Mikami T, Okayasu I (1999) The role of the Epstein-Barr virus in the oncogenesis of EBV(+) gastric carcinomas. Virchows Arch 434:17–22

Takasaka N, Tajima M, Okinaga K, Satoh Y, Hoshikawa Y, Katsumoto T, Kurata T, Sairenji T (1998) Productive infection of Epstein-Barr virus (EBV) in EBV-genome-positive epithelial cell lines (GT38 and GT39) derived from gastric tissues. Virology 247:152–159

Takeuchi H, Kobayashi R, Hasegawa M, Hirai K (1997) Detection of latent Epstein-Barr virus (EBV) DNA in paraffin sections of nasopharyngeal carcinomas expressing no EBV-encoded small RNAs using in situ PCR. Arch Virol 142:1743–1756

Tao Q, Srivastava G, Chan ACL, Ho FCS (1995) Epstein-Barr-virus-infected nasopharyngeal intraepithelial lymphocytes. Lancet 345:1309–1310

Tashiro Y, Arikawa J, Itoh T, Tokunaga M (1998) Clinico-pathological findings of Epstein-Barr virus-related gastric cancer. In: Osato T, Takada K, Tokunaga M (eds) Epstein-Barr virus and human cancer. Gann Monograph in Cancer Res No. 45, Karger, Basel Tokyo, pp 87–97

Timens W, Boes A, Vos H, Poppema S (1991) Tissue distribution of the C3d/EBV-receptor: CD21 monoclonal antibodies reactive with a variety of epithelial cells, medullary thymocytes, and peripheral T-cells. Histochemistry 95:605–611

Tokunaga M, Land CE, Uemura Y, Tokudome T, Tanaka S, Sato E (1993) Epstein-Barr virus in gastric carcinoma. Am J Pathol 143:1250–1254

Tomei LD, Noyes I, Blocker D, Holliday J, Glaser R (1987) Phorbol ester and Epstein-Barr virus dependent transformation of normal primary human skin epithelial cells. Nature 329:73–75

Uemura Y, Tokunaga M, Arikawa J, Yamamoto N, Hamasaki Y, Tanaka S, Sato E, Land CE (1994) A unique morphology of Epstein-Barr virus-related early gastric carcinoma. Cancer Epidemiol Biomarkers Prev 3:607–611

Wang D, Liebowitz D, Kieff E (1985) An EBV membrane protein expressed in immortalized lymphocytes transforms established rodent cells. Cell 43:831–840

Wang X, Hutt-Fletcher LM (1998) Epstein-Barr virus lacking glycoprotein gp42 can bind to B cells but is not able to infect. J Virol 72:158–163

Wang X, Kenyon WJ, Li Q, Mullberg J, Hutt-Fletcher LM (1998) Epstein-Barr virus uses different complexes of glycoproteins gH and gL to infect B lymphocytes and epithelial cells. J Virol 72:5552–5558

Watanabe H, Enjoji M, Imai T (1976) Gastric carcinoma with lymphoid stroma. Its morphologic characteristics and prognostic correlations. Cancer 38:232–243

Webster-Cyriaque J, Raab-Traub N (1998) Transcription of Epstein-Barr virus latent cycle genes in oral hairy leukoplakia. Virology 248:53–65

Wei MX, Ooka T (1989) A transforming function of the BARF1 gene encoded by Epstein-Barr virus. EMBO J 8:2897–2903

Wei MX, de Turenne-Tessier M, Decaussin G, Benet G, Ooka T (1997) Establishment of a monkey kidney epithelial cell line with the BARF1 open reading frame from Epstein-Barr virus. Oncogene 14:3073–3081

Weiss LM, Movahed LA, Butler AE, Swanson SA, Friesson HF, Cooper PH, Colby TV (1989) Analysis of lymphoepithelioma and lymphoepithelioma-like carcinomas for Epstein-Barr virus by in situ hybridisation. Am J Surg Pathol 13:625–631

Westphal EM, Mauser A, Swenson J, Davis MG, Talarico CL, Kenney SC (1999) Induction of lytic Epstein-Barr virus (EBV) infection in EBV-associated malignancies using adenovirus vectors in vitro and in vivo. Cancer Res 59:1485–1491

Yanai H, Takada K, Shimizu N, Mizugaki Y, Tada M, Okita K (1997) Epstein-Barr virus infection in non-carcinomatous gastric epithelium. J Pathol 183:293–298

Yoshiyama H, Imai S, Shimizu N, Takada K (1997) Epstein-Barr virus infection of human gastric carcinoma cells: Implication of the existence of a new virus receptor different from CD21. J Virol 71:5688–5691

Yoshizaki T, Sato H, Furukawa M, Pagano JS (1998) The expression of matrix metalloproteinase 9 is enhanced by Epstein-Barr virus latent membrane protein 1. Proc Natl Acad Sci USA 95:3621–3626

Young LS, Sixbey JW, Clark D, Rickinson AB (1986) Epstein-Barr virus receptor on human pharyngeal epithelia. Lancet i:240–242

Young LS, Dawson CW, Clark D, Rupani H, Busson P, Tursz T, Johnson A, Rickinson AB (1988) Epstein-Barr virus gene expression in nasopharyngeal carcinoma. J Gen Virol 69:1051–1065

Young LS, Dawson CW, Brown KW, Rickinson AB (1989) Identification of a human epithelial cell surface protein sharing an epitope with the C3d/Epstein-Barr virus receptor molecule of B lymphocytes. Int J Cancer 43:786–794

Zeng Y, Miao XC, Jaio B, Li HY, Ni HY, Ito Y (1984) Epstein-Barr virus activation in Raji cells with ether extracts of soil from different areas in China. Cancer Lett 23:53–59

Zong YS, Zhang CQ, Zhang F, Ruan JB, Chen MY, Feng KT, et al. (1993) Infiltrating lymphocytes and accessory cells in nasopharyngeal carcinoma. Jpn J Cancer Res 84:900–905

Characterization of EBV-Infected Epithelial Cell Lines from Gastric Cancer-Bearing Tissues

T. Sairenji[1,6], M. Tajima[2], M. Kanamori[1], N. Takasaka[1], X. Gao[1], M. Murakami[1], K. Okinaga[3], Y. Satoh[1], Y. Hoshikawa[1], H. Ito[5], Y. Miyazawa[4], and T. Kurata[6]

1	Introduction	185
2	"Spontaneous" Establishment of EBV-Positive Gastric Epithelial Cell Lines	186
2.1	GT38 and GT39 Cell Lines	187
2.2	Chromosomal Analysis	187
3	EBV Infection in GT38 and GT39 Cell Lines	187
3.1	Latency Type	187
3.2	Genotype of EBV	189
3.3	Spontaneous EBV Reactivation	189
3.4	EBV Reactivation with TPA	189
3.5	Production of Infectious Virus	190
4	Nitric Oxide Downregulates EBV Reactivation	190
4.1	iNOS Gene Expression in GT38 and GT39 Cell Lines	191
4.2	EBV Reactivation by a Competitive Inhibitor of NOS, L-NMMA	191
4.3	Effect of TPA on EBV Reactivation and iNOS mRNA Expression	191
4.4	Effect of L-NMMA and SNAP, a NO Donor on TPA-Induced EBV Reactivation	192
5	Loss of EBV DNA Copy in GT38 and GT39 Cell Lines	192
6	Oncogenic Potential	194
6.1	Cell Growth in Soft Agar	194
6.2	Tumorigenesis in SCID Mice	194
7	Conclusions	194
	References	196

1 Introduction

The long-term goal of our study is to explore how Epstein-Barr virus (EBV) associates with the pathogenesis of gastric carcinoma. The first goal is to establish

---

[1] Department of Biosignaling, School of Life Science, Faculty of Medicine, Tottori University, Yonago 683-8503, Japan
[2] Central Clinical Laboratory, Teikyo University School of Medicine, Tokyo 173-0003, Japan
[3] Department of Second Surgery, Teikyo University School of Medicine, Tokyo 173-0003, Japan
[4] Department of Clinical Pathology, Teikyo University School of Medicine, Tokyo 173-0003, Japan
[5] First Department of Pathology, Faculty of Medicine, Tottori University, Yonago 683-8503, Japan
[6] Department of Pathology, National Institute of Infectious Diseases, Tokyo 162-0052, Japan

EBV-positive epithelial cell lines from EBV-infected gastric carcinoma tissues and to characterize the cell lines and EBV infection in the cells. EBV, a ubiquitous human herpesvirus with oncogenic potential, is predominantly associated with the infection of two target tissues in vivo: (1) B lymphocytes, where the infection is largely nonproductive, and (2) the epithelium, in which virus replication occurs (RICKINSON and KIEFF 1996). Both target tissues are susceptible to EBV-associated malignant change, leading to tumors of B-cell origin, such as Burkitt's lymphoma, or of epithelial-cell origin, such as nasopharyngeal carcinoma (NPC) (PATHMANATHAN et al. 1995). Recently, EBV has also emerged as an etiologic agent implicated in gastric carcinoma (MIN et al. 1991; SHIBATA et al. 1991; SHIBATA and WEISS 1992; TOKUNAGA et al. 1993a,b; FUKAYAMA et al. 1994; IMAI et al. 1994; OHFUJI et al. 1996; IWASAKI et al. 1998). EBV has been found in most cases of rare gastric lymphoepithelioma-like carcinomas (MIN et al. 1991; SHIBATA et al. 1991; OHFUJI et al. 1996) and a small but significant proportion of common gastric adenocarcinomas (SHIBATA and WEISS 1992; TOKUNAGA et al. 1993a,b). EBV infection was found in approximately 7% of Japanese gastric carcinomas (TOKUNAGA et al. 1993a,b; FUKAYAMA et al. 1994; IMAI et al. 1994). The world distribution of gastric carcinomas with EBV infection is shown from 4% to 18% of total gastric carcinoma in different countries (TASHIRO et al. 1998).

The interaction of EBV with epithelial cells and the role of the virus in the etiology of EBV-associated NPC and gastric carcinoma are poorly understood. The major reason is that the EBV-infected epithelial cell line which is available for the study in vitro had not been established, while numerous EBV-infected B-cell lines have been established and used for the study on the bases of cellular and molecular biology to the etiological role of EBV. Recently Cheung et al. reported the establishment of NPC cell line (C666-1) consistently harboring EBV (CHEUNG et al. 1999). To understand the role of EBV infection in the epithelial cells, we have been trying to establish EBV-positive epithelial cell lines from EBV-infected gastric carcinomas. Two EBV-infected epithelial cell lines, GT38 and GT39 have been established from gastric tissues bearing carcinoma by Tajima, one of the co-authors of this paper (TAJIMA et al. 1998). This review describes, according to the character of cell lines, EBV latency, reactivation, effect of 12-0-tetradecanoylphorbol 13-acetate (TPA) on the cell growth and EBV reactivation, regulation of EBV latency by nitric oxide (NO) and tumorigenesis of the cell lines in severe combined immunodeficient (SCID) mouse.

2 "Spontaneous" Establishment of EBV-Positive Gastric Epithelial Cell Lines

There was no report to the spontaneous establishment of EBV-positive epithelial cell lines from gastric carcinomas, while many epithelial cell lines have been established from them (SEKIGUCHI and SUZUKI 1994).

2.1 GT38 and GT39 Cell Lines

GT38 and GT39 cell lines were established spontaneously from gastric tissues of two male patients bearing adenocarcinoma. These cell lines were derived from noncancerous portions, but not from the cancerous portions in each tissue after lengthy cultivation in vitro. The cell lines grew as adherent cells and were epithelioid cells in morphology. They were positive for cytokeratin, an epithelial marker, but were negative for lymphocyte-related markers such as CD3 and CD19. We tested the clonality of cell lines. It has been reported that all EBV-carrying gastric carcinoma cells have individual single clonotypes of EBV DNA (FUKAYAMA et al. 1994; IMAI et al. 1994; IWASAKI et al. 1998). The clonality of EBV genomes indicated, by extension, cellular monoclonality (RAAB-TRAUB and FLYNN 1986). Hybridization with a *Xho*I fragment identified a single EBV DNA restriction enzyme fragment in each of the cell lines of GT38 and GT39. It implied that these cell lines may be clonal proliferations of a single EBV-infected cell in agreement with Raab-Traub and Flynn.

2.2 Chromosomal Analysis

Chromosomal abnormalities are observed in all types of gastric carcinoma cell lines without specific patterns (SEKIGUCHI and SUZUKI 1994). Preliminary karyotypic analysis of GT38 and GT39 cells demonstrated a modal number 44–49 (Fig. 1). The 3n nor 4n was not seen. This is interesting; however, we could not explain the exact abnormalities by the obstruction of enormous chromosomal aberrations in the karyotypes. It remains to be studied why they had such aberrations and whether the aberrations are related to the persistent and productive EBV infection in the cell lines.

3 EBV Infection in GT38 and GT39 Cell Lines

3.1 Latency Type

Different patterns of viral gene expression have been identified in EBV-carrying cells (RICKINSON and KIEFF 1996). Expression of all the latent viral genes detected in lymphoblastoid cell lines is a feature of the immunoblastic lymphomas of immunosuppressed patients. This form of latency is commonly defined as latency III. The EBV-nuclear antigen (EBNA) was detected by immunofluorescence in all GT38 and GT39 cells (TAKASAKA et al. 1998). The EBNA1, EBNA2, and latent membrane protein (LMP)1 were detected by immunoblotting in both cell lines. The EBV infection on these cell lines was referred to latency III classification by RICKINSON and KIEFF (1996). The latency III is also determined by the analysis of

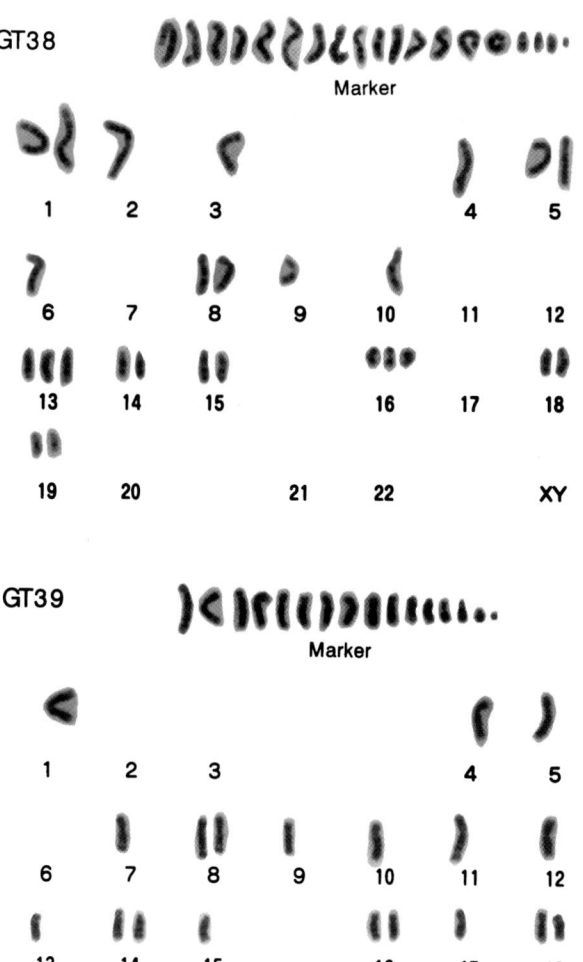

Fig. 1. Karyotype of G-banded metaphase chromosomes of GT38 and GT39 cells

transcriptional promoters Cp and Wp for EBNA genes. Transcripts initiated from Cp and Wp were detected in GT38 and GT39 cell lines (MURAKAMI et al. 2000). The latency III pattern in these cell lines is very interesting, because it is characteristic of the immunoblastic lymphomas of B-cell type, and of EBV-transformed lymphoblastoid cell lines (LCL) from nonmalignant B cells (KLEIN 1998). The EBV-associated gastric carcinoma occurs in immunocompetent individuals who can eliminate latency III cells by the cytotoxic T-cell mediated immunity. The latency type of GT38 and GT39 was clearly different from the latency I of EBV-associated gastric carcinoma cells (FUKAYAMA et al. 1994; IMAI et al. 1994; IWASAKI et al. 1998) and of the gastric carcinoma cell lines infected with EBV in vitro

(YOSHIYAMA et al. 1997; IMAI et al. 1998) and of the immortalized primary gastric epithelial cells with EBV infection in vitro (NISHIKAWA et al. 1999).

3.2 Genotype of EBV

The EBV genotype determined by differences in the sequence of the EBNA2 region was type 1 in both GT38 and GT39 cells (TAKASAKA et al. 1998). The type 1 virus was observed in EBV-associated gastric carcinoma (FUKAYAMA et al. 1994) and was the predominant type in throat washing of normal donors in Japan (KUNIMOTO et al. 1992; IKUTA et al. 2000).

3.3 Spontaneous EBV Reactivation

These cell lines had spontaneous EBV reactivation in the short-term passaged cells after establishment (TAKASAKA et al. 1998). The EBV immediate-early BZLF1 replication activator (ZEBRA) protein, the early antigen diffuse component (EA-D), and the EBV major envelope protein (gp350/220) were detected usually in 3%–5% of both cell lines. The infectious virus was detected in the culture fluids. However, the spontaneous EBV reactivation has disappeared in the cell lines for long-term passage in vitro as mentioned below.

3.4 EBV Reactivation with TPA

TPA or n-butyrate can induce EBV reactivation (ZUR HAUSEN et al. 1978; LUKA et al. 1979). To analyze the reactivation, GT38 and GT39 cells were treated with these chemicals (TAKASAKA et al. 1998). TPA and/or n-butyrate clearly induced EBV reactivation in GT38 and GT39 cells, although TPA for EBV reactivation was more efficient than n-butyrate. This was different from B-cell lines in which the EBV reactivation was induced more efficiently by n-butyrate than TPA (SAIRENJI and HINUMA 1980). The differences between TPA and n-butyrate may reflect the mechanisms leading to the disruption of EBV latency in B cells and epithelial cells (ZALANI et al. 1996). TPA is a tumor promoter that promotes cell growth (CASTAGNA et al. 1982) and reactivates EBV in the cells. We have characterized the effect of TPA on cell growth and EBV reactivation for GT38, GT39, and a Burkitt's lymphoma cell line Raji in detail (KANAMORI et al. 2000). The mode of actions of TPA in GT38 and GT39 cells was clearly distinguished from Raji in terms of the effect of cell growth and EBV reactivation. The inhibition of cell growth of GT38, GT39, and Raji was induced with high concentrations (0.5–20 ng/ml) of TPA, dose-dependently in liquid medium. These cell lines formed colonies in soft agar and were analyzed for the effect of TPA in the condition. The colony formation of GT38 and GT39 cells was enhanced with low concentrations (0.01–0.05 ng/ml) of TPA, but was inhibited with the high concentrations (>0.5 ng/ml). On the other hand, the colony formation of Raji cells was enhanced with the concentrations (0.01–0.5 ng/ml); however, it was inhibited with 5 ng/ml of TPA. The latent EBV was reactivated with the high concentrations of TPA as shown by the expression of ZEBRA. The effective concentrations of TPA for cell growth inhibition and EBV reactivation were much higher in Raji cells than GT38 and GT39 cells. These results demonstrated that TPA affects differentially for the stimulation and inhibition of cell growth, and EBV reactivation is not induced by the stimulation of cell growth but by the inhibition of cell growth.

3.5 Production of Infectious Virus

The virus particles were found in the nuclei of the TPA-treated GT38 cells by an electron microscope (TAKASAKA et al. 1998). We analyzed the production of infectious virus. The production of the transforming virus was demonstrated by measuring the ability of the virus to transform cord lymphocytes in both cell lines. The titers were $10^{4.5}$ and $10^{4.0}$ 50% transforming doses (TD_{50})/ml in GT38 and GT39 cells, respectively. The titer was increased 5–10 times in the cells treated with TPA or TPA plus n-butyrate. The production of EA-inducing virus was assayed by measuring the ability of EA to superinfect Raji cells. The titers were estimated at $10^{4.6}$ and $10^{4.3}$ EA-inducing units/ml in the culture fluids of GT38 and GT39 cells, respectively. The EBV with transforming and EA-inducing ability was observed in NPC-KT cells of a NPC hybrid cell line by fusing primary NPC epithelial cells with an epithelial cell line derived from a human adenoid tissue (TAKIMOTO et al. 1985). Zhang et al. reported that only an EA inducing virus was rescued from a NPC tumor cell line by treatment with iododeoxyuridine (1990). No cell line that produces a transforming and EA-inducing virus has been reported in B-cell lines, while the P3HR-1 cell line produces EA-inducing virus, and the parental Jijoye line is known to produce a transforming virus (SAIRENJI and HINUMA 1980). The P3HR-1 virus is probably uniquely efficient in inducing lytic infection in Raji cells because of the presence of defective virus in P3HR-1 virus (COUNTRYMAN and MILLER 1985). The strains of EBV on GT38 and GT39 cells appear to be unique in that they have transforming and EA-inducing properties. The infection of GT38 and GT39 may represent the EBV infection of gastric epithelial cells, because the other gastric epithelial cell line, GTC-4, had also the same type of EBV production (data not shown). These viruses could serve as new prototype viruses derived from gastric epithelial cells, and as in vitro models for studying the system of EBV infection for epithelial cells.

4 Nitric Oxide Downregulates EBV Reactivation

The spontaneous EBV reactivation has been lost in the GT38 and GT39 cell lines passaged in vitro long-term as mentioned above. We wondered why the spontaneous reactivation disappeared in these cell lines and hypothesized that a cellular factor(s) may regulate the spontaneous EBV reactivation. NO is an important signaling molecule regulating a wide array of biological activities in neural, vascular, and immune cells (NATHAN 1992). INF-γ-induced NO production by macrophages has been implicated in resistance to intracellular pathogens, such as parasites, fungi, mycobacteria, and viruses, such as vaccinia virus (HARRIS et al. 1995), Friend leukemia virus (AKARID et al. 1995), herpes simplex virus type 1 (CROEN 1993), and murine hepatitis virus type 3 (POPE et al. 1998). NO has been shown to inhibit viral reactivation in EBV-infected B cells (MANNICK et al. 1994) and DNA replication in EBV superinfected Raji cells (KAWANISHI 1995). These observations lead us to postulate that NO might be a controlling factor in EBV reactivation.

4.1 iNOS Gene Expression in GT38 and GT39 Cell Lines

In this study we found that a constitutive, low level of inducible NO synthase (iNOS) messenger (m)RNA was expressed in GT38 and GT39 cell lines, by analysis with the reverse transcriptase polymerase chain reaction (RT-PCR) (GAO et al. 1999). We detected iNOS mRNA in both of them, although the iNOS mRNA expression level was much higher in GT38 cells than that in GT39 cells. The results are consistent with our idea that spontaneous EBV reactivation is controlled by endogenous NO.

4.2 EBV Reactivation by a Competitive Inhibitor of NOS, L-NMMA

We analyzed the effects of endogenous NO on EBV reactivation in GT38 and GT39 cells. N^G-monomethyl-L-arginine (L-NMMA), a specific competitive inhibitor of NOS, was used to examine the reactivation of EBV in GT38 and GT39 cells. The ZEBRA was induced in a time-dependent manner after exposure to L-NMMA. These results showed that latent EBV can be reactivated in these cells by L-NMMA, which blocks NO synthesis, supporting the conclusion that endogenous NO can have an inhibiting effect on EBV reactivation in these cells.

4.3 Effect of TPA on EBV Reactivation and iNOS mRNA Expression

The BZLF1 immediate-early transactivator initiates the switch between latent and productive infection in B cells (COUNTRYMAN and MILLER 1985; CHEVALLIER-GRECO et al. 1986; TAKADA et al. 1986). We examined the induction of BZLF1 gene by TPA in the GT38 and GT39 cells. The mRNAs of BZLF1 and BRLF1 genes were expressed in a time-dependent manner. To investigate whether iNOS mRNA expression could be affected by TPA, we further examined the levels of iNOS mRNA in the cells treated with TPA. TPA led to a marked reduction of iNOS mRNA expression. The reduction of iNOS mRNA coincided with the appearance of BZLF1 and BRLF1 transcripts. These results suggested that TPA induced EBV reactivation through inhibiting iNOS activation.

4.4 Effect of L-NMMA and SNAP, a NO Donor on TPA-Induced EBV Reactivation

We analyzed whether L-NMMA could enhance TPA-induced EBV reactivation. The TPA-induced BZLF1 and BRLF1 levels were increased more by treatment with L-NMMA. L-NMMA increased TPA-induced ZEBRA expression, indicating that the interaction between L-NMMA and TPA in EBV reactivation is probably through a common pathway. We examined whether exogenous NO also has a similar effect in inhibiting EBV reactivation. The effect of NO donor S-nitroso-N-acetylpenicillamine (SNAP) on EBV reactivation was tested. SNAP generates NO in an aqueous solution, resulting in the formation of peroxynitrite (GILAD et al. 1997). SNAP inhibited the TPA-induced BZLF1 and BRLF1 expression. These results demonstrated that reactivation of EBV in GT38 and GT39 cells could be inhibited by exogenous NO.

5 Loss of EBV DNA Copy in GT38 and GT39 Cell Lines

We have reported that GT38 and GT39 cell lines were EBV producers expressing lytic antigens and producing infectious virus spontaneously. However, they have been changed to nonproducer cells in the long-term passage in vitro. We demonstrated that the EBV reactivation could be suppressed by endogenous NO in GT38 and GT39 cells. We hypothesized that the EBV genome number in the cells may be reduced by the inhibition of EBV reactivation. It has been reported that EBV DNA was lost from NPC cell lines HNE-1 and HONE-1 (GLASER et al. 1989; YAO 1990) and from a Burkitt's lymphoma Akata cell line (SHIMIZU et al. 1994). The copy number of EBV DNA was analyzed quantitatively by fluorescent in situ hybridization (FISH) using EBV DNA BamHI-W fragments (M. Kanamori, manuscript in preparation). The short-term cultured cells, after their establishment, had more EBV DNA copies compared with the long-term cultured cells for more than 2 years after establishment. EBV DNA-negative cells were observed in a small proportion of the long-term cultured cells (Fig. 2), but not in the short-term cultured cells. These results demonstrated that EBV DNA is not stably maintained in the cells and is lost spontaneously in the long-term-passaged cells.

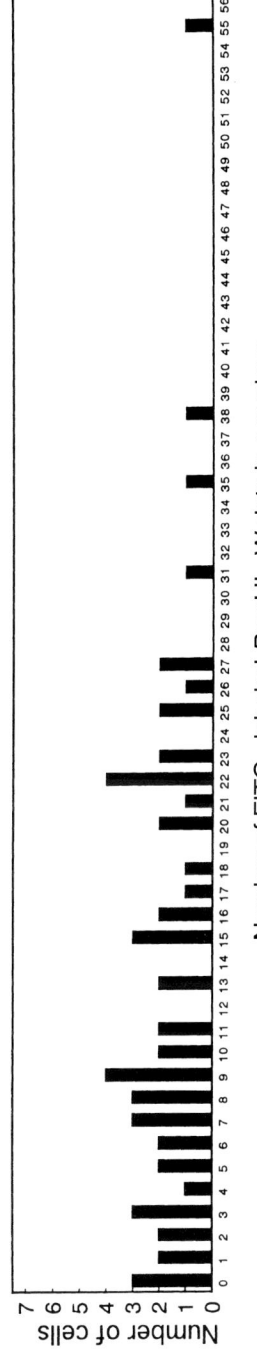

Fig. 2. EBV DNA copy numbers in GT39 cells. The FITC-labeled BamHI-W dots were counted for 50–100 interphase nuclei followed by FISH. One dot equals one copy of EBV DNA

6 Oncogenic Potential

6.1 Cell Growth in Soft Agar

One of the markers of tumor cells is anchorage-independent growth in soft agar (MacPherson and Montagnier 1964). The growth of GT38 and GT39 cells was tested by a colony formation procedure in soft agar. The colony formation was observed in these cell lines (Kanamori et al. 2000). The colony-forming efficiencies of GT38 and GT39 cells were 13.8% and 7.3%, respectively, under the same growth condition. This result indicated that GT38 and GT39 cell lines have the character of tumor cells.

6.2 Tumorigenesis in SCID Mice

Most cultured cells of stomach cancer origin are tumorigenic in athymic nude mice (Sekiguchi and Suzuki 1994). To study the tumorigenesis of GT38 and GT39, we inoculated these cells under the skin of severe combined immunodeficiency (SCID) mice. The development of tumors in SCID mice was observed approximately 2 months after the inoculation of GT39 or GT38 cells (Murakami et al. 2000). Both tumors were diagnosed as undifferentiated carcinomas. EBV-encoded small RNA was detected in the tumor cells by in situ hybridization. These tumors were solid carcinoma and were clearly different from EBV-positive B-cell lymphoma in the SCID mouse. This is the first evidence of the tumorigenesis of EBV-positive epithelial cell lines derived from gastric tissues in the SCID mouse, while there have been many discoveries on EBV-infected B cells in SCID mice (Rowe et al. 1991; Katano et al. 1996). However, we do not know the exact origin of GT38 and GT39 cell lines because they have been established from noncancerous portions of gastric tissues bearing carcinoma and after lengthy cultivation in vitro (Tajima et al. 1998). Recently, it has been reported that a transplantable EBV-associated gastric carcinoma was propagated in SCID mice (Iwasaki et al. 1998). The EBV-infected tumor tissues were trimmed and inoculated subcutaneously into the backs of SCID mice. The tumor grew in the SCID mice. It was a poorly differentiated adenocarcinoma and transplantable from SCID mouse to SCID mouse. The pattern of latency gene expression of EBV was the same as the original gastric carcinoma. However, the cell line could not grow in vitro (M. Fukayama, personal communication).

7 Conclusions

We characterized two EBV-positive epithelial cell lines, GT38 and GT39, established from gastric tissues bearing carcinomas. These cell lines had intensive chromosome abnormality and character of tumor cells as shown by the colony-forming ability in soft agar and tumorigenesis in SCID mice. However, we question the origin of these cell lines, because they were not established from cancerous portions but from noncancerous portions after lengthy cultivation in vitro. Also, the EBV latency was not type I, which is seen in EBV-infected gastric carcinoma cells, but type III, which is common in LCL. These questions will be cleared by the establishment of real EBV-infected gastric carcinoma cell lines which have the same genetic markers or tumor-associated proteins, such as alpha-fetoprotein and carcinoembryonic antigen of gastric carcinoma. In spite of such questions, these cell lines are useful to understand the EBV infection on gastric epithelial cell lines. We observed the spontaneous EBV reactivation and a lot of production of EA-inducing virus on GT38 and GT39 cells cultured for the short-term after establishment of the cell lines, although the spontaneous reactivation disappeared in the cells passaged for long-term in vitro. The EA-inducing virus may be associated with the spontaneous EBV reactivation and/or cell transformation in vitro. Therefore, we have been trying to isolate the virus.

We demonstrated that all epithelial cell lines GT38, GT39, and GTC-4 have high sensitivity to TPA. The cell-growth inhibition and EBV reactivation in the epithelial cell lines were induced with lower concentrations of TPA than those of B-cell lines. It indicates the difference of intracellular signaling pathways between epithelial cell lines and B-cell lines. It may be related that EBV replication occurs in some epithelial cells; although, the infection in B cells is largely nonproductive. Our preliminary data show that the productive infection was clearly detected in EBV-infected gastric carcinomas (Y. Hoshikawa et al., manuscript in preparation).

We found that constitutive iNOS mRNA expression in GT38 and GT39 cells can downregulate EBV reactivation in the cells. The iNOS mRNA expression paralleled the reduction of spontaneous EBV reactivation in these cell lines. Now we are trying to get the EBV genome free clones in order to understand the biological role of EBV in the epithelial cells. The role of NO on EBV reactivation and loss of EBV genome should be expanded in EBV-infected gastric carcinoma cells.

We have been encouraged to establish more EBV-infected epithelial cell lines from gastric carcinoma tissues by the establishment of GT38 and GT39 cell lines.

Acknowledgements. This work was supported by Grants-in-Aid Scientific Research on Priority Areas from the Ministry of Education, Science, Sports and Culture of Japan; the Japan Health Science Foundation; the Second-Term Comprehensive Ten-Year Strategy for Cancer Control from the Ministry of Health and Welfare of Japan. We thank Drs. Nanao Kamada and Tomoko Takahashi for chromosomal analysis. We thank Ms. Sayuri Matsumoto for typing and Ms. Tomoko Sairenji for reading this paper.

References

Akarid K, Sinet M, Desforges B, Gougerot-Pocidalo MA (1995) Inhibitory effect of nitric oxide on the replication of a murine retrovirus in vitro and in vivo. J Virol 69:7001–7005

Castagna M, Takai Y, Kaibuchi K, Sano K, Kikkawa U, Nishizuka Y (1982) Direct activation of calcium-activated, phospholipid-dependent protein kinase by tumor-promoting phorbol esters. J Biol Chem 257:7847–7851

Cheung ST, Huang DP, Hui AB, Lo KW, Ko CW, Tsang YS, Wong N, Whiteney BM, Lee JCK (1999) Nasopharyngeal carcinoma cell line (C666-1) consistently harbouring Epstein-Barr virus. Int J Cancer 83:121–126

Chevallier-Greco A, Manet E, Chavrier P, Mosnier C, Daillie J, Sergeant A (1986) Both Epstein-Barr virus (EBV)-encoded trans-acting factors, EB1 and EB2, are required to activate transcription from an EBV early promoter. Embo J 5:3243–3249

Countryman J, Miller G (1985) Activation of expression of latent Epstein-Barr herpesvirus after gene transfer with a small cloned subfragment of heterogeneous viral DNA. Proc Natl Acad Sci USA 82:4085–4089

Croen KD (1993) Evidence for antiviral effect of nitric oxide: Inhibition of herpes simplex virus type 1 replication. J Clin Invest 91:2446–2452

Fukayama M, Hayashi Y, Iwasaki Y, Chong J, Ooba T, Takizawa T, Koike M, Mizutani S, Miyaki M, Hirai K (1994) Epstein-Barr virus-associated gastric carcinoma and Epstein-Barr virus infection of the stomach. Lab Invest 71:73–81

Gao X, Tajima M, Sairenji T (1999) Nitric oxide down-regulates Epstein-Barr virus reactivation in epithelial cell lines. Virology 258:375–381

Gilad E, Zingarelli B, Salzman AL, Szabo C (1997) Protection by inhibition of poly (ADP-ribose) synthetase against oxidant injury in cardiac myoblasts in vitro. J Mol Cell Cardiol 29:2585–2597

Glaser R, Zhang HY, Yao KT, Zhu HC, Wang FX, Li GY, Wen DS, Li YP (1989) Two epithelial tumor cell lines (HNE-1 and HONE-1) latently infected with Epstein-Barr virus that were derived from nasopharyngeal carcinomas. Proc Natl Acad Sci USA 86:9524–9528

Harris N, Buller RM, Karupiah G (1995) Gamma interferon-induced, nitric oxide-mediated inhibition of vaccinia virus replication. J Virol 69:910–915

zur Hausen H, O'Neill FJ, Freese UK, Hecker E (1978) Persisting oncogenic herpesvirus induced by the tumour promotor TPA. Nature 272:373–375

Ikuta K, Satoh Y, Hoshikawa Y, Sairenji T (2000) Detection of Epstein-Barr virus in salivas and throat washings in healthy adults and children. Microbes and Infection 2:115–120

Imai S, Koizumi S, Sugiura M, Tokunaga M, Uemura Y, Yamamoto N, Tanaka S, Sato E, Osato T (1994) Gastric carcinoma: monoclonal epithelial malignant cells expressing Epstein-Barr virus latent infection protein. Proc Natl Acad Sci USA 91:9131–9135

Imai S, Nishikawa J, Takada K (1998) Cell-to-cell contact as an efficient mode of Epstein-Barr virus infection of diverse human epithelial cells. J Virol 72:4371–4378

Iwasaki Y, Chong JM, Hayashi Y, Ikeno R, Arai K, Kitamura M, Koike M, Hirai K, Fukayama M (1998) Establishment and characterization of a human Epstein-Barr virus-associated gastric carcinoma in SCID mice. J Virol 72:8321–8326

Kanamori M, Tajima M, Satoh Y, Hoshikawa Y, Miyazawa Y, Okinaga K, Kurata T, Sairenji T (2000) Differential effect of TPA on cell growth and Epstein-Barr virus reactivation in epithelial cell lines derived from gastric tissues and B cell line Raji. Virus Genes 20:117–125

Katano H, Morishita Y, Cui LX, Watanabe T, Hirai K, Mori S (1996) Expression of latent membrane protein 1 in clinically isolated cases and animal models of AIDS-associated non-Hodgkin's lymphomas. Pathol Int 46:568–574

Kawanishi M (1995) Nitric oxide inhibits Epstein-Barr virus DNA replication and activation of latent EBV. Intervirology 38:206–213

Klein G (1998) EBV and B cell lymphomas. In: Medveczky PG, Friedman H, Bendinelli M (eds) Herpesvirus and immunity. Plenum Press, New York and London

Kunimoto M, Tamura S, Tabata T, Yoshie O (1992) One-step typing of Epstein-Barr virus by polymerase chain reaction: predominance of type 1 virus in Japan. J Gen Virol 73:455–461

Luka J, Kallin B, Klein G (1979) Induction of the Epstein-Barr virus (EBV) cycle in latently infected cells by n-butyrate. Virology 94:228–231

MacPherson I, Montagnier L (1964) Agar suspension culture for the selective assay of cells transformed by polyoma virus. Virology 23:291–297
Mannick JB, Asano K, Izumi K, Kieff E, Stamler JS (1994) Nitric oxide produced by human B lymphocytes inhibits apoptosis and Epstein-Barr virus reactivation. Cell 79:1137–1146
Min KW, Holmquist S, Peiper SC, O'Leary TJ (1991) Poorly differentiated adenocarcinoma with lymphoid stroma (lymphoepithelioma-like carcinomas) of the stomach: Report of three cases with Epstein-Barr virus genome demonstrated by the polymerase chain reaction. Am J Clin Pathol 96: 219–227
Murakami M, Hoshikawa Y, Satoh Y, Ito H, Tajima M, Okinaga K, Miyazawa Y, Kurata T, Sairenji T (2000) Tumorigenesis of Epstein-Barr virus-positive epithelial cell lines derived from gastric tissues in the SCID mouse. Virology 277:20–26
Nathan C (1992) Nitric oxide as a secretory product of mammalian cells. FASEB J 6:3051–3064
Nishikawa J, Imai S, Oda T, Kojima T, Okita K, Takada K (1999) Epstein-Barr virus promotes epithelial cell growth in the absence of EBNA2 and LMP1 expression. J Virol 73:1286–1292
Ohfuji S, Osaki M, Tsujitani S, Ikeguchi M, Sairenji T, Ito H (1996) Low frequency of apoptosis in Epstein-Barr virus-associated gastric carcinoma with lymphoid stroma. Int J Cancer 68:710–715
Pathmanathan R, Prasad U, Sadler R, Flynn K, Raab-Traub N (1995) Clonal proliferations of cells infected with Epstein-Barr virus in preinvasive lesions related to nasopharyngeal carcinoma. N Engl J Med 333:693–698
Pope M, Marsden PA, Cole E, Sloan S, Fung LS, Ning Q, Ding JW, Leibowiz JL, Phillips MJ, Levey GA (1998) Resistance to murine hepatitis virus strain 3 is dependent on production of nitric oxide. J Virol 72:7084–7090
Raab-Traub N, Flynn K (1986) The structure of the termini of the Epstein-Barr virus as a marker of clonal cellular proliferation. Cell 47:883–889
Rickinson AB, Kieff E (1996) Epstein-Barr virus. In: Fields BN, Knipe DM, Howley PM, et al. (eds) Fields Virology, 3rd edn. Raven Press, New York
Rowe M, Young LS, Crocker J, Stokes H, Henderson S, Rickinson AB (1991) Epstein-Barr virus (EBV)-associated lymphoproliferative disease in the SCID mouse model: Implications for the pathogenesis of EBV-positive lymphomas in man. J Exp Med 173:147–158
Sairenji T, Hinuma Y (1980) Re-evaluation of a transforming strain of Epstein-Barr virus from the Burkitt lymphoma cell line, Jijoye. Int J Cancer 26:337–342
Sekiguchi M, Suzuki T (1994) Gastric tumor cell lines. In: Hay RJ, Park J-G, Gazdar AF (eds) Atlas of human tumor cell lines. Academic Press, London
Shibata D, Tokunaga M, Uemura Y, Sato E, Tanaka S, Weiss LM (1991) Association of Epstein-Barr virus with undifferentiated gastric carcinomas with intense lymphoid infiltration. Lymphoepithelioma-like carcinoma. Am J Pathol 139:469–474
Shibata D, Weiss LM (1992) Epstein-Barr virus-associated gastric adenocarcinoma. Am J Pathol 140:769–774
Shimizu N, Tanabe Tochikura A, Kuroiwa Y, Takada K (1994) Isolation of Epstein-Barr virus (EBV)-negative cell clones from the EBV-positive Burkitt's lymphoma (BL) line Akata: Malignant phenotypes of BL cells are dependent on EBV. J Virol 68:6069–6073
Tajima M, Komuro M, Okinaga K (1998) Establishment of Epstein-Barr virus-positive human gastric epithelial cell lines. Jpn J Cancer Res 89:262–268
Takada K, Shimizu N, Sakuma S, Ono Y (1986) *Trans* activation of the latent Epstein-Barr virus (EBV) genome after transfection of the EBV DNA fragment. J Virol 57:1016–1022
Takasaka N, Tajima M, Okinaga K, Satoh Y, Hoshikawa Y, Katsumoto T, Kurata T, Sairenji T (1998) Productive infection of Epstein-Barr virus (EBV) in EBV-genome-positive epithelial cell lines (GT38 and GT39) derived from gastric tissues. Virology 247:152–159
Takimoto T, Ogura H, Sato H, Umeda R, Hatano M (1985) Isolation of transforming and early antigen-inducing Epstein-Barr virus from nasopharyngeal carcinoma hybrid cells (NPC-KT). J Natl Cancer Inst 74:57–60
Tashiro Y, Arikawa J, Itoh T, Tokunaga M (1998) Clinico-pathological findings of Epstein-Barr virus-related gastric cancer. In: Osato T, Takada K, Tokunaga M (eds) Epstein-Barr virus and human cancer. Monograph on cancer research No. 45, Japan Scientific Society Press, Karger, Basel Freiburg Paris London New York New Delhi Bangkok Singapore Tokyo Sydney
Tokunaga M, Land CE, Uemura Y, Tokudome T, Tanaka S, Sato E (1993a) Epstein-Barr virus in gastric carcinoma. Am J Pathol 143:1250–1254

Tokunaga M, Uemura Y, Tokudome T, Ishidate T, Masuda H, Okazaki E, Kaneko K, Naoe S, Ito M, Okamura A, Shimada A, Sato E, Land CE (1993b) Epstein-Barr virus related gastric cancer in Japan: a molecular patho-epidemiological study. Acta Pathol Jpn 43:574–581

Yao KT, Zhang HY, Zhu HC, Wang FX, Li GY, Wen DS, Li YP, Tsai CH, Glaser R (1990) Establishment and characterization of two epithelial tumor cell lines (HNE-1 and HONE-1) latently infected with Epstein-Barr virus and derived from nasopharyngeal carcinomas. Int J Cancer 45: 83–89

Yoshiyama H, Imai S, Shimizu N, Takada K (1997) Epstein-Barr virus infection of human gastric carcinoma cells: Implication of the existence of a new virus receptor different from CD21. J Virol 71:5688–5691

Zalani S, Holley Guthrie E, Kenney S (1996) Epstein-Barr viral latency is disrupted by the immediate-early BRLF1 protein through a cell-specific mechanism. Proc Natl Acad Sci USA 93:9194–9199

Zhang HY, Yao K, Zhu HC, Glaser R (1990) Expression of the Epstein-Barr virus genome in a nasopharyngeal carcinoma epithelial tumor cell line. Int J Cancer 46:944–949

IV
Animal Model and New Therapeutic Approach to Malignancy

A New Animal Model for Epstein-Barr Virus Pathogenesis

F. Wang

1	Introduction.	201
2	Lymphocryptoviruses Naturally Infecting Old World Nonhuman Primates	202
2.1	Serologic Studies	202
2.2	LCV-Infected B Cells	203
2.3	Nomenclature	203
3	Similarities Between Natural Infection with EBV and Simian LCV	204
3.1	LCV-Associated Malignancies	204
4	Molecular Comparisons Between EBV and Simian LCV	205
4.1	Hybridization Studies	205
4.2	Lytic Infection Genes	206
4.3	Latent Infection Genes	207
4.4	A Species-Restricted Block to LCV-Induced B-Cell Immortalization	207
4.5	Two Types of Rhesus LCV	208
4.6	Nonconserved and Conserved Molecular Strategies for Immune Evasion	208
5	Experimental Infection of Naïve Rhesus Macaques with Rhesus LCV	209
5.1	Identifying Rhesus LCV-Naïve Hosts	209
5.2	Experimental Rhesus LCV Inoculation	210
5.3	Acute Clinical and Laboratory Responses to Experimental Infection	210
5.4	Persistent Infection After Experimental Inoculation	211
6	Future Directions for the Rhesus LCV Animal Model	212
6.1	Molecular Pathogenesis of Acute and Persistent Infection	212
6.2	Immune Responses and Immune Evasion	213
6.3	Pathogenesis of Epithelial Cell Infection	213
6.4	LCV-Associated Tumorigenesis	214
6.5	Vaccine Development	214
7	Conclusion	215
	References	215

1 Introduction

Animal models have not been widely used for studying Epstein-Barr virus (EBV) pathogenesis. In large part this has been due to the lack of good animal model

---

Harvard Medical School, Brigham and Women's Hospital, Channing Laboratory, 181 Longwood Avenue, Boston, MA 02115, USA

systems which reproduce the many different aspects of EBV infection in humans. Many aspects of EBV pathogenesis would benefit from an authentic animal model system, including studies about infectious mononucleosis, immune responses that control acute and persistent infection, mechanisms for immune evasion and persistent infection, natural reservoirs for infection in hematologic and epithelial cell compartments, transition to hematologic and epithelial cell malignancies, and vaccine development.

Murine models using xenogeneic transplantation of human peripheral blood lymphocytes (PBLs) into severe combined immunodeficiency (SCID) mice have been used for studying EBV-associated lymphomagenesis (MOSIER 1996). However, in this system EBV-infected human cells grow uncontrolled, and the murine host is not actually infected. In addition, the reconstitution of the human immune response is not complete. EBV infection of New World primates, such as cotton-top tamarins, has been used to develop potential vaccine strategies (EPSTEIN et al. 1985). In this model, host cells are infected with EBV and can develop malignant tumors. However, the natural route of transmission is bypassed by parenteral virus injection, and in most cases there is no homeostasis established between the host immune response and viral evasion resulting in the lack of persistent viral infection. More recently, infection of nonhuman primates with lymphocryptoviruses (LCVs) closely related to EBV has been developed for study of EBV pathogenesis (MOGHADDAM et al. 1997). Recent molecular characterization of these Old World nonhuman primate LCVs has demonstrated the close and well-conserved relationship between the simian and human viruses. The similar biologic responses to LCV infection in human and nonhuman hosts suggests that this animal model system can provide an authentic animal model for studying EBV pathogenesis.

2 Lymphocryptoviruses Naturally Infecting Old World Nonhuman Primates

2.1 Serologic Studies

Studies beginning in the early 1970s recognized that Old World nonhuman primates, such as rhesus monkeys, were infected with lymphotropic herpesviruses (LCVs) closely related to EBV (FRANK et al. 1976; reviewed in ABLASHI et al. 1979). Using EBV-infected human cell lines, cross-reactive antibodies to viral capsid antigens, early antigens, and nuclear antigens could be routinely found in Old World monkeys and great apes, but not prosimians nor New World monkeys (LANDON and MALAN 1971; DUNKEL et al. 1972; KALTER et al. 1972; KALTER et al. 1973; DILLNER et al. 1987). Evidence for LCV infection has now been identified in species from all families and virtually all subfamilies of Old World primates.

2.2 LCV-Infected B Cells

Soon after the serologic studies, investigators identified that LCV-infected cell lines could be derived from malignancies as well as the peripheral blood of healthy, seropositive animals, similar to the EBV-infected cell lines derived from humans (LANDON et al. 1968; STEVENS et al. 1970; O'GARA et al. 1971; SCHABLE et al. 1974; FRANK et al. 1976; RABIN et al. 1977; RANGAN et al. 1986). LCV-infected B-cell lines have now been derived from tumors or the peripheral blood of several Old World primates including rhesus monkeys (*Macaca mulatta*; RANGAN et al. 1986; CHO et al. 1999), cynomolgus monkeys (*Macaca fascicularis*; HEBERLING et al. 1981; FEICHTINGER et al. 1990), stumptailed monkeys (*Macaca arctoides*; LAPIN et al. 1985), African green monkeys (*Cercopithecus aethiops*; BOCKER et al. 1980), baboons (*Papio* sp.; AGRBA et al. 1975; FALK et al. 1976), chimpanzees (*Pan troglodytes*; LANDON et al. 1968; GERBER et al. 1977), orangutan (*Pongo pygmaeus*; RASHEED et al. 1977; RABIN et al. 1978), gorilla (*Gorilla gorilla*; NEUBAUER et al. 1979).

These cell lines have phenotypic characteristics similar to human EBV-infected lymphoblastoid cell lines, are derived from B lymphocytes, can contain herpesvirus particles as seen by electron microscopy, and express viral antigens that can be detected by immune animal and human sera (FALK et al. 1974; FRANK et al. 1976; ABLASHI et al. 1979; DILLNER et al. 1987; LI et al. 1993). Cell-free virus derived from some cell lines can also immortalize na B lymphocytes from the autologous species as well as other primate species, further demonstrating the conserved growth transforming potential in these related viruses (FRANK et al. 1976; ABLASHI et al. 1979; RANGAN et al. 1986).

2.3 Nomenclature

Historically, these viruses have acquired various names using nonstandard nomenclature such as rhesus EBV (rhesus macaque), cyno EBV (cynomolgous macaque), herpesvirus papio (baboon), and herpesvirus pan (chimpanzee). Unfortunately, these colloquial names are often neither correct nor very specific. Epstein-Barr virus is the name applied to the LCVs found in humans and the simian LCV described subsequently, although similar to EBV, they are genetically and biologically distinct. Many Old World nonhuman primates are infected with alpha- and betaherpesviruses as well as gammaherpesviruses, so that names such as herpesvirus papio are not very specific. Official nomenclature recommended by the Study Group for the International Committee on the Taxonomy of Viruses (http://www.ncbi.nlm.nih.gov/ICTV/) uses a sequential number in order of discovery combined with the family or subfamily of the natural host, e.g., cercopithicine herpesvirus 15 for "rhesus EBV" as the 15th herpesvirus found to naturally infect species in the family Ceropithicidae. This system recognizes the unique identity of these various virus species, but fails to provide a very useful and easily recognizable name. Thus, we propose using the combination of the host species and LCV for vernacular usage, e.g., rhesus LCV, baboon LCV, and chimpanzee LCV.

3 Similarities Between Natural Infection with EBV and Simian LCV

Like humans, nearly all Old World nonhuman primates in a given species are infected with LCV by adulthood. Similarly, newborn animals are often LCV seropositive due to maternal antibody transplacental transfer. When newborn rhesus macaques are followed in captivity, the majority become seronegative by 6 months of age due to the loss of maternal antibodies. In conventional breeding colonies, 60%–70% will become LCV seropositive by 1 year of age due to natural virus infection (F. Wang, unpublished observations). If animals are hand reared, they are more likely to remain LCV seronegative (RAO et al. 2000). This indicates that simian LCV infection, like EBV, is transmitted horizontally through close contact with persistently infected hosts. Since LCVs can be detected in the oral secretions of naturally infected animals (MOGHADDAM et al. 1997), transmission is most likely via the oral exchange of virus containing saliva, as in humans.

Once infected, simian LCV establishes asymptomatic persistent infection similar to EBV infection in humans. Simian LCV establishes a persistent lytic infection in the oropharynx, which can be detected by PCR amplification of DNA from oral secretions (CHO et al. 1999). Persistent simian LCV infection in the peripheral blood mononuclear cells can be detected by the spontaneous outgrowth of LCV immortalized B-cell lines in vitro (RABIN et al. 1978) or by reverse transcriptase polymerase chain reaction (RT-PCR) amplification of the small nonpolyadenylated RNAs, or EBER homologs, expressed during latent B-cell infection (RAO et al. 2000).

3.1 LCV-Associated Malignancies

Simian LCV infection also has the potential for inducing B-cell malignancies in immunosuppressed hosts. In the late 1960s, an epidemic of lymphoma in over 30 baboons was reported at the Institute of Experimental Pathology, USSR Academy of Medical Sciences in Sukhumi (LAPIN 1974). An EBV-like agent with herpesvirus particles, a viral capsid antigen that cross-reacted with EBV-immune sera, and nucleic acid homology to human EBV was demonstrated in cell lines established from lymphomatous baboons. This agent was very closely related to baboon LCV isolated from a normal healthy baboon in the USA. The precipitating cause of the outbreak was unclear, but it was epidemiologically associated with introduction of baboons that had been inoculated with human leukemic blood. Interestingly, a primate retrovirus was later isolated from a lymphomatous baboon in the Sukhumi colony suggesting that a simian immunodeficiency virus (SIV) outbreak may have contributed to the malignant epidemic (GOLDBERG et al. 1974).

In the late 1960s, an outbreak of lymphoma in 40 rhesus monkeys was also reported at the National Center for Primate Biology at Davis, California (STEVENS et al. 1970). The tumors were histologically characterized as poorly differentiated neoplastic cells with a number of histological features resembling Burkitt's lymphoma. Epidemiologic studies showed that the outbreak was temporally associated with several unusual events at the primate center including the inoculation of human Burkitt's lymphoma tissue into several chimpanzees, a fatal outbreak of herpes simplex in owl monkeys, and the presence of splenectomized malaria-infected and treated animals.

More recent work with SIV-infected primates suggests a direct analogy with EBV-induced lymphomas in AIDS patients. FEICHTINGER et al. (1990) reported that malignant B-cell lymphomas, containing DNA which cross-hybridized with human EBV, appeared 5–15 months after SIV infection of cynomolgus macaques. Tumors formed at a very high frequency in LCV seropositive animals infected with SIV (62%; 8 of 13 LCV-seropositive animals developed tumors). These animals were naturally infected with LCV prior to experimental infection with SIV, thus the development of LCV-positive B-cell lymphomas in these SIV-infected animals is similar to the EBV-positive B-cell lymphomas which arise in adult AIDS patients.

It is less clear whether simian LCVs are associated with B-cell malignancies in hosts without overt immunosuppression, i.e., similar to the EBV association with Burkitt's lymphoma and Hodgkin's disease. It is also unclear whether simian LCV infection is associated with malignancies in the epithelial cell compartment, i.e., similar to the EBV association with nasopharyngeal and gastric carcinomas. The failure to recognize an association between simian LCV infection and these types of malignancies in nonhuman primates may be due to the relatively low frequency of disease and small captive population. However, it is interesting to note that simian LCV infection has been detected in epithelial skin and esophageal lesion in SIV-immunosuppressed animals suggesting that LCV can cause epithelial cell disease in both human and nonhuman hosts (BASKIN et al. 1995; M. Simon, personal communication).

4 Molecular Comparisons Between EBV and Simian LCV

4.1 Hybridization Studies

Hybridization studies performed in the 1980s indicated that EBV and baboon LCV DNA share ~40% homology, have similar densities (1.717–1.718), are colinear, and share a similar structural format with tandem repeats at both ends and tandem direct repeats separating long and short unique regions (HELLER et al. 1981; HELLER and KIEFF 1981). Chimpanzee, baboon, and orangutan LCV also showed 30%–50% nucleotide homology with EBV (FRANK et al. 1976; GERBER et al. 1976; HELLER et al. 1982). This level of homology is similar to that between herpes simplex type 1 and type 2 viruses (HELLER and KIEFF 1981).

4.2 Lytic Infection Genes

More recently cosmid clones from the baboon and rhesus LCV have been derived, and selected genes have been sequenced. In addition, our laboratory has initiated an effort to derive the complete nucleotide sequence for the rhesus LCV genome. From these more detailed studies, it appears that the baboon and rhesus LCV genome encode for an identical repertoire of lytic and latent infection genes. Overall, the lytic genes are more well-conserved with 50%–90% amino acid identity to the EBV homologs (Table 1). This is not surprising given the well-conserved nature of many lytic infection genes among all herpesviruses, e.g., BALF2, the major DNA binding protein. Several lytic infection genes are unique to LCV, and homologs are not found even in gamma 2 herpesviruses, e.g., gp350, viral interleukin (vIL)-10, BARF1, BALF1 (MOGHADDAM et al. 1998; P. Rivailler et al., unpublished observations). These genes are probably not essential for lytic replication or B-cell immortalization in vitro, but are likely to play an important role during virus replication in vivo. The preservation of these genes in the rhesus LCV with a high degree of homology underscores the similar biology and pathogenesis conserved through the evolution of these human and nonhuman LCV.

Table 1. Homology of simian LCV lytic and latent infection genes to EBV[a]

	Rhesus LCV (%)	Baboon LCV (%)	Description
Lytic gene			
BARF1	75.7	ND	Lytic oncogene; CSF1R homolog
BALF1	84.1	ND	bcl2 Homolog
BALF2	94.7	ND	Major DNA-binding protein
BALF3	82.9	ND	
BCRF1	84.7	79.4	IL-10 homolog
BZLF1	71.3	ND	Zebra transactivator
BFRF3	69.4	ND	Viral capsid antigen
gp350	49.2	ND	Major membrane glycoprotein
Latent gene			
LMP1	36.5	ND	
	23.3	23.8	Carboxy terminus
LMP2A	54.7	51.1	
	31.4	29.8	First exon only
LMP2B	27.0	ND	Nucleotide homology, first exon only
EBNA-1	46.4	44.1	
EBNA-2A	26.5	26.3	
EBNA-2B	31.5	ND	Type 2 EBV and rhesus LCV EBNA-2
EBNA-3A	29.4	ND	
EBNA-3B	29.1	ND	
EBNA-3C	31.0	ND	
EBNA-LP	53.3	ND	
EBER1	70.1	71.9	Nucleotide homology
EBER2	41.7	44.7	Nucleotide homology

ND, sequence not determined yet.
[a] Homology is expressed as % amino acid similarity, or nucleotide homology where noted, as determined by performing clustal alignments with the DNAstar program.

4.3 Latent Infection Genes

Somewhat surprisingly, the latent infection genes are not as well-conserved (Table 1). Epstein-Barr nuclear antigen (EBNA)-1 is one of the most well-conserved, and this likely reflects the essential role for episomal maintenance during latent LCV infection (YATES et al. 1996; BLAKE et al. 1999). Most other latent infection genes are much less well-conserved, but all appear to function similar to the EBV homologs. For example, the baboon and rhesus LCV EBNA-2, EBNA-3A, -3B and -3C can all interact with RBP-Jκ, and the simian EBNA-2 and EBNA-3C homologs can act as transcriptional transactivators in human cells despite having relatively low overall amino acid homology with the EBV homologs (LING and HAYWARD 1995; PENG et al. 2000; Jiang et al., in press). The baboon and rhesus LCV latent membrane protein (LMP)-1s have conserved Transformation effector site (TES)-1 and TES-2 sites that can interact with TNF receptor associated factors (TRAFs) and induce nuclear factor (NF)-κB activity in human cells, despite otherwise dramatic sequence divergence in the carboxy terminal cytoplasmic domain (FRANKEN et al. 1996). The baboon and rhesus LCV LMP2As are most divergent in the first exon, but have conserved the immunoreceptor tyrosine-based activation motif (ITAM) and the ability to interact with src kinases in human B cells (FRANKEN et al. 1995; RIVAILLER et al. 1999). This sequence divergence associated with conserved function suggests that these are relatively new genes in evolution, arising from a relatively diverse gene pool, with convergence of domains important for gene function.

4.4 A Species-Restricted Block to LCV-Induced B-Cell Immortalization

Despite this functional similarity and interchangeable nature for in vitro assays, there is a species-restricted block to B-cell immortalization (MOGHADDAM et al. 1998). Thus, EBV can efficiently immortalize human B cells and B cells from closely related Old World nonhuman primates such as chimpanzees. However, EBV cannot efficiently immortalize B cells from more distant Old World species, such as rhesus macaques. Similarly, baboon LCV can immortalize B cells from other Old World species in the same cercopithecine family, e.g., rhesus monkeys, but cannot efficiently immortalize human B cells.

Studies clearly demonstrate that this species-specific block is beyond the point of virus binding and penetration (MOGHADDAM et al. 1998). In fact, baboon LCV can infect, persist, and replicate in human B cells, but only in the presence of EBV co-infection. The simplest hypothesis is that one or more latent infection genes perform some yet-to-be-described function essential for B-cell immortalization, which is species specific. The EBNA-3 genes may be one locus for this species restriction since chimeric EBV where the EBNA-3 genes had been replaced with the rhesus LCV EBNA-3 locus were able to initiate human B-cell growth, but unable to sustain B-cell immortalization (Jiang et al., in press). Other preliminary experiments suggest that multiple loci may contribute to the species restricted block, but the precise genetic loci responsible remain to be determined. This species restriction may identify molecular interactions that are important for B-cell immortalization using a comparative approach with the EBV and simian LCV genes.

These species differences may also explain why investigators were generally unsuccessful at infecting Old World nonhuman primates with EBV in the 1980s. Pre-existing natural LCV infection in Old World nonhuman primates may also have resulted in some cross-reactive immunity that may have blocked infection with EBV. These factors suggest that careful screening for natural LCV infection and the autologous virus species must be used for an experimental model in Old World nonhuman primates. The molecular similarities between rhesus LCV and EBV suggest that use of rhesus LCV will provide a valid and accurate model for EBV infection.

4.5 Two Types of Rhesus LCV

The discovery of two types of rhesus LCV infection resembling type 1 and type 2 EBV further underscores the similarities between the simian LCV and EBV. These two types of rhesus LCV were identified by polymorphisms in the EBNA-2 gene from two naturally occurring rhesus LCV isolates (CHO et al. 1999). The two rhesus LCV EBNA-2 alleles not only demonstrated similar degrees of sequence divergence as the EBV-1 and EBV-2 EBNA-2s, but the domains with the most divergence were positionally conserved in the rhesus LCV and EBV EBNA-2s. It is still unclear what biologic advantage is provided by the two different EBV types, but it appears that the same biologic advantage has also resulted in the evolution of two rhesus LCV types. Furthermore, the observation that co-infection with both rhesus LCV types can occur in immunocompetent animals is consistent with the increasing evidence that some humans can be naturally co-infected with two types of EBV (SIXBEY et al. 1989; APOLLONI and SCULLEY 1994; BROOKS et al. 2000).

4.6 Nonconserved and Conserved Molecular Strategies for Immune Evasion

The only significant difference discovered to date between the simian LCV and EBV has been associated with the putative immune evasion function associated with the glycine-alanine repeat (GAR) repeats in the EBV EBNA-1. The GAR domain in EBV EBNA-1 has been shown to inhibit antigen presentation of cytotoxic T-cell epitopes in *cis* (LEVITSKAYA et al. 1995). The GAR is hypothesized to prevent molecules from entering the class I antigen presentation by protecting GAR-containing molecules from proteosomal degradation, and this mechanism may contribute to the apparent lack of robust EBNA-1 specific CTL activity in EBV infected individuals.

The baboon and rhesus LCV EBNA-1 both contain similar GAR domains; however, in both instances the repeats are somewhat simpler and significantly fewer in number compared to EBV EBNA-1 (YATES et al. 1996; BLAKE et al. 1999). When these simian LCV GAR domains were used to replace the GAR domain in EBV EBNA-1, they failed to reproduce the effect of the EBV GAR by failing to inhibit antigen presentation (BLAKE et al. 1999). When cytotoxic T-lymphocyte (CTL) epitopes were cloned into the rhesus LCV EBNA-1s, the simian LCV GAR domain still failed to inhibit antigen presentation in the context of the native protein. Since this system also used rhesus monkey antigen-presenting cells and CTLs, this ruled out any potential species-restricted effect (BLAKE et al. 1999). Thus, despite the dramatic conservation of function among the various simian LCV latent infection genes studied to date, the protection of antigen presentation by the EBV EBNA-1 GAR has not been conserved in other Old World nonhuman primate LCV. This suggests that this is not a strategy for immune evasion common to all LCVs.

It is interesting to note that a functional homolog for the type 1 latency Qp EBNA-1 promoter has been conserved in both rhesus and baboon LCV (RUF et al. 1999). Thus, the downregulation of latent gene expression (ROWE et al. 1987) due to a Cp/Wp to Qp switch does appear to be a common strategy for immune evasion in human and nonhuman LCV.

5 Experimental Infection of Naïve Rhesus Macaques with Rhesus LCV

The biologic and molecular similarities between the rhesus LCV and EBV suggest that experimental infection with the native virus in na rhesus macaques will provide a valid model for EBV infection in humans. The species restriction for LCV-induced B-cell immortalization suggests that using EBV to infect rhesus macaques is unlikely to reproduce the natural LCV infection seen in macaques and humans. Theoretically, it might be possible to engineer chimeric EBV capable of successfully infecting rhesus monkeys, similar to the chimeric SIV containing the HIV envelope gene. Chimeric LCV/EBV may be especially useful for evaluating the efficacy of potential therapeutic agents against EBV in an animal model system. However, the loci responsible for the species restriction remain to be defined, and studies to date suggesting the possible involvement of multiple loci may make chimeric LCV a difficult proposition (Jiang et al., in press).

5.1 Identifying Rhesus LCV-Naïve Hosts

Experimental infection requires the use of rhesus LCV-naive animals. The group housing used at most primate centers results in ubiquitous natural LCV infection at even earlier ages than in humans. Thus, by 1 year of age, most rhesus macaques raised in conventional colonies are LCV seropositive (Wang, unpublished observations). Fortunately, there have been widespread efforts to raise specific pathogen-free colonies in order to reduce the adverse effects of simian retrovirus D infection to rhesus macaques and to reduce the risk of transmitting herpesvirus B infection to human care-takers (DESROSIERS 1997). Typically, these colonies are initiated by hand-rearing abandoned newborns. Animals are serologically screened on a regular basis to ensure that they are free of retrovirus D and herpes B infection. Naive animals are then housed together. When sufficient animals have been raised, breeding colonies can be established to naturally maintain the specific-pathogen-free (SPF) colony. Although these efforts have not typically screened for LCV infection, we have found that these procedures can be successful for raising LCV-naive rhesus macaques (RAO et al. 2000). However, the inadvertent introduction of rhesus LCV seropositive animals can result in relatively rapid spread of LCV infection in the na colony. Therefore, vigilant serologic screening for LCV infection should be used. Recent cloning of the VCA p18 (BFRF3) homolog from rhesus LCV indicates that immunodominant epitopes are present in the carboxy terminus as for EBV, and enzyme immunoassays with the rhesus LCV VCA peptide are a sensitive assay for both type 1 and type 2 rhesus LCV infection (RAO et al. 2000).

5.2 Experimental Rhesus LCV Inoculation

We have chosen non-traumatic oral instillation as our route of inoculation in order to most closely mimic natural transmission (MOGHADDAM et al. 1997). Generally, 10×6 transforming units in 10–20cc of tissue culture media are used to bathe the oropharynx, and variable amounts of virus can be swallowed or expectorated making the virus dose difficult to titrate by this route. However, this relatively imprecise mode of inoculation is useful since we do not know the site for initial EBV/LCV infection. The concept that EBV first infects epithelial cells has been recently challenged (ANAGNOSTOPOULOS et al. 1995), and the observation that EBV infection can be associated with the gastric epithelium and gastric carcinomas (ROWLANDS et al. 1993; GULLEY et al. 1996; ARIKAWA et al. 1997) raise the possibility that other parts of the gastro-intestinal tract beyond the oropharynx may be involved.

5.3 Acute Clinical and Laboratory Responses to Experimental Infection

It is difficult to evaluate subjective symptoms of acute infections in rhesus macaques, e.g., fatigue, malaise. Fevers are also difficult to measure without implanted devices since anesthesia alone may elevate body temperature. Physical exam can detect lymphadenopathy and splenomegaly in some, but not all experimentally infected animals (MOGHADDAM et al. 1997). It remains to be determined whether development of these physical signs in response to rhesus LCV infection has some degree of age-dependence similar to the age dependent association of the infectious mononucleosis syndrome with adolescent humans.

Laboratory studies often provide simpler and more objective measurements of acute infection. Atypical lymphocytosis is seen in many animals, but the degree can be quite variable, similar to the physical findings (MOGHADDAM et al. 1997). More recently, we have developed specific assays for viral load and immune activation to study the responses during acute infection. Using real-time quantitative PCR, we can detect viral DNA in the peripheral blood mononuclear cells of experimentally infected animals 7–14 days after oral inoculation. This initial viral load generally peaks around 14–21 days and slowly resolves over the next 10–12 weeks when it often becomes undetectable by DNA PCR (RAO et al. 1999).

Evidence for immune activation can be detected as early as 3 days after oral inoculation by the appearance of T cells expressing intracellular cytokines, such as IL-2, gamma interferon, and IL-10 (KAUR et al. 1999). Intracellular IL-10 is prominent in both T- and B-cell populations, and sometime has a biphasic pattern of appearance. The proportion of cellular and viral encoded IL-10 remains to be determined, but the high expression levels of cellular and viral IL-10 in infectious mononucleosis patients supports a prominent role for this cytokine during acute LCV infection (TAGA et al. 1995). Preliminary evidence for activated CTL activity against rhesus LCV-infected cells has been detected in the first few weeks after infection similar to that described in infectious mononucleosis patients (KAUR et al. 1999). Humoral responses to viral capsid antigen and EBNA-2 develop relatively early after oral inoculation (MOGHADDAM et al. 1997; RAO et al. 2000). Thus, experimental rhesus LCV infection by oral inoculation induces a similar clinical and laboratory picture as EBV infection in humans, and experimental infection allows for detailed study of the early events initiating primary infection as well as those contributing to the resolution of primary infection.

5.4 Persistent Infection After Experimental Inoculation

Persistent infection can be detected in both the peripheral blood B-cell compartment and oropharynx of experimentally infected animals. Persistent B-cell infection can be detected by the recovery of rhesus LCV-infected B cells after their spontaneous outgrowth in tissue culture, as in humans (MOGHADDAM et al. 1997). Furthermore, persistent infection can be detected in experimentally infected animals by RT-PCR amplification for the rhesus LCV EBER homologs in peripheral blood mononuclear cells (RAO et al. 2000). Persistent infection detected by EBER RT-PCR has been detected in experimentally infected animals as long as 3 years after infection. EBER RT-PCR can detect persistent infection in approximately 95% of naturally infected animals; however, a semiquantitative EBER assay showed a very broad range of expression levels, e.g., 4 logs. Thus, this is a useful qualitative assay for persistent infection, but other assays need to be developed in order to determine whether there is any subtle, quantitative difference in persistent infection after experimental versus natural infection. Persistent infection in the oropharynx can also be detected in experimentally and naturally infected animals by PCR amplification of DNA extracts from oral washes (MOGHADDAM et al. 1997; CHO et al. 1999).

Thus, these experiments demonstrate that rhesus macaques can be successfully infected by experimental inoculation with rhesus LCV. The animals develop an acute syndrome, which provides a unique window into the early events immediately after viral inoculation, and experimental infection results in an acute and persistent infection that is similar to that of EBV infection in humans. The rhesus LCV animal model can be useful for addressing a number of questions.

6 Future Directions for the Rhesus LCV Animal Model

6.1 Molecular Pathogenesis of Acute and Persistent Infection

Genetic EBV studies have now identified many viral genes which appear to be 'nonessential' for virus replication and B-cell immortalization in vitro. Most, if not all, of these genes have been conserved in the rhesus LCV genome suggesting they play an essential role for viral pathogenesis in vivo, even though they are apparently unnecessary in vitro. Laboratory studies can provide some clue as to the function for some of these genes, but how they affect acute or persistent LCV infection in vivo remains unknown.

For example, functional attributes have been assigned from in vitro experiments to 'nonessential' latent infection genes such as LMP2A/B, EBNA-3B, and EBERs (SWAMINATHAN et al. 1992; TOMKINSON and KIEFF 1992; LONGNECKER et al. 1993), and one can speculate how they may contribute to viral pathogenesis in vivo. LMP2A may promote latency by inhibiting viral replication, but recent experiments suggest that LMP2A might also contribute a positive effect to B-cell growth and persistence in vivo (CALDWELL et al. 1998). This hypothesis can be directly tested by constructing LMP2A rhesus LCV mutants and asking whether experimental infection results in excessive lytic infection or a failure to develop persistence. The mystery surrounding the role of LMP2B is further compounded by the strong conservation in simian LCV and the apparent lack of recognizable functional domains (RIVAILLER et al. 1999). One might predict that the EBERs might play an important role during acute infection due to their potential ability to block interferon-induced responses (ROSA et al. 1981). The presumed EBNA-3B function as a transcriptional regulator suggests that it may be important for inducing cell genes that modulate cell survival or the immune response in vivo (LE ROUX et al. 1994).

LCVs also encode for a number of lytic infection genes which appear to be 'nonessential' for virus replication in vitro (MARCHINI et al. 1991; SWAMINATHAN et al. 1993; ROBERTSON et al. 1996; COHEN and LEKSTROM 1999). Many of these genes are cell gene homologs and have also been conserved in the simian LCV suggesting an essential role for viral pathogenesis in vivo. These genes include two bcl2 homologs (BHRF1 and BALF1), a CSF1 receptor homolog (BARF1), and a viral IL-10 homolog (BCRF1) (P. Rivailler et al., unpublished observations). The capture of a vIL-10 homolog is particularly intriguing since it has been tightly conserved in the rhesus LCV, and only a few viruses are known to encode for a viral IL-10. EBV IL-10 can be detected in the serum of infectious mononucleosis patients along with high titers of cellular IL-10, so the unique requirement for vIL-10 during acute infection is not evident (TAGA et al. 1995). It is difficult to attribute a significant role for vIL-10 in the immune evasion of latently infected B cells, since vIL-10 is a lytic infection gene. Perhaps, vIL-10 is important for persistent viral infection in the oropharynx where local immunosuppression may be required to mask the antigenic viral proteins produced during replication. These types of questions can be directly addressed by genetically mutating these genes in the rhesus LCV and assessing the impact in vivo after experimental infection.

6.2 Immune Responses and Immune Evasion

The animal model will also be useful for identifying host immune responses important for control of acute infection and to test the relevance of viral strategies for immune evasion and persistent infection. The EBNA-3 genes are immunodominant targets for the CTL response in humans (KHANNA et al. 1992; STEVEN et al. 1996), however, it is unknown whether these CTLs are protective and whether EBNA-3 epitopes would represent an effective target for potential vaccination. A conserved immunodominant response to the EBNA-3s in rhesus macaques would argue for a protective role versus a bias due to the human MHC background. Similarly, it is unknown whether the vigorous and persistent response to lytic infection genes might represent an important component of the immune response important for control of virus infection (CALLAN et al. 1998; STEVEN et al. 1997).

Conservation of the latency type 1 Qp promoter suggests that downregulation of latent gene expression is an important strategy of immune evasion common to all LCV. This hypothesis can be directly tested by deleting the Qp and asking whether the virus is still able to establish persistent infection. In this way, the virus may become attenuated, while still evoking a strong immune response, important characteristics for a potential live viral vaccine.

6.3 Pathogenesis of Epithelial Cell Infection

The initial site of EBV infection in humans remains controversial. The classic paradigm is that EBV first infects epithelial cells in the oropharynx and then infects B cells that migrate close to the mucosal surface. There is no question that EBV can infect epithelial cells in immunosuppressed patients, e.g., oral hairy leukoplakia in AIDS patients, or during malignant transformation, e.g., nasopharyngeal carcinoma. However, whether EBV naturally infects epithelial cells in immunocompetent hosts has been challenged by some investigators who fail to find evidence for EBV infection in tonsillar epithelial cells from patients with infectious mononucleosis (ANAGNOSTOPOULOS et al. 1995).

Naturally or experimentally infected rhesus macaques can be used to ask whether epithelial cells are infected during acute infection or if epithelial cells a site for persistent infection. Although many clinical trials have shown that acyclovir is ineffective when used to treat patients presenting with infectious mononucleosis (WANG 1999), the ability to treat rhesus macaques with acyclovir prior to infection would test whether viral replication is required to establish primary infection. The association of EBV infection in some gastric carcinomas also raises the possibility that other regions of the gastrointestinal tract beyond the oropharynx may be important for LCV infection.

6.4 LCV-Associated Tumorigenesis

Providing a model for LCV-induced tumors is also an important goal. Simian LCV are clearly associated with the development of B-cell lymphomas, but the ability to reproduce these tumors experimentally represents a significant challenge. The high frequency of LCV-infected B-cell tumors in cynomolgous macaques with natural LCV infection followed by experimental SIV infection is encouraging (FEICHTINGER et al. 1990). In fact, this frequency is probably much higher than in humans, since only a fraction of EBV seropositive adult AIDS patients develop lymphomas. But, even the reported incidence and rate of tumor development in the SIV-infected cynomolgous macaques may not be experimentally feasible for an expensive animal model system. In addition, the incidence of tumors in these SIV-infected cynomolgous macaques appears higher than other centers using SIV-infected rhesus macaques, and the reason for this difference, e.g., species, viral strain, etc., is not obvious.

Therefore, we have recently begun exploring the risk of tumorigenesis when primary LCV infection is initiated after Simian-Human Immunodeficiency Virus (SHIV)-induced immunosuppression. The markedly higher risk of lymphoproliferative disease in pediatric transplants, who acquire primary EBV infection in the post-transplant period, provide precedence for an increased risk of tumorigenesis (Ho et al. 1988). SHIV infection is particularly advantageous since $CD4^+$ T cells decrease rapidly and reproducibly after experimental inoculation (REIMANN et al. 1996). Our preliminary experiments suggest that acute rhesus LCV infection is initially controlled, perhaps by some innate immune response unaffected by the SHIV infection, and uncontrolled LCV viral loads can develop later, which may be a primary event or secondary to opportunistic infection. The potential for a tumorigenesis model remains to be developed, but this system also provides interesting insight into the interaction of SIV/LCV interactions and the role of the innate immune response.

6.5 Vaccine Development

A proposed EBV vaccine candidate currently under development is based on the membrane antigen, gp350. While a gp350 vaccine can induce serum-neutralizing antibodies (Gu et al. 1995), it is still uncertain how successful this vaccine strategy will be for preventing infectious mononucleosis or establishment of persistent infection. It also remains unknown whether the induction of mucosal immunity or cellular immunity against lytic or latent infection antigens may be important for an effective EBV vaccine. The membrane antigen in rhesus LCV shows 50% amino acid identity with EBV gp350 (MOGHADDAM et al. 1998), and the rhesus LCV animal model can be used to address how successful these different vaccine strategies may be for protecting against an oral virus challenge.

7 Conclusion

The rhesus LCV animal model holds great promise as an experimental model system for studying EBV pathogenesis. Recent work has demonstrated the strong evolutionary conservation between the rhesus LCV and EBV genomes with highly conserved genetic repertoires and molecular signaling pathways. Modern breeding practices and dedicated resources for developing SPF colonies make experiments with rhesus LCV-naive animals feasible and provide important infrastructure for these biologic studies. Experimental infection of rhesus macaques with rhesus LCV reproduces the natural mode of transmission and provides a model system for studying the earliest phases of acute infection, the establishment of persistent infection, the development of host immune responses important for controlling infection, and viral strategies for immune evasion. The impending completion of the rhesus LCV genome sequence will be an important step in the development of genetic systems for mutating rhesus LCV and for subsequent molecular pathogenesis studies in vivo. Co-infection with SIV and rhesus LCV provides a potential model for LCV-induced B-cell lymphomas and AIDS-associated malignancies. The rhesus LCV animal model will also contribute to the development of potential EBV therapies through a better understanding of LCV pathogenesis and a model system for testing novel vaccines and therapeutics.

References

Ablashi DV, Gerber P, Easton J (1979) Oncogenic herpesviruses of nonhuman primates. Comp Immunol Microbiol Infect Dis 2:229–241

Agrba VZ, Yakovleva LA, Lapin BA, Sangulija IA, Timanovskaya VV, Markarjan DS, Chuvirov GN, Salmanova EA (1975) The establishment of continuous lymphoblastoid suspension cell cultures from hematopoietic organs of baboon (Papio hamadryas) with malignant lymphoma. Exp Pathol (Jena) 10:318–332

Anagnostopoulos I, Hummel M, Kreschel C, Stein H (1995) Morphology immunophenotype and distribution of latently and/or productively Epstein-Barr virus-infected cells in acute infectious mononucleosis: implications for the interindividual infection route of Epstein-Barr virus. Blood 85:744–750

Apolloni A, Sculley TB (1994) Detection of A-type and B-type Epstein-Barr virus in throat washings and lymphocytes. Virology 202:978–981

Arikawa J, Tokunaga M, Satoh E, Tanaka S, Land CE (1997) Morphological characteristics of Epstein-Barr virus-related early gastric carcinoma: a case-control study. Pathol Int 47:360–367

Baskin GB, Roberts ED, Kuebler D, Martin LN, Blauw B, Heeney J, Zurcher C (1995) Squamous epithelial proliferative lesions associated with rhesus Epstein-Barr virus in simian immunodeficiency virus-infected rhesus monkeys. J Infect Dis 172:535–539

Blake NW, Moghaddam A, Rao P, Kaur A, Glickman R, Cho YG, Marchini A, Haigh T, Johnson RP, Rickinson AB, Wang F (1999) Inhibition of antigen presentation by the glycine/alanine repeat domain is not conserved in simian homologs of Epstein-Barr virus nuclear antigen 1. J Virol 73:7381–7389

Bocker JF, Tiedemann KH, Bornkamm GW, Zur HH (1980) Characterization of an EBV-like virus from African green monkey lymphoblasts. Virology 101:291–295

Brooks JM, Croom-Carter DS, Leese AM, Tierney RJ, Habeshaw G, Rickinson AB (2000) Cytotoxic T-lymphocyte responses to a polymorphic Epstein-Barr virus epitope identify healthy carriers with coresident viral strains. J Virol 74:1801–1809

Caldwell RG, Wilson JB, Anderson SJ, Longnecker R (1998) Epstein-Barr virus LMP2A drives B cell development and survival in the absence of normal B cell receptor signals. Immunity 9:405–411

Callan MF, Tan L, Annels N, Ogg GS, Wilson JD, O'Callaghan CA, Steven N, McMichael AJ, Rickinson AB (1998) Direct visualization of antigen-specific CD8+ T cells during the primary immune response to Epstein-Barr virus in vivo. J Exp Med 187:1395–1402

Cho YG, Gordadze AV, Ling PD, Wang F (1999) Evolution of two types of rhesus lymphocryptovirus similar to type 1 and type 2 Epstein-Barr virus. J Virol 73:9206–9212

Cohen JI, Lekstrom K (1999) Epstein-Barr virus BARF1 protein is dispensable for B-cell transformation and inhibits alpha interferon secretion from mononuclear cells. J Virol 73:7627–7632

Desrosiers RC (1997) The value of specific pathogen-free rhesus monkey breeding colonies for AIDS research. AIDS Res Hum Retroviruses 13:5–6

Dillner J, Rabin H, Letvin N, Henle W, Henle G, Klein G (1987) Nuclear DNA-binding proteins determined by the Epstein-Barr virus-related simian lymphotropic herpesviruses H. gorilla, H. pan, H. pongo and H. papio. J Gen Virol 68:1587–1596

Dunkel VC, Pry TW, Henle G, Henle W (1972) Immunofluorescence tests for antibodies to Epstein-Barr virus with sera of lower primates. J Natl Cancer Inst 49:435–440

Epstein MA, Morgan AJ, Finerty S, Randle BJ, Kirkwood JK (1985) Protection of cottontop tamarins against Epstein-Barr virus-induced malignant lymphoma by a prototype subunit vaccine. Nature 318:287–289

Falk L, Deinhardt F, Nonoyama M, Wolfe LG, Bergholz C (1976) Properties of a baboon lymphotropic herpesvirus related to Epstein-Barr virus. Int J Cancer 18:798–807

Falk L, Wolfe L, Deinhardt F, Paciga J, Dombos L, Klein G, Henle W, Henle G (1974) Epstein-Barr virus: transformation of non-human primate lymphocytes in vitro. Int J Cancer 13:363–376

Feichtinger H, Putkonen P, Parravicini C, Li ST, Kaya EE, Bottiger D, Biberfeld P (1990) Malignant lymphomas in cynomolgus monkeys infected with simian immunodeficiency virus. Am J Pathol 137:1311–1315

Frank A, Andiman WA, Miller G (1976) Epstein-Barr virus and nonhuman primates: natural and experimental infection. Adv Cancer Res 23:171–201

Franken M, Annis B, Ali AN, Wang F (1995) 5′ Coding and regulatory region sequence divergence with conserved function of the Epstein-Barr virus LMP2A homolog in herpesvirus papio. J Virol 69:8011–8019

Franken M, Devergne O, Rosenzweig M, Annis B, Kieff E, Wang F (1996) Comparative analysis identifies conserved tumor necrosis factor receptor-associated factor 3 binding sites in the human and simian Epstein-Barr virus oncogene LMP1. J Virol 70:7819–7826

Gerber P, Kalter SS, Schidlovsky G, Peterson WJ, Daniel MD (1977) Biologic and antigenic characteristics of Epstein-Barr virus-related herpesviruses of chimpanzees and baboons. Int J Cancer 20:448–459

Gerber P, Pritchett RF, Kieff ED (1976) Antigens and DNA of a chimpanzee agent related to Epstein-Barr virus. J Virol 19:1090–1099

Goldberg RJ, Scolnick EM, Parks WP, Yakovleva LA, Lapin BA (1974) Isolation of a primate type-C virus from a lymphomatous baboon. Int J Cancer 14:722–730

Gu SY, Huang TM, Ruan L, Miao YH, Lu H, Chu CM, Motz M, Wolf H (1995) First EBV vaccine trial in humans using recombinant vaccinia virus expressing the major membrane antigen. Dev Biol Stand 84:171–177

Gulley ML, Pulitzer DR, Eagan PA, Schneider BG (1996) Epstein-Barr virus infection is an early event in gastric carcinogenesis and is independent of bcl-2 expression and p53 accumulation. Hum Pathol 27:20–27

Heberling RL, Bieber CP, Kalter SS (1981) Establishment of a lymphoblastoid cell line from a lymphomous cynomolgus monkey. In: Yohn DS, Blakeslee JR (eds) Advances in comparative leukemia research. Elsevier, Amsterdam, pp 385–386

Heller M, Gerber P, Kieff E (1982) DNA of herpesvirus pan a third member of the Epstein-Barr virus-Herpesvirus papio group. J Virol 41:931–939

Heller M, Gerber P, Kieff E (1981) Herpesvirus papio DNA is similar in organization to Epstein-Barr virus DNA. J Virol 37:698–709

Heller M, Kieff E (1981) Colinearity between the DNAs of Epstein-Barr virus and herpesvirus papio. J Virol 37:821–826

Ho M, Jaffe R, Miller G, Breinig MK, Dummer JS, Makowka L, Atchison RW, Karrer F, Nalesnik MA, Starzl TE (1988) The frequency of Epstein-Barr virus infection and associated lymphoproliferative syndrome after transplantation and its manifestations in children. Transplantation 45:719–727

Jiang H, Cho YG, Wang F (2000) Structural, functional, and genetic comparisons of Epstein-Barr virus nuclear antigen 3A, 3B, and 3C homologues encoded by the rhesus lymphocryptovirus. J Virol 74(13):5921–5932

Kalter SS, Heberling RL, Ratner JJ (1972) EBV antibody in sera of non-human primates. Nature 238:353–354

Kalter SS, Herberling RL, Ratner JJ (1973) EBV antibody in monkeys and apes. Bibl Haematol 39: 871–875

Kaur A, Rao P, Cho Y, Hale C, Johnson RP, Wang F (1999) Host responses to lymphocryptovirus infection in rhesus macaques – an animal model for EBV infection. In: 24th International Herpesvirus Workshop

Khanna R, Burrows SR, Kurilla MG, Jacob CA, Misko IS, Sculley TB, Kieff E, Moss DJ (1992) Localization of Epstein-Barr virus cytotoxic T cell epitopes using recombinant vaccinia: implications for vaccine development. J Exp Med 176:169–176

Landon JC, Ellis LB, Zeve VH, Fabrizio DP (1968) Herpes-type virus in cultured leukocytes from chimpanzees. J Natl Cancer Inst 40:181–192

Landon JC, Malan LB (1971) Seroepidemiologic studies of Epstein-Barr virus antibody in monkeys. J Natl Cancer Inst 46:881–884

Lapin BA (1974) The epidemiologic and genetic aspect of an outbreak of leukemia among baboons of the Sukhumi monkey colony in Dutcher and Chieco-Bianchi. Unifying concepts of leukemia. Biblphyl Haematol 39:263–268

Lapin BA, Timanovskaya VV, Yakovleva LA (1985) Herpesvirus HVMA: a new representative in the group of the EBV-like B- lymphotropic herpesviruses of primates. Hamatol Bluttransfus 29:312–313

Le Roux A, Kerdiles B, Walls D, Dedieu JF, Perricaudet M (1994) The Epstein-Barr virus determined nuclear antigens EBNA-3A -3B and -3C repress EBNA-2-mediated transactivation of the viral terminal protein 1 gene promoter. Virology 205:596–602

Levitskaya J, Coram M, Levitsky V, Imreh S, Steigerwald-Mullen PM, Klein G, Kurilla MG, Masucci MG (1995) Inhibition of antigen processing by the internal repeat region of the Epstein-Barr virus nuclear antigen-1. Nature 375:685–688

Li SL, Feichtinger H, Kaaya E, Migliorini P, Putkonen P, Biberfeld G, Middeldorp JM, Biberfeld P, Ernberg I (1993) Expression of Epstein-Barr-virus-related nuclear antigens and B-cell markers in lymphomas of SIV-immunosuppressed monkeys. Int J Cancer 55:609–615

Ling PD, Hayward SD (1995) Contribution of conserved amino acids in mediating the interaction between EBNA2 and CBF1/RBPJk. J Virol 69:1944–1950

Longnecker R, Miller CL, Tomkinson B, Miao XQ, Kieff E (1993) Deletion of DNA encoding the first five transmembrane domains of Epstein-Barr virus latent membrane proteins 2A and 2B. J Virol 67:5068–5074

Marchini A, Tomkinson B, Cohen JI, Kieff E (1991) BHRF1 the Epstein-Barr virus gene with homology to Bcl2 is dispensable for B-lymphocyte transformation and virus replication. J Virol 65:5991–6000

Moghaddam A, Koch J, Annis B, Wang F (1998) Infection of human B lymphocytes with lymphocryptoviruses related to Epstein-Barr virus. J Virol 72:3205–3212

Moghaddam A, Rosenzweig M, Lee-Parritz D, Annis B, Johnson RP, Wang F (1997) An animal model for acute and persistent Epstein-Barr virus infection. Science 276:2030–2033

Mosier DE (1996) Viral pathogenesis in hu-PBL-SCID mice. Semin Immunol 8:255–262

Neubauer RH, Rabin H, Strnad BC, Nonoyama M, Nelson RW (1979) Establishment of a lymphoblastoid cell line and isolation of an Epstein-Barr-related virus of gorilla origin. J Virol 31:845–848

O'Gara RW, Adamson RH, Kelly MG, Dalgard DW (1971) Neoplasms of the hematopoietic system in nonhuman primates: report of one spontaneous tumor and two leukemias induced by procarbazine. J Natl Cancer Inst 46:1121–1130

Peng R, Gordadze AV, Fuentes Panana EM, Wang F, Zong J, Hayward GS, Tan J, Ling PD (2000) Sequence and functional analysis of EBNA-LP and EBNA2 proteins from nonhuman primate lymphocryptoviruses. J Virol 74:379–389

Rabin H, Neubauer RH, Hopkins R ed, Rasheed S (1978) In vitro lymphocyte transformation by Epstein-Barr virus (EBV)-like viruses isolated from Old-World non-human primates. Iarc Sci Publ 553–557

Rabin H, Neubauer RH, Hopkins RF, Dzhikidze EK, Shevtsova ZV, Lapin BA (1977) Transforming activity and antigenicity of an Epstein-Barr-like virus from lymphoblastoid cell lines of baboons with lymphoid disease. Intervirology 8:240–249

Rabin H, Neubauer RH, Hopkins RFd, Nonoyama M (1978) Further characterization of a herpesvirus-positive orang-utan cell line and comparative aspects of in vitro transformation with lymphotropic old world primate herpesviruses. Int J Cancer 21:762–767

Rangan SR, Martin LN, Bozelka BE, Wang N, Gormus BJ (1986) Epstein-Barr virus-related herpesvirus from a rhesus monkey (Macaca mulatta) with malignant lymphoma. Int J Cancer 38:425–432

Rao P, Jiang H, Kaur A, Wang F (1999) Acute and persistent viral loads using real-time PCR after experimental lymphocryptovirus infection of rhesus monkeys. 24th International Herpesvirus Workshop Abstract 10.010

Rao P, Jiang H, Wang F (2000) Cloning of the rhesus lymphocryptovirus viral capsid antigen and Epstein-Barr virus-encoded small RNA homologues and use in diagnosis of acute and persistent infections. J Clin Microbiol 38(9):3219–3225

Rasheed S, Rongey RW, Bruszweski J, Nelson-Rees WA, Rabin H, Neubauer RH, Esra G, Gardner MB (1977) Establishment of a cell line with associated Epstein-Barr-like virus from a leukemic orangutan. Science 198:407–409

Reimann KA, Li JT, Veazey R, Halloran M, Park IW, Karlsson GB, Sodroski J, Letvin NL (1996) A chimeric simian/human immunodeficiency virus expressing a primary patient human immunodeficiency virus type 1 isolate env causes an AIDS-like disease after in vivo passage in rhesus monkeys. J Virol 70:6922–6928

Rivailler P, Quink C, Wang F (1999) Strong selective pressure for evolution of an Epstein-Barr virus LMP2B homolog in the rhesus lymphocryptovirus. J Virol 73:8867–8872

Robertson ES, Ooka T, Kieff ED (1996) Epstein-Barr virus vectors for gene delivery to B lymphocytes. Proc Natl Acad Sci USA 93:11334–11340

Rosa MD, Gottlieb E, Lerner MR, Steitz JA (1981) Striking similarities are exhibited by two small Epstein-Barr virus-encoded ribonucleic acids and the adenovirus-associated ribonucleic acids VAI and VAII. Mol Cell Biol 1:785–796

Rowe M, Rowe DT, Gregory CD, Young LS, Farrell PJ, Rupani H, Rickinson AB (1987) Differences in B cell growth phenotype reflect novel patterns of Epstein-Barr virus latent gene expression in Burkitt's lymphoma cells. Embo J 6:2743–2751

Rowlands DC, Ito M, Mangham DC, Reynolds G, Herbst H, Hallissey MT, Fielding JW, Newbold KM, Jones EL, Young LS, et al. (1993) Epstein-Barr virus and carcinomas: rare association of the virus with gastric adenocarcinomas. Br J Cancer 68:1014–1019

Ruf IK, Moghaddam A, Wang F, Sample J (1999) Mechanisms that regulate Epstein-Barr virus EBNA-1 gene transcription during restricted latency are conserved among lymphocryptoviruses of Old World primates. J Virol 73:1980–1989

Schable CA, Murphy BL, Berquist KR, Gravelle CR, Maynard JE (1974) Inability to detect hepatitis B virus or specific antigens in transformed chimpanzee lymphocytes. Infect Immun 10:1443–1444

Sixbey JW, Shirley P, Chesney PJ, Buntin DM, Resnick L (1989) Detection of a second widespread strain of Epstein-Barr virus. Lancet 2:761–765

Steven NM, Annels NE, Kumar A, Leese AM, Kurilla MG, Rickinson AB (1997) Immediate early and early lytic cycle proteins are frequent targets of the Epstein-Barr virus-induced cytotoxic T cell response. J Exp Med 185:1605–1617

Steven NM, Leese AM, Annels NE, Lee SP, Rickinson AB (1996) Epitope focusing in the primary cytotoxic T cell response to Epstein-Barr virus and its relationship to T cell memory. J Exp Med 184:1801–1813

Stevens DA, Pry TW, Blackham EA, Manaker RA (1970) Comparison of antigens from human and chimpanzee herpes-type virus-infected hemic cell lines. Proc Soc Exp Biol Med 133:678–683

Swaminathan S, Hesselton R, Sullivan J, Kieff E (1993) Epstein-Barr virus recombinants with specifically mutated BCRF1 genes. J Virol 67:7406–7413

Swaminathan S, Huneycutt BS, Reiss CS, Kieff E (1992) Epstein-Barr virus-encoded small RNAs (EBERs) do not modulate interferon effects in infected lymphocytes. J Virol 66:5133–5136

Taga H, Taga K, Wang F, Chretien J, Tosato G (1995) Human and viral interleukin-10 in acute Epstein-Barr virus-induced infectious mononucleosis. J Infect Dis 171:1347–1350

Tomkinson B, Kieff E (1992) Use of second-site homologous recombination to demonstrate that Epstein-Barr virus nuclear protein 3B is not important for lymphocyte infection or growth transformation in vitro. J Virol 66:2893–2903

Wang F (1999) Epstein-Barr Virus and Human Herpesvirus-8. In: Yu VL, Merigan TC, Barriere SL (eds) Antimicrobial Therapy and Vaccines. Williams & Wilkins, Baltimore, pp 1215–1219

Yates JL, Camiolo SM, Ali S, Ying A (1996) Comparison of the EBNA1 proteins of Epstein-Barr virus and herpesvirus papio in sequence and function. Virology 222:1–13

Adoptive Immunotherapy of EBV-Associated Malignancies with EBV-Specific Cytotoxic T-Cell Lines

C.M. ROONEY, L.K. AGUILAR, M.H. HULS, M.K. BRENNER, and H.E. HESLOP

1	Introduction	221
2	EBV-Specific CTLs for the Treatment of Lymphoproliferative Disease in Stem Cell Recipients	222
2.1	CTL Activation and Expansion	223
2.2	Prophylactic Use of EBV-Specific CTLs	224
2.2.1	Safety	225
2.2.2	Persistence	225
2.2.3	Immunological Efficacy	225
2.2.4	Virological Efficacy	225
2.3	CTLs as Treatment	225
2.3.1	Pitfalls Associated with the Use of CTLs as Treatment	226
3	Hodgkin's Disease	226
References		228

1 Introduction

The goal of immunotherapy is to overcome the immune response deficits of the host or the immune stimulatory deficits of the tumor and activate an effective tumor-specific immune response. The cytotoxic T-lymphocyte (CTL) arm of the cellular immune response is thought to be the most important defense against tumors and virus-infected cells. CTLs recognize short peptides derived from viral antigens that are carried to the infected cell surface in association with major histocompatibility (MHC) molecules (see Fig. 1). Epstein-Barr virus (EBV)-associated malignancies express a range of antigens against which to target CTLs. For immunotherapy, CTLs may be activated and expanded in vivo or ex vivo. In vivo strategies involve immunization with DNA, tumor vaccines, or antigen- or peptide-loaded dendritic cells. Ex vivo strategies involve exposing T cells to tumor or viral antigens expressed on antigen-presenting cells (APCs) and expanding them in T-cell growth factors in vitro. Although the ex vivo approach may be more costly in the time, effort, and expertise required to grow CTLs for patient infusion, it may be the only option in

---

Center for Cell and Gene Therapy, Baylor College of Medicine, 6621 Fannin St, Houston, TX 77030, USA

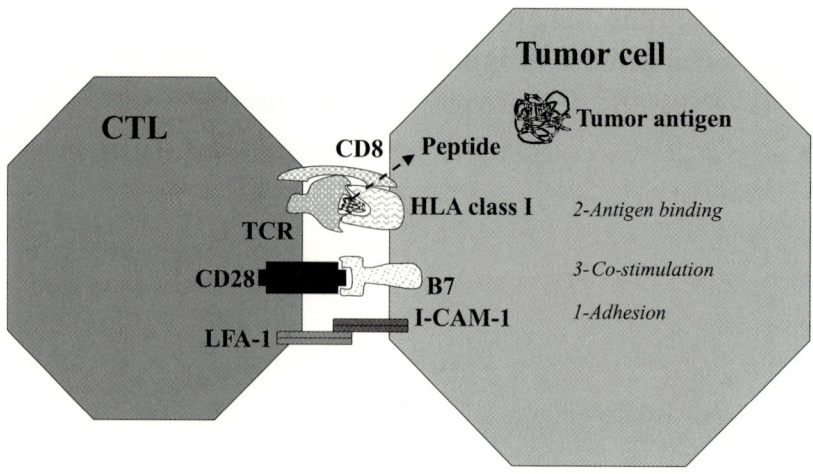

Fig. 1. Requirements for reactivation of memory CTL in vitro. First, adhesion molecules are important for bringing the T cell into contact with the antigen-presenting cell (APC). Second, the T-cell receptor (TCR) must bind its specific antigenic peptide in the groove of the appropriate HLA molecule. Third, a costimulatory signal, which can be provided by CD28 on the T-cell surface binding to B7 on the APC; without costimulation, the CTL may be anergized

cases where the patient is immunosuppressed or where the tumor secretes inhibitory factors. In these cases, it is often possible to activate and expand antigen-specific CTL precursors in a culture environment that is conducive to CTL growth, even though the same cells cannot expand and function in vivo. There are additional advantages to the ex vivo approach. The phenotype and function of the CTLs can be determined prior to infusion, anti-tumor activity can be ensured, and anti-host activity can be excluded. CTL numbers can be controlled and repeat infusions given if required. Finally, transfer of marker genes into CTL lines allows the function and persistence of CTLs in vivo to be followed, transfer of suicide genes allows the in vivo destruction of the CTLs should they prove toxic, and transfer of functional genes may improve the activity of CTLs once infused (Hwu et al. 1993; Heslop et al. 1996; Bonini et al. 1997). We have used the ex vivo approach for the prevention and treatment of EBV-related malignancies in stem cell recipients, because they are unable to mount an immune response in vivo, and in patients with relapsed EBV-positive Hodgkin's disease, because the tumor secretes inhibitory factors.

2 EBV-Specific CTLs for the Treatment of Lymphoproliferative Disease in Stem Cell Recipients

EBV-associated lymphoproliferative disease (EBV-LPD) in stem cell recipients has provided an excellent model system in which to test the biological efficacy of ex vivo-expanded, adoptively transferred antigen-specific CTL lines. EBV-LPD occurs in up to 20% of recipients of T cell-depleted stem cells from HLA-mismatched or unrelated donors (AGUILAR et al. 1999). No anti-viral agents are reproducibly effective for EBV-LPD. Immunotherapy with unmanipulated donor leukocytes is associated with a high incidence of disease progression and survivors have a high incidence of graft-vs-host disease (GvHD) (O'REILLY et al. 1996; LUCAS et al. 1998). To avoid the GvHD and the immunopathology associated with the treatment of active disease, we have treated patients with selectively activated and expanded, EBV-specific CTLs. EBV-specific CTL lines are readily established from the majority of stem cell donors, since over 90% are persistently-infected with EBV and therefore carry a high frequency of circulating EBV-specific CTL precursors. EBV-transformed B-lymphoblastoid cell lines (LCLs) are easy to establish from normal donors, they reproducibly activate EBV-specific CTL lines from seropositive donors and provide a continuous source of excellent APC.

2.1 CTL Activation and Expansion

Figure 2 shows our method of generating EBV-specific CTL lines. The first step is the generation of EBV-transformed B-cell lines for use as APCs. Peripheral blood mononuclear cells (5×10^6) are resuspended in concentrated supernatant from the B95-8 virus producer line and cultured in the presence of cyclosporin A. The LCL is

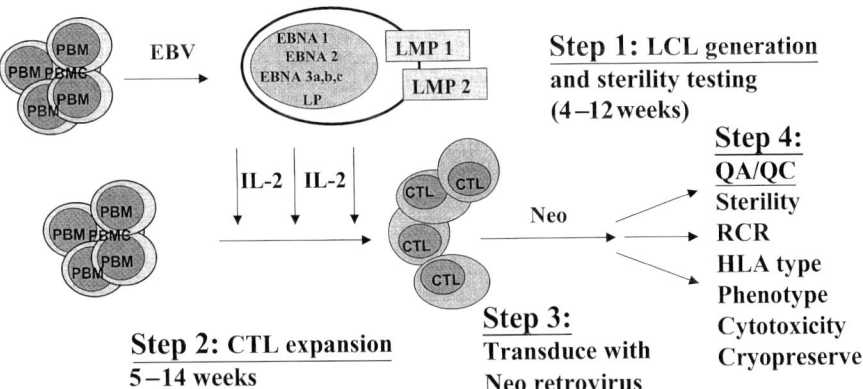

Fig. 2. Generation of EBV-specific CTL lines. The first step is the generation of an EBV transformed B-cell line (LCL) from the donor. Peripheral blood mononuclear cells are infected with concentrated virus stock from the B95-8 cell line in the presence of cyclosporin A, an inhibitory of T-cell activation. From day 14 the cells are expanded in the presence of acyclovir to inhibit the release of infectious virus. Once the LCL is established and sterility testing is complete, it is irradiated and co-cultured with fresh or frozen PBMC from the same donor. This stimulation is repeated on day 10 and then weekly until the patient dose is reached. IL-2 at 40–100units/ml is added twice weekly from day 14. Retrovirus transduction is performed less than 3 days before cryopreservation of the CTL line. The CTL product is tested for sterility, identity (HLA typing), phenotype, and cytotoxic specificity before it is released for patient infusion. The test for replication competent retrovirus (RCR) is performed, but is not a release criterion

usually established within about 4 weeks, when it can be sterility tested and cryopreserved before being used to initiate CTL lines. The LCLs are maintained in 100µM acyclovir to prevent the production of infectious virus.

For CTL generation, fresh or frozen donor peripheral blood mononuclear cells (PBMCs) are cultured with the irradiated autologous LCL at a responder:stimulator ratio of 40:1 in 2ml wells. They are restimulated on day 9 and interleukin (IL)-2 is added on day 14. Thereafter, the cultures are restimulated weekly with the irradiated LCL (4:1 ratio) and twice weekly with IL-2 (40–100units/ml). When sufficient cell numbers for patient treatment are obtained, the cells are safety tested (see Table 1) and frozen using a controlled rate freezer. EBV-specific CTL lines were established from all but three of over 150 normal donors. Two of these were seronegative and one had recently seroconverted.

CTL lines established in this way are variable in phenotype and cytotoxic activity. Both CD4 and CD8 numbers ranged from 2% to 98%, with widely varying numbers of $CD56^+$ lymphokine-activated killer cells. The release criteria require first that the line should contain less than 1% B cells, ensuring that there is no LCL growth in the line, and second that there should be significant killing of the autologous LCL with less than 10% killing of patient-derived phytohemagglutinin (PHA) blasts (grown from blood obtained prior to transplant), a measure of potential GvHD activity.

2.2 Prophylactic Use of EBV-Specific CTLs

Since 1993, over 60 recipients of T cell-depleted bone marrow from HLA-mismatched or unrelated donors have received infusions of donor-derived, EBV-specific CTL lines, as prophylaxis for EBV-LPD (HESLOP et al. 1994; ROONEY et al. 1998). The first 26 patients received CTL that had been genetically marked with a retrovirus vector carrying the neomycin-resistance gene. Marking efficiencies of 0.5–10% allowed us to track the in vivo persistence of the CTLs and to determine their involvement in any toxicity. An initial dose escalation study revealed that low numbers of CTLs were biologically effective and so currently, patients

Table 1. Release criteria required by CTLs for infusion

Test	Result required
Viability	>70%
Sterility	
Fungus, bacteria	No growth after 7 days
Mycoplasma	Negative
Endotoxin	<5EU/ml
Identity	Donor HLA class I type
Phenotype	<2% B lymphocytes
Cytotoxic specificity	<10% killing of patient-derived lymphoblasts at an E:T ratio of 20:1

EU, endotoxin unit.

receive one dose of $2 \times 10^7 \text{CTL}/\text{m}^2$. The target date of infusion is day 45, at which time GvHD should be apparent if it is to occur.

2.2.1 Safety

Used prophylactically, CTLs are safe and effective. The main anticipated toxicity was GvHD. No patient developed de novo GvHD, although previously occurring GvHD reactivated in two. Both responded to low-dose steroids. No other toxicities could be associated with CTL infusions.

2.2.2 Persistence

Amplification of marker DNA from EBV-specific CTLs reactivated from patients at serial times after infusion showed that the infused CTL lines expanded up to 4 logs in vivo and then persisted for up to 6 years after infusion (ROONEY et al. 1995b, 1998; HESLOP et al. 1996). Marking studies also showed that the CTLs could expand in vivo in response to EBV reactivation; a rise in EBV DNA levels in two patients over one year after infusion was accompanied by reappearance of the marker gene.

2.2.3 Immunological Efficacy

Prior to CTL infusion, we were unable to detect EBV-specific CTL precursors. Within 1 week of infusion, low level activity could be detected, and within 4 weeks the levels of immunity were within the range seen in normal individuals, where they remained (HESLOP et al. 1996).

2.2.4 Virological Efficacy

High EBV DNA levels, as measured by amplification of EBV DNA in peripheral blood, are frequently accompanied by fevers and lymphadenopathy and are associated with an extremely high risk for EBV-LPD (ROONEY et al. 1995a). About 20% of our patients developed a high virus load after bone marrow transplantation (BMT). Infusion of CTLs uniformly resulted in a dramatic drop in virus load, to low or undetectable levels. The most important result of the study was that none of the patients who received prophylactic CTL developed EBV-LPD, in contrast with about 12% of controls.

2.3 CTLs as Treatment

Further evidence for anti-tumor effects came from four patients who had not received CTLs and developed frank lymphoma. These four received CTLs as treatment and three achieved complete remission. One case, in which tumor biopsies could be studied, clearly demonstrated that EBV-specific CTL home to tumor sites, then accumulate or expand to effect lymphoma regression. Comparison of tumor biopsy material taken pre- and 10 days post-CTL infusion showed that $CD20^+$ tumor B cells were replaced almost entirely by $CD3^+$ T cells. In situ PCR demonstrated *neo*-marked CTLs in the tumor tissue at levels over 100-fold seen in peripheral blood (ROONEY et al. 1998).

2.3.1 Pitfalls Associated with the Use of CTLs as Treatment

Two of the patients who received CTLs as treatment also illustrated two of the pitfalls that can result from treating bulky disease. First, if the tumor occurs in a sensitive anatomical location, the inflammatory response can be damaging. Second, mutation of important CTL epitopes becomes increasingly likely with tumor size. The second patient, who presented with airway obstruction due to bulky disease in the nasopharynx, developed increased swelling after CTL infusion, requiring intubation and tracheotomy. He also developed ulceration of the soft palate and gut, illustrating the damage that a CTL-mediated inflammatory response can generate. The patient eventually recovered and is now well over 3 years after CTL infusion. The third patient also presented with bulky disease involving the nasopharynx and lung (GOTTSCHALK et al. 1998). She received CTLs but the tumor progressed and she died of respiratory failure 24 days after CTL infusion. EBV-transformed B-cell lines grew rapidly from patient peripheral blood before and 7 days after CTL infusion. Comparative analysis of tumor virus DNA with DNA from the patient-derived B-cell lines and B95-8 DNA revealed a deletion in the tumor's Epstein-Barr virus nuclear antigen (EBNA) 3B gene that removed two immunodominant HLA-A11-restricted epitopes. Analysis of the donor CTL line showed that it also was dominated by these two EBNA 3B epitopes. No other epitope specificity could be identified from peptides predicted from the literature to sensitize targets to killing through the donor HLA class I allotype (RICKINSON and MOSS 1997).

Thus, while the use of EBV-specific CTL lines is a safe and effective strategy to prevent EBV-associated disease after stem cell transplant, CTLs as therapy should be used with caution. Inflammatory responses can be damaging because of the aggressive nature of the tumor and its preference for respiratory tissues. Escape mutants may also be a problem, since even polyclonal EBV-specific CTL lines may be specific for only a few viral epitopes, and with a large tumor burden there is ample opportunity for mutation.

3 Hodgkin's Disease

EBV-positive Hodgkin's disease is another candidate for EBV-specific CTL, since it provides viral antigens as targets. A number of problems can be anticipated to complicate the use of CTLs to treat Hodgkin's disease (see Fig. 3). First, the tumor

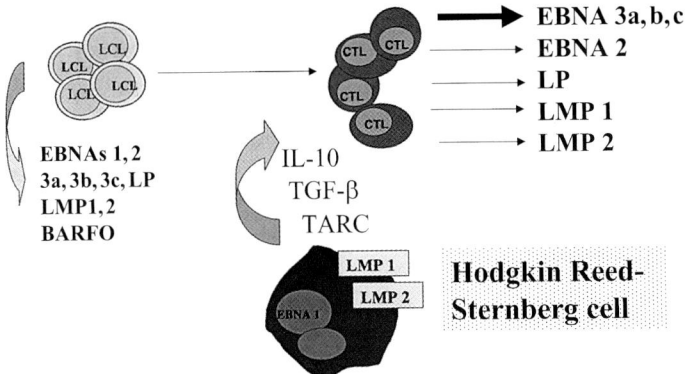

Fig. 3. Immune evasion strategies of Hodgkin's Reed-Sternberg cells. CTL epitopes in EBNA 3a, 3b, or 3c dominate the CTL response to EBV when CTL lines are activated by in vitro culture with LCLs. Reed-Sternberg cells do not express the EBNA 3 proteins and therefore are not susceptible to the bulk of circulating EBV-specific CTLs. They also secrete chemokines and cytokines that inhibit the recruitment, activation, and function of CTLs

cells downregulate the immunodominant latency-associated proteins that are expressed in post-transplant lymphomas, Hodgkin's lymphoma cells express only EBNA1, which is not processed for presentation by HLA class I molecules, and latent membrane protein LMP1 and LMP2 (LEVITSKAYA et al. 1997). Therefore, if LCLs are used to activate CTLs, the major activity will likely be directed against the immunodominant EBNAs 3A, 3B, and 3C proteins that are not expressed in Hodgkin's Reed-Sternberg cells. Second, the tumor cells secrete cytokines and chemokines that are inhibitory to CTLs. IL-10 is an anti-inflammatory cytokine that prevents the activation of professional APC, which may be important in cross-priming the immune response to viral proteins. Transforming growth factor (TGF)-β is directly inhibitory to CTLs, and thymus and activation-associated cytokine (TARC) is a chemokine that specifically recruits TH2 cells that promote antibody responses and inhibit CTL responses. Third, since these patients have received multiple rounds of chemotherapy and irradiation as well as autologous stem cell rescue, the immune system is usually in poor shape when blood is received. Lymphocyte yields are low and they respond poorly to in vitro stimulus. Even the LCL may take up to 4 months to establish.

To ensure the successful generation of CTL lines, after initial specific activation, CTLs are expanded in a powerful mitogenic cocktail of irradiated allogeneic PBMCs as well as irradiated LCLs, high dose IL-2 and anti-CD3 antibody (SMITH et al. 1995). Nevertheless, some patients have died before completion of the CTL line. Seven patients have received autologous EBV-specific CTL lines on a dose escalation study, and had temporary improvements in clinical and immunological parameters. These included increases in EBV-specific CTL precursor frequency, reduction in high virus load, resolution of type B symptoms and mixed tumor responses (ROSKROW et al. 1998). Study of these patients revealed that their CTL lines did have anti-tumor specificity, despite the fact that LCLs were used as APC (see Fig. 3). Comparative polymerase chain reaction (PCR) and in situ hybridization to the neo marker gene showed CTLs in a malignant pleural effusion in one patient, and in a mediastinal tumor in a second (ROSKROW et al. 1998). Further, tetramer studies demonstrated the presence of LMP2-specific CTLs in the patient lines, albeit with a low frequency (L.K. Aguilar, unpublished observations). Current improvements include the generation of lines in which the majority of CTLs are specific for the limited range of viral antigens expressed in Reed-Sternberg cells. Dendritic cells, transduced with viral vectors that express individual EBV proteins, are effective antigen-presenting cells, even in na or non-responder individuals (NAIR et al. 1993; CHOUDHURY et al. 1998). Further developments should involve the genetic modification of antigen-specific CTLs with genes that facilitate CTL function in a TH2 environment.

Acknowledgements. This work was supported in part by NIH grants CA61384, CA74126, and Cancer Center Support CORE Grant 21765, the American Lebanese Syrian Associated Charities (ALSAC), and the Department of Pediatrics, Baylor College of Medicine. We thank Belinda Rossitter for editing the manuscript.

References

Aguilar LK, Rooney CM, Heslop HE (1999) Lymphoproliferative disorders involving Epstein-Barr virus after hemopoietic stem cell transplantation. Curr Opin Oncol 11:96–101
Bonini C, Ferrari G, Verzeletti S, Servida P, Zappone E, Ruggieri L, Ponzoni M, Rossini S, Malvilio F, Traversari C, Bordignon C (1997) HSV-TK gene transfer into donor lymphocytes for control of allogeneic graft versus leukemia. Science 276:1719–1724
Choudhury A, Toubert A, Sutaria S, Charron D, Champlin RE, Claxton DF (1998) Human leukemia-derived dendritic cells: ex-vivo development of specific antileukemic cytotoxicity. Crit Rev Immunol 18:121–131
Gottschalk S, Ng CYC, Perez M, Brenner MK, Heslop HE, Rooney CM (2000) Mutation in EBV produces immunoblastic lymphoma unresponsive to CTL immunotherapy. Blood (in press)
Heslop HE, Brenner MK, Rooney CM, Krance RA, Roberts WM, Rochester R, Smith CA, Turner V, Sixbey J, Moen R, Boyett JM (1994) Administration of neomycin-resistance-gene-marked EBV-specific cytotoxic T lymphocytes to recipients of mismatched-related or phenotypically similar unrelated donor marrow grafts. Hum Gene Ther 5:381–397
Heslop HE, Ng CYC, Li C, Smith CA, Loftin SK, Krance RA, Brenner MK, Rooney CM (1996) Long-term restoration of immunity against Epstein-Barr virus infection by adoptive transfer of gene-modified virus-specific T lymphocytes. Nature Medicine 2:551–555
Hwu P, Shafer GE, Treisman J, Schindler DG, Gross G, Cowherd R, Rosenberg SA, Eshhar Z (1993) Lysis of ovarian cancer cells by human lymphocytes redirected with a chimeric gene composed of an antibody variable region and the Fc receptor gamma chain. J Exp Med 178:361–366
Levitskaya J, Sharipo A, Leonchiks A, Ciechanover A, Masucci MG (1997) Inhibition of ubiquitin/proteasome-dependent protein degradation by the Gly-Ala repeat domain of the Epstein-Barr virus nuclear antigen 1. Proc Natl Acad Sci 94(23):12616–12621. 11–11–0997
Lucas KG, Burton RL, Zimmerman SE, Wang J, Cornetta KG, Robertson KA, Lee CH, Emanuel DJ (1998) Semiquantitative Epstein-Barr virus (EBV) polymerase chain reaction for the determination of patients at risk for EBV-induced lymphoproliferative disease after stem cell transplantation. Blood 91:3654–3661
Nair S, Babu JS, Dunham RG, Kanda P, Burke RL, Rouse BL (1993) Induction of primary, antiviral cytotoxic and proliferative responses with antigens administered via dendritic cells. J Virol 67:4062–4069

O'Reilly RJ, Lacerda JF, Lucas KG, Rosenfield NS, Small TN, Papadopoulos EB (1996) Adoptive cell therapy with donor lymphocytes for EBV-associated lymphomas developing after allogeneic marrow transplants. In: DeVita VT, Hellman S, Rosenberg SA (eds) Important Advances in Oncology 1996. Lippincott-Raven, Philadelphia, pp 149–166

Rickinson AB, Moss DJ (1997) Human cytotoxic T lymphocyte responses to Epstein-Barr virus infection. Annu Rev Immunol 15:405–431

Rooney CM, Loftin SK, Holladay MS, Brenner MK, Krance RA, Heslop HE (1995a) Early identification of Epstein-Barr virus-associated post-transplant lymphoproliferative disease. Br J Haematol 89:98–103

Rooney CM, Smith CA, Ng CYC, Loftin SK, Sixbey JW, Gan Y-J, Srivastava D-K, Bowman LC, Krance RA, Brenner MK, Heslop HE (1998) Infusion of cytotoxic T cells for the prevention and treatment of Epstein-Barr virus-induced lymphoma in allogeneic transplant recipients. Blood 92:1549–1555

Rooney CM, Smith CA, Ng C, Loftin SK, Li C, Krance RA, Brenner MK, Heslop HE (1995b) Use of gene-modified virus-specific T lymphocytes to control Epstein-Barr virus-related lymphoproliferation. Lancet 345:9–13

Roskrow MA, Suzuki N, Gan Y-J, Sixbey JW, Ng CYC, Kimbrough S, Hudson MM, Brenner MK, Heslop HE, Rooney CM (1998) EBV-specific cytotoxic T lymphocytes for the treatment of patients with EBV positive relapsed Hodgkin's disease. Blood 91:2925–2934

Smith CA, Ng CYC, Heslop HE, Holladay MS, Richardson S, Turner EV, Loftin SK, Li C, Brenner MK, Rooney CM (1995) Production of genetically modified EBV-specific cytotoxic T cells for adoptive transfer to patients at high risk of EBV-associated lymphoproliferative disease. J Hematother 4:73–79

Subject Index

A
AGS 56
AIDS lymphoma 121–133
– anaplastic large-cell lymphomas (ALCL) 123, 130
– *Burkitt's* lymphoma (BL) 123, 164
– diffuse large-cell lymphoma (DLCL) 123
Akata cells 35–47
animal model 201–215
antigen-presenting cells 221
apoptosis 99, 144, 170
assembly 61
atrophic gastritis 98
atrophy 97
autonomous replicating sequences (ARS) 14, 15

B
bacteriophage T7 76
BALF4 52
BARF1 169
BBRF3 59
B-cell immortalization 207
Bcl-2 127, 155, 172
bcl-2 99, 149
BDLF 51
BDLF3 51
BILF1 51
BILF2 51
BKRF2 54
BLLF1 52
BLRF1 58
BmHI A rightward transcripts (BARTs) 168
BMRF2 51
body-cavity-based lymphomas (BCBL) 117, 129
breast carcinoma 162
Burkitt's lymphoma (BL) 40, 104, 123, 141, 164
BXLF2 54
BZLF1 69, 166
BZLF2 54

C
CD21 52, 163
CD44 99
CD45 123
CD138 123
c-fes/fps 126
chronic pyothorax 109
cis-acting elements 5
clonal analysis 92
c-myc 124, 141
complement receptor type 2 (CR2) 52, 163
CR2 52, 163
cyclin D 130
cytotoxic T lymphocyte (CTL) 106, 127, 176, 221
– epitopes 114
– persistence 222

D
diffuse large-cell lymphoma 123
dipyrimidine site 117
disassembly 61
DNA
– helicase 6
– polymerase δ 75
DNA-ISH 97
DNA-looping/-linking 7
DS-independent replication 24

E
E. coli DNA *Pol* III holoenzyme 75
E2 protein 7
EBP2 8
EBV
– DNA levels 235
– genotype 189
– life cycle 66–68
– lytic replication origin 68, 69
– reactivation 189
– receptor 163
– replication enzymes 65–83
EBVaGC 91
EBV-associated lymphoproliferative disease 222

EBV-determined nuclear antigens (EBNAs) 106, 168
- EBNA1 1, 5, 153–160
- EBNA2 35–47, 111, 127
- EBNA4 114
- – epitope 113
EBV-encoded small RNA (EBER) 104, 126, 141–149, 168
- EBER-ISH 92
EBV-positive gastric epithelial cell lines 186, 187
EBV-specific CTL lines 223
egress 51
entry 51
enzyme-linked immunosorbent assay (ELISA) 132
epithelial cell infection 213
Epstein-Barr virus (*see* EBV)
extrachromosomal replicon 3

F
field cancerization 95

G
gastric carcinoma 91–100, 162, 185–195
- advanced 92
- early 92
gastritis 97
- atrophic 98
gB 53
gene
- expression 153
- therapy 100
genetic pathways 98
gH 53
gH/gL/gp42 53–57
gL 54
gN/gM 53, 58–61
gp25 54
gp42 54, 167
gp78 52
gp85 53, 167
gp110 53
gp125 53
gp150 51
gp220 52
gp350 52
gp350/220 52, 163
G-protein-coupled receptor (GPCR) 130
GT38 cell lines 187–194

H
H. pylori 98, 100
helicase primase 79
hepatocellular carcinoma 162
HHV-8 122, 128

highly active anti-retroviral therapy (HAART) 122
HIV 121
HLA allele 113
HLA-A 113
HLA-A2 106
HLA-DP 167
HLA-DQ 167
HLA-DR 55, 167
Hodgkin's disease 104, 107, 108, 123, 126
human herpes virus 8 117, 122, 128
- serology 132

I
IL-6 112, 128, 131
IL-10 114
immunoblastic lymphoma (IBL) 123
immunofluorescence assay (IFA) 132
immunosuppressive cytokine 113
immunotherapy 221–228
in situ hybridization 104
interferon-induced protein kinase 148
intestinal metaplasia 97

K
Kaposi's sarcoma 128
Kaposi's sarcoma-associated herpesvirus (KSHV) 117, 122
KT tumor 99

L
large noncleaved cell lymphoma (LNCCL) 123
latency 153
- type III 127
latency-associated nuclear antigen (LANA) 129
latent infection 129
- genes 207
latent membrane proteins (LMPs) 106, 168
- LMP1 111, 127, 168, 170
- LMP2A 168
- LMP2B 168
LCV-associated malignancies 204, 214
lethal midline granuloma 104
leukocyte common antigen 123
licensing of replication 9
load 51
lymphocryptoviruses (LCV) 202, 203
lymphocytic infiltration 97
lymphoepithelioma 91
lymphoma (see also AIDS lymphoma, *Burkitt's* lymphoma) 104–117
lytic
- infection genes 206
- phase 81

M

malignant
- lymphoma   122
- phenotype   149

marker genes   222
MCM complex   8
MHC class I   106
microsatellite   98
multiple carcinomas   95

N

nasal natural killer-cell lymphoma   104
nasopharyngeal carcinoma (NPC)   104, 161
nitric oxide   190
non-*Hodgkin's* lymphoma (NHL)   104, 121

O

OPL-1   111
OPL-2   111
oral hairy leukoplakia   161
origin recognition complex (ORC)   18, 22
oriLyt   68–70
oriP   5

P

P32/TAP/gClq-R   8
p53   99, 107, 172
PCNA   75
PCR-RFLP   98
PCR-SSCP   117
penetration   61
persistent generalized lymphadenopathy (PGL)   124
phage T4   75
PKR   148
Pol
- accessory subunit   72
- catalytic subunit   71, 72

polymeric IgA (pIgA)   163

primary effusion lymphoma (PEL)   117, 122
pyothorax-associated lymphoma (PAL)   108, 111

R

Rch1/importin-α 8
recombinant virus   164
Reed-Sternberg cells   107, 126
Rep*   6
replication
- forks   82
- licensing factors   20

replicative intermediates   3
revised European-American lymphoma
   classification   130
RF-C   75
rhesus LCV   204
RPA   8

S

SCID mice   99, 127, 194
simian LCV   204, 205, 208
single-strand conformation polymorphism
   (SSCP) analysis   107
Sp1   7
SVKCR2   56
syndecan-1   123

T

T-cell NHL   123
trans-acting viral gene   5
transforming growth factor (TGF)-β 114
translocation   124
tropism   51, 61

V

vIRF   130

Y

yeast one- and two-hybrid assays   8

Current Topics in Microbiology and Immunology

Volumes published since 1989 (and still available)

Vol. 215: **Shinnick, Thomas M. (Ed.):** Tuberculosis. 1996. 46 figs. XI, 307 pp. ISBN 3-540-60985-7

Vol. 216: **Rietschel, Ernst Th.; Wagner, Hermann (Eds.):** Pathology of Septic Shock. 1996. 34 figs. X, 321 pp. ISBN 3-540-61026-X

Vol. 217: **Jessberger, Rolf; Lieber, Michael R. (Eds.):** Molecular Analysis of DNA Rearrangements in the Immune System. 1996. 43 figs. IX, 224 pp. ISBN 3-540-61037-5

Vol. 218: **Berns, Kenneth I.; Giraud, Catherine (Eds.):** Adeno-Associated Virus (AAV) Vectors in Gene Therapy. 1996. 38 figs. IX,173 pp. ISBN 3-540-61076-6

Vol. 219: **Gross, Uwe (Ed.):** Toxoplasma gondii. 1996. 31 figs. XI, 274 pp. ISBN 3-540-61300-5

Vol. 220: **Rauscher, Frank J. III; Vogt, Peter K. (Eds.):** Chromosomal Translocations and Oncogenic Transcription Factors. 1997. 28 figs. XI, 166 pp. ISBN 3-540-61402-8

Vol. 221: **Kastan, Michael B. (Ed.):** Genetic Instability and Tumorigenesis. 1997. 12 figs.VII, 180 pp. ISBN 3-540-61518-0

Vol. 222: **Olding, Lars B. (Ed.):** Reproductive Immunology. 1997. 17 figs. XII, 219 pp. ISBN 3-540-61888-0

Vol. 223: **Tracy, S.; Chapman, N. M.; Mahy, B. W. J. (Eds.):** The Coxsackie B Viruses. 1997. 37 figs. VIII, 336 pp. ISBN 3-540-62390-6

Vol. 224: **Potter, Michael; Melchers, Fritz (Eds.):** C-Myc in B-Cell Neoplasia. 1997. 94 figs. XII, 291 pp. ISBN 3-540-62892-4

Vol. 225: **Vogt, Peter K.; Mahan, Michael J. (Eds.):** Bacterial Infection: Close Encounters at the Host Pathogen Interface. 1998. 15 figs. IX, 169 pp. ISBN 3-540-63260-3

Vol. 226: **Koprowski, Hilary; Weiner, David B. (Eds.):** DNA Vaccination/Genetic Vaccination. 1998. 31 figs. XVIII, 198 pp. ISBN 3-540-63392-8

Vol. 227: **Vogt, Peter K.; Reed, Steven I. (Eds.):** Cyclin Dependent Kinase (CDK) Inhibitors. 1998. 15 figs. XII, 169 pp. ISBN 3-540-63429-0

Vol. 228: **Pawson, Anthony I. (Ed.):** Protein Modules in Signal Transduction. 1998. 42 figs. IX, 368 pp. ISBN 3-540-63396-0

Vol. 229: **Kelsoe, Garnett; Flajnik, Martin (Eds.):** Somatic Diversification of Immune Responses. 1998. 38 figs. IX, 221 pp. ISBN 3-540-63608-0

Vol. 230: **Kärre, Klas; Colonna, Marco (Eds.):** Specificity, Function, and Development of NK Cells. 1998. 22 figs. IX, 248 pp. ISBN 3-540-63941-1

Vol. 231: **Holzmann, Bernhard; Wagner, Hermann (Eds.):** Leukocyte Integrins in the Immune System and Malignant Disease. 1998. 40 figs. XIII, 189 pp. ISBN 3-540-63609-9

Vol. 232: **Whitton, J. Lindsay (Ed.):** Antigen Presentation. 1998. 11 figs. IX, 244 pp. ISBN 3-540-63813-X

Vol. 233/I: **Tyler, Kenneth L.; Oldstone, Michael B. A. (Eds.):** Reoviruses I. 1998. 29 figs. XVIII, 223 pp. ISBN 3-540-63946-2

Vol. 233/II: **Tyler, Kenneth L.; Oldstone, Michael B. A. (Eds.):** Reoviruses II. 1998. 45 figs. XVI, 187 pp. ISBN 3-540-63947-0

Vol. 234: **Frankel, Arthur E. (Ed.):** Clinical Applications of Immunotoxins. 1999. 16 figs. IX, 122 pp. ISBN 3-540-64097-5

Vol. 235: **Klenk, Hans-Dieter (Ed.):** Marburg and Ebola Viruses. 1999. 34 figs. XI, 225 pp. ISBN 3-540-64729-5

Vol. 236: **Kraehenbuhl, Jean-Pierre; Neutra, Marian R. (Eds.):** Defense of Mucosal Surfaces: Pathogenesis, Immunity and Vaccines. 1999. 30 figs. IX, 296 pp. ISBN 3-540-64730-9

Vol. 237: **Claesson-Welsh, Lena (Ed.):** Vascular Growth Factors and Angiogenesis. 1999. 36 figs. X, 189 pp. ISBN 3-540-64731-7

Vol. 238: **Coffman, Robert L.; Romagnani, Sergio (Eds.):** Redirection of Th1 and Th2 Responses. 1999. 6 figs. IX, 148 pp. ISBN 3-540-65048-2

Vol. 239: **Vogt, Peter K.; Jackson, Andrew O. (Eds.):** Satellites and Defective Viral RNAs. 1999. 39 figs. XVI, 179 pp. ISBN 3-540-65049-0

Vol. 240: **Hammond, John; McGarvey, Peter; Yusibov, Vidadi (Eds.):** Plant Biotechnology. 1999. 12 figs. XII, 196 pp. ISBN 3-540-65104-7

Vol. 241: **Westblom, Tore U.; Czinn, Steven J.; Nedrud, John G. (Eds.):** Gastroduodenal Disease and Helicobacter pylori. 1999. 35 figs. XI, 313 pp. ISBN 3-540-65084-9

Vol. 242: **Hagedorn, Curt H.; Rice, Charles M. (Eds.):** The Hepatitis C Viruses. 2000. 47 figs. IX, 379 pp. ISBN 3-540-65358-9

Vol. 243: **Famulok, Michael; Winnacker, Ernst-L.; Wong, Chi-Huey (Eds.):** Combinatorial Chemistry in Biology. 1999. 48 figs. IX, 189 pp. ISBN 3-540-65704-5

Vol. 244: **Daëron, Marc; Vivier, Eric (Eds.):** Immunoreceptor Tyrosine-Based Inhibition Motifs. 1999. 20 figs. VIII, 179 pp. ISBN 3-540-65789-4

Vol. 245/I: **Justement, Louis B.; Siminovitch, Katherine A. (Eds.):** Signal Transduction and the Coordination of B Lymphocyte Development and Function I. 2000. 22 figs. XVI, 274 pp. ISBN 3-540-66002-X

Vol. 245/II: **Justement, Louis B.; Siminovitch, Katherine A. (Eds.):** Signal Transduction on the Coordination of B Lymphocyte Development and Function II. 2000. 13 figs. XV, 172 pp. ISBN 3-540-66003-8

Vol. 246: **Melchers, Fritz; Potter, Michael (Eds.):** Mechanisms of B Cell Neoplasia 1998. 1999. 111 figs. XXIX, 415 pp. ISBN 3-540-65759-2

Vol. 247: **Wagner, Hermann (Ed.):** Immunobiology of Bacterial CpG-DNA. 2000. 34 figs. IX, 246 pp. ISBN 3-540-66400-9

Vol. 248: **du Pasquier, Louis; Litman, Gary W. (Eds.):** Origin and Evolution of the Vertebrate Immune System. 2000. 81 figs. IX, 324 pp. ISBN 3-540-66414-9

Vol. 249: **Jones, Peter A.; Vogt, Peter K. (Eds.):** DNA Methylation and Cancer. 2000. 16 figs. IX, 169 pp. ISBN 3-540-66608-7

Vol. 250: **Aktories, Klaus; Wilkins, Tracy, D. (Eds.):** Clostridium difficile. 2000. 20 figs. IX, 143 pp. ISBN 3-540-67291-5

Vol. 251: **Melchers, Fritz (Ed.):** Lymphoid Organogenesis. 2000. 62 figs. XII, 215 pp. ISBN 3-540-67569-8

Vol. 252: **Potter, Michael; Melchers, Fritz (Eds.):** B1 Lymphocytes in B Cell Neoplasia. 2000. XIII, 326 pp. ISBN 3-540-67567-1

Vol. 253: **Gosztonyi, Georg (Ed.):** The Mechanisms of Neuronal Damage in Virus Infections of the Nervous System. 2001. approx. XVI, 270 pp. ISBN 3-540-67617-1

Vol. 254: **Privalsky, Martin L. (Ed.):** Transcriptional Corepressors. 2001. 25 figs. XIV, 190 pp. ISBN 3-540-67569-8

Vol. 255: **Hirai, Kanji (Ed.):** Marek's Disease. 2001. 22 figs. XII, 294 pp. ISBN 3-540-67798-4

Vol. 256: **Schmaljohn, connie S.; Nichol, Stuart T. (Eds.):** Hantaviruses . 2001, 24 figs. XI, 196 pp. ISBN 3-540-41045-7

Vol. 257: **van der Goot, Gisou (Ed.):** Pore-Forming Toxins, 2001. 19 figs. IX, 166 pp. ISBN 3-540-41386-3

Printing (Computer to Film): Saladruck, Berlin
Binding: Stürtz AG, Würzburg